Reducing Burglary

Andromachi Tseloni • Rebecca Thompson
Nick Tilley

Reducing Burglary

 Springer

Andromachi Tseloni
Quantitative and Spatial Criminology
School of Social Sciences
Nottingham Trent University
Nottingham, UK

Rebecca Thompson
Quantitative and Spatial Criminology
School of Social Sciences
Nottingham Trent University
Nottingham, UK

Nick Tilley
Jill Dando Institute
Department of Security and Crime Science
University College London
London, UK

ISBN 978-3-319-99941-8   ISBN 978-3-319-99942-5   (eBook)
https://doi.org/10.1007/978-3-319-99942-5

Library of Congress Control Number: 2018955907

This Springer imprint is published by the registered company Springer Nature Switzerland AG
The registered company address is: Gewerbestrasse 11, 6330 Cham, Switzerland

Acknowledgements

Chapters 4, 5, 6, 8 and 9 of this book are based on the past work undertaken with funding from the Economic and Social Research Council, Secondary Data Analysis Initiative Phase 1 (ES/K003771/1 and ES/K003771/2). The authors are indebted to the project's Advisory Committee (http://www4.ntu.ac.uk/app_research/soc/document_uploads/178982.pdf) for their support for the duration of the research project and contribution to this work. Crime survey data sets used in this project are cited as follows: Home Office, Research, Development and Statistics Directorate, TNS-BMRB. (2012). *British Crime Survey, 1992–2011.* [data collection]. UK Data Service. Retrieved from https://beta.ukdataservice.ac.uk/datacatalogue/series/series?id=200009; and Office for National Statistics. (2013). *Crime Survey for England and Wales, 2011–2012.* [data collection]. UK Data Service. Retrieved from https://beta.ukdataservice.ac.uk/datacatalogue/series/series?id=200009. Any errors or omissions are entirely the authors' responsibility.

Contents

1 **Introduction** 1
   Andromachi Tseloni, Rebecca Thompson, and Nick Tilley
   1.1 Domestic Burglary: Definition, Data Sources and Counts 2
   1.2 The Distribution of Burglary 4
   1.3 Repeat Victimisation 12
   1.4 Burglary Trends 14
   1.5 The Impact of Burglary 15
   1.6 Responding to Burglary 15
   1.7 Outline of the Remainder of the Book 16
   References 17

2 **A Short History of the England and Wales National
  Burglary Security Initiatives** 21
   Gloria Laycock and Nick Tilley
   2.1 Introduction 21
   2.2 Background 23
   2.3 Programmes 25
        2.3.1 Neighbourhood Watch 25
        2.3.2 Property Marking 26
        2.3.3 Publicity Schemes 26
        2.3.4 The Kirkholt Burglary Prevention Project 27
        2.3.5 Safer Cities 28
        2.3.6 Huddersfield and the 'Olympic' Model 31
        2.3.7 Crime Reduction Programme 32
        2.3.8 Design Against Crime 33
        2.3.9 Alley Gating 33
        2.3.10 Estate Action, Single Regeneration Budget
             and Priority Estates 34
        2.3.11 Improved Street Lighting 35
   2.4 The Vexed Question of Displacement 36

   2.5 Lessons Learned 37
       2.5.1 What Worked?.. 37
       2.5.2 What Didn't Work.................................... 39
       2.5.3 The Importance of Context......................... 39
       2.5.4 Strategy 40
   2.6 Conclusions.. 41
   References 41

3 Domestic Burglary: Burglar Responses to Target Attractiveness 45
   Rachel Armitage
   3.1 Introduction... 45
   3.2 Reducing Burglary Through Secured by Design 46
       3.2.1 Place-Based Crime Reduction 46
       3.2.2 Crime Prevention through Environmental
             Design (CPTED)................................... 47
       3.2.3 Secured by Design (SBD): Development,
             Management and Implementation.................... 47
       3.2.4 Consideration for Crime Prevention Within
             the Planning System 48
       3.2.5 Evaluating the Effectiveness of Secured
             by Design (SBD)................................... 51
       3.2.6 The Principles of Secured by Design (SBD)
             and Their Individual Impact on Crime 53
   3.3 Accounting for Burglar Perceptions 55
       3.3.1 Methodology....................................... 55
       3.3.2 Limitations of the Research 56
   3.4 Burglar Accounts of Target Attractiveness: Research Findings 58
       3.4.1 What Makes a Suitable Target?...................... 58
       3.4.2 What Makes an Unsuitable Target? 59
       3.4.3 Surveillance....................................... 61
       3.4.4 Movement Control 63
       3.4.5 Defensible Space................................... 66
       3.4.6 Physical Security................................... 67
       3.4.7 Management and Maintenance....................... 70
   3.5 What Can Secured by Design Learn from Burglar Accounts?..... 71
   References 73

4 Which Security Devices Reduce Burglary? 77
   Rebecca Thompson, Andromachi Tseloni, Nick Tilley,
   Graham Farrell, and Ken Pease
   4.1 Introduction... 77
   4.2 Previous Work on Security Availability 78
   4.3 Previous Work on Security Device Effectiveness
       Against Burglary... 80
       4.3.1 Victimisation Survey Data 80
       4.3.2 Offender Interviews 82
       4.3.3 Large-Scale Initiatives 84

4.4   Data and Methods..   85
      4.4.1   Why Examine Attempted Burglary and Burglary
              with Entry Separately?...........................   86
4.5   Results...   88
      4.5.1   FAVOR-able Cues: Accessibility and Occupancy   88
      4.5.2   Which Security Devices Deter and Which Thwart?.......   90
4.6   Discussion and Conclusion................................   93
Appendix A   97
A.1   Introduction..   97
A.2   Crime Survey for England and Wales Sample Selection   97
A.3   Crime Survey for England and Wales Questionnaire Structure   97
A.4   Limitations   98
      A.4.1   Security Information Is Not Available
              for All Burglary Victims   98
      A.4.2   Victims of Both Attempted Burglary and Burglary
              with Entry Are Excluded........................... 100
A.5   Alternative Deter/Thwart Calculations...................... 100
A.6   More Information 102
References 102

5   **Household- and Area-Level Differences in Burglary Risk
    and Security Availability over Time**......................... 107
    Andromachi Tseloni and Rebecca Thompson
5.1   Introduction.. 107
5.2   Theoretical Framework................................... 109
5.3   Previous Research Evidence.............................. 111
      5.3.1   Burglary Risks in Context.......................... 111
      5.3.2   Security Availability in Context 112
      5.3.3   Security Availability *and* Burglary Risk in Context 113
      5.3.4   Who Has Benefited the Most (or, Conversely,
              Drew Negligible Benefits) from the Reduction
              in Burglary Risk and the Increase
              in Security Availability? 114
      5.3.5   Limitations of Previous Research.................... 116
5.4   Effective Security Availability and Burglary Risks
      During the Crime Drop................................... 117
      5.4.1   Data and Methodology 117
      5.4.2   Burglary Risk and Effective Security Correlation
              During the Crime Drop............................ 119
5.5   Effective Security Availability and Burglary Risks
      in Context over the Period of the Crime Drop 121
      5.5.1   General Remarks, Population Groups
              and Their (National Average) Burglary Risks............ 121
      5.5.2   Effective Security and Burglary Risk Across
              Ethnic Groups................................... 127
      5.5.3   Effective Security and Burglary Risk with Respect
              to Household Composition......................... 129

	5.5.4	Effective Security and Burglary Risk with Respect to Household Tenure	133
	5.5.5	Effective Security and Burglary Risk with Respect to Annual Household Income	135
	5.5.6	Effective Security and Burglary Risk with Respect to Household Car Ownership	138
	5.5.7	Effective Security and Burglary Risk by Area of Residence	139
5.6	Security-Driven Burglary Drop and Distributive Justice		141
5.7	How Can Crime Prevention Redress the Uneven Burglary Drop and Reignite Overall Falls?		144
Appendix B			146
B.1	Data and Methodology		146
	B.1.1	Variables	146
	B.1.2	Data and Sample Sizes	147
	B.1.3	Statistical Model and Modelling Strategy	151
B.2	The Correlation of Burglary Risk and Effective Security Availability Nationally, 1993–2011/2012		156
B.3	Estimated Bivariate Logit Regression Models of Burglary Risk and WIDE Security Availability During the Crime Drop		157
References			161

6 An Evaluation of a Research-Informed Target Hardening Initiative 165
James Hunter and Andromachi Tseloni

6.1	Introduction		165
6.2	Evaluation of Burglary Reduction Initiatives		167
	6.2.1	Theoretical Underpinnings: Repeat and Near Repeat Victimisation	167
	6.2.2	Key Methodological Issues in the Evaluation of Burglary Reduction Initiatives	167
6.3	Project Context		169
	6.3.1	The City of Nottingham	169
	6.3.2	Burglary Profile of Nottingham	170
	6.3.3	Nottingham Crime and Drugs Partnership (NCDP)	170
6.4	The Nottingham Pilot Burglary Target Hardening Initiative		171
	6.4.1	Project Inception and Operational Framework	171
	6.4.2	Research-Informed Project Aims and Protocol	171
	6.4.3	Selection of Participating Areas	173
	6.4.4	Pilot Process: Planning, Implementation, Security Cost and Evaluation	175
6.5	Evaluation		180
	6.5.1	Pilot Data	180
	6.5.2	Evaluation Data: Did It Work?	183
6.6	Discussion and Conclusion		185

Appendix C 186
C.1 The Nottingham Crime and Drugs Partnership (NCDP) 186
    C.1.1 Statement 186
    C.1.2 For Recognition 187
    C.1.3 History 187
C.2 Selected Protocol and Home Security Assessment Templates 188
References 191

7 The Role of Security Devices Against Burglaries: Findings from the French Victimisation Survey 195
Amandine Sourd and Vincent Delbecque
7.1 Introduction... 195
7.2 Source, Contextual Data and Modelling...................... 198
    7.2.1 Source 198
    7.2.2 Defining the Three Stages of the Burglary 199
    7.2.3 Security Features and Information Regarding the Presence of Someone in the Housing Unit 202
    7.2.4 Environmental Factors and Lifestyle 204
    7.2.5 Modelling 205
7.3 Results... 206
    7.3.1 The Role of Security Devices 206
    7.3.2 Analysis of Combinations of Devices................. 211
    7.3.3 The Specific Case of Repeat Victimisations 214
7.4 Discussion.. 217
Appendix D 220
References 221

8 The Role of Security in Causing Drops in Domestic Burglary 223
Nick Tilley, Graham Farrell, Andromachi Tseloni,
and Rebecca Thompson
8.1 Introduction... 223
8.2 A Comprehensive Theory of the Crime Drop.................. 226
    8.2.1 Seventeen Propositions and Four Tests................. 226
    8.2.2 The Security Hypothesis............................ 226
8.3 Testing the Security Hypothesis for the Burglary Drop: A Data Signatures Approach............................... 229
8.4 Security-Led Burglary Drop in England and Wales 230
    8.4.1 Signature 1: There Would Be an Overall Increase in the Level of Security of Dwellings.................. 230
    8.4.2 Signature 2: There Would Be a Reduction in the Proportion of Dwellings Unprotected by Security Measures 231
    8.4.3 Signature 3: Dwellings with More Security Would Generally Be Less Vulnerable to Burglary than Those with Less Security 232

        8.4.4   Signature 4: The Use of More Effective Security Devices
                and Combinations Will Grow More than the Use of Less
                Effective Security Devices and Combinations   233
        8.4.5   Signature 5: The Protection Conferred by the Presence
                of Security Devices Would Increase over Time   233
        8.4.6   Signature 6: There Will Be No Downward Trend
                in Burglary Amongst Properties with No Security   236
        8.4.7   Signature 7: There Would Be a Greater Fall
                in Burglary with Forced Entry Where the Offender
                Has to Overcome Security Devices, than in Unforced
                Entry Where This Is Not Necessary   236
   8.5 Discussion   237
   8.6 The Curious Case of Burglar Alarms   239
   8.7 The Importance of Design and Detailed Understanding   241
   8.8 Conclusion   242
   References   242

9 **From Project to Practice: Utilising Research Evidence
   in the Prevention of Crime**   245
   Rebecca Thompson and Kate Algate
   9.1 Introduction   245
   9.2 Context   246
   9.3 The Project   247
   9.4 Key Factors   248
        9.4.1   Relationships   249
        9.4.2   Communication   251
   9.5 Challenges in Exchanging Knowledge and Facilitating Impact   254
        9.5.1   Articulating the Potential Practical Benefits
                of Involvement   254
        9.5.2   How to Trace and Document Impact?   256
   9.6 Discussion   257
   9.7 Conclusion   258
   References   259

10 **Conclusions: Reducing Burglary – Summing Up**   265
   Andromachi Tseloni, Rebecca Thompson, and Nick Tilley
   10.1 Burglary Trends and Patterns   266
   10.2 Which Security Devices Work and How?   267
   10.3 Burglary Prevention Lessons   269
   10.4 Future Opportunities   270
   References   272

Index   273

About the Authors

Kate Algate is a Third Sector Chief Executive, currently leading Coventry Citizens Advice. She was the Inaugural Chief Executive for the Neighbourhood and Home Watch Network (England and Wales) from its inception in 2010 through to 2017 when it became a CIO and rebranded as the Neighbourhood Watch Network. Kate has a postgraduate diploma in Policing and Social Conflict from the University of Leicester, and her passion throughout her career has been crime prevention and community participation. Kate has worked on a number of national crime prevention and community engagement programmes across England and Wales and was part of the Home Office Advisory Panel for the Modern Crime Prevention Strategy launched in March 2016.

Rachel Armitage is a Professor of Criminology at the University of Huddersfield. She founded the highly successful multidisciplinary institute, the Secure Societies Institute (SSI), which she directed between 2014 and 2018. Professor Armitage's research focuses upon the role of design (place, space, products and systems) in influencing both antisocial and prosocial behaviour. She has conducted research on the subject of Crime Prevention through Environmental Design (CPTED) for over 20 years. Her work has been referenced in local, national and international planning policy and guidance, and she aims to ensure that consideration for crime prevention is on the agenda of all agencies involved in planning and developing the built environment. Details of Rachel's research can be found at: https://pure.hud.ac.uk/admin/workspace.xhtml?uid=6

Vincent Delbecque has a PhD in Economics from Paris Nanterre University. He was Deputy Head of Statistics and Head of Criminology Studies at the French National Observatory of Crime and Criminal Justice (ONDRP). He has carried out research on burglary in France and has been involved in the development of the French crime and victimisation survey (Cadre de Vie et Sécurité) at the national and international level. He was Professor of Quantitative Methods in Criminology at the National Conservatory of Arts and Crafts (Conservatoire National des Arts et Métiers, CNAM) Paris.

Graham Farrell is a Professor in the Centre for Criminal Justice at the School of Law, University of Leeds. He has published books and over 100 research papers, mostly in the area of crime science, particularly situational crime prevention and crime analysis, repeat victimisation, policing and illicit drug control. He has published widely on the security hypothesis that identified the role of security in the international crime drop.

James Hunter has a PhD in Social Geography from the University of Glamorgan and is Principal Lecturer in Public Policy at Nottingham Trent University. His research interests focus on how place shapes the geography of crime and public service provision. Recent research has explored equity and the crime drop, and spatial justice in policing provision in England and Wales. He has also recently developed a community engagement area classification for all neighbourhoods across England that is designed to provide police forces and officers with a bespoke policy tool that can help shape their community engagement strategies and targeting of initiatives.

Gloria Laycock has a BSc and PhD in Psychology from University College London (UCL). She established and headed the Police Research Group in the UK Home Office and was founding Director of the UCL Jill Dando Institute. She has carried out research and development in prisons, policing and crime prevention and has acted as a consultant and trainer on policing matters around the world. She is currently UCL Professor of Crime Science. She was awarded an OBE in the Queen's Birthday Honours 2008 for services to crime policy.

Ken Pease is a chartered forensic psychologist and Professor of Policing at the University of Derby. His current interests are the use of Bayesian statistics in evaluating police work and the renewed relevance of personal construct theory in the digital age.

Amandine Sourd is a Research Officer at the French National Observatory of Crime and Criminal Justice (ONDRP) since 2015. She graduated from the Faculty of Human and Social Sciences – Sorbonne of Paris Descartes University. Following her research interest in household victimisation, she has carried out research on burglary in France. Her recent work focuses on human trafficking, gender and domestic violence.

Rebecca Thompson is a Senior Lecturer in Criminology at Nottingham Trent University. Her research focuses upon household burglary, antisocial behaviour and police-academic collaboration. Much of this involves working with external partners, for example the East Midlands Policing Academic Collaboration (EMPAC). Rebecca graduated with a PhD in Criminology in 2014. Prior to this, she worked for a police force in the United Kingdom (with a specific remit around crime reduction and community safety) before being awarded a Vice-Chancellor's Scholarship to undertake doctoral study. Since her PhD, she has held research and teaching positions at a number of academic institutions.

Nick Tilley is a Professor in the UCL Department of Security and Crime Science, Emeritus Professor of Sociology at Nottingham Trent University and an Adjunct Professor at the Griffith Criminology Institute in Brisbane. His academic work has been devoted to developing and delivering theoretically informed applied social science. Specific interests lie in evaluation methodology, the international crime drop, problem-oriented policing and situational crime prevention, about all of which he has published extensively. He was awarded an OBE for Services to Policing and Crime Reduction in 2005 and elected a Fellow of the Academy of Social Sciences (FAcSS) in 2009.

Andromachi Tseloni is a Professor of Quantitative Criminology at Nottingham Trent University. She has a BA (Hons.) and an MA from Athens University of Economics and Business and a PhD in Econometrics and Social Statistics from the University of Manchester and has held posts in a number of universities in Greece, the United Kingdom and the United States. Her research focuses upon the individual and environmental factors that shape victimisation risk and repetition. This research has explored victimisation inequalities and the role that security has played in the crime drop across a number of offence types.

List of Figures

Fig. 1.1   Sample size CSEW 1981–2017   4

Fig. 1.2   Variations in prevalence rates of household crime,
            England and Wales, year ending September 2016
            (Source: Office for National Statistics CSEW table at:
            https://www.ons.gov.uk/peoplepopulationandcommunity/
            crimeandjustice/adhocs/006558csewperception
            andasbdatabypoliceforceareayearendingseptember2016.
            Accessed 13 June 2018).   7

Fig. 1.3   Variations in incidence rates of domestic burglary per 1000
            population, England and Wales, year ending September 2016,
            recorded crime data (Source: Office for National Statistics
            table at: https://www.ons.gov.uk/peoplepopulationand
            community/crimeandjustice/datasets/policeforceareadatatables.
            Accessed 13 June 2018).   8

Fig. 1.4   Variations in recorded domestic burglary incidence rates
            per 1000 households for 371 Crime and Disorder Reduction
            Partnership areas in 2007–2008. Note: Data drawn or
            downloaded from https://data.gov.uk/dataset/crime-
            england-wales-2008-2009@2012-06-27T16:12:50.324579.
            Accessed 13 June 2018   9

Fig. 1.5   Variations in recorded domestic burglary incidence rates per
            1000 households for 40 Crime and Disorder Reduction
            Partnership areas within the East Midlands region
            in 2007–2008. Note: Data drawn or downloaded from
            https://data.gov.uk/dataset/crime-england-wales-2008-2009
            @2012-06-27T16:12:50.324579. Accessed 13 June 2018   9

Fig. 1.6   Variations in recorded domestic burglary incidence rates across
            census output areas within the CDRP with the highest domestic
            burglary rate in the East Midlands in 2003/2004. Note: Data
            and analysis by Home Office research team, Government
            Office for the East Midlands, unpublished   10

Fig. 1.7   The ratio of observed to expected repeat burglaries in Victoria,
              Australia (Source: Sagovsky and Johnson 2007, pp.1–26)   13
Fig. 1.8   Domestic burglary incidence rate per 1000 households trend
              in England and Wales 1981–2016, CSEW. Note: The figures
              up to 2000 refer to calendar years and those after that
              to financial years (see also Fig. 1.1) (Source: ONS 2017).   14

Fig. 2.1   Trends in Crime Survey for England and Wales and police
              recorded burglary, year ending December 1981 to year ending
              March 2017. Notes: new Home Office counting rules were
              introduced in April 1988; the National Crime Recording
              Standard was introduced in April 2002; police recorded
              crimes up till 1997 refer to the calendar year and from 1999
              to the years ended 31 March; crime surveys were not
              conducted for 1982, 1984–1986, 1988–1990, 1992, 1994,
              1996, 1998 and 2000–2001; and the figures for these are
              interpolated (Source: Crime Survey for England and Wales,
              Office for National Statistics and Police Recorded Crime)   22

Fig. 3.1   Illustration of 16 images   57

Fig. 4.1   Security Protection Factors against burglary with entry
              for the ten security device combinations with the highest
              SPFs (data taken from the 2008/2009–2011/2012 sweeps
              of the Crime Survey for England and Wales) (significant
              in burglary with entry, p-value <0.05) (capped at 60)   89
Fig. 4.2   Security Protection Factors against attempted burglary for
              the ten security device combinations with the highest SPFs
              (data taken from the 2008/2009–2011/2012 sweeps
              of the Crime Survey for England and Wales)
              (significant in attempted burglary, p-value <0.05)   89

Fig. 5.1   National average correlation between burglary risk and the
              availability of effective security combinations, WD, EWD
              and WIDE, over the period of the crime drop
              (1996–2011/2012 CSEW data). Note: The y-axis values
              refer to the national average (unconditional) correlation
              estimated from joint logit empty models of burglary risk and
              availability of respective WD, EWD and WIDE effective
              security combinations from five aggregated CSEW data sets,
              1994–1996, 1998–2000, 2001/2002–2004/2005, 2005/2006–
              2007/2008 and 2008/2009–2011/2012. The first two models for
              each security combination refer to years 1993–1996 and 1997–
              2000, respectively. The in-between years' correlation estimates
              have been interpolated from the values given by the models of
              adjacent periods   120

Fig. 5.2   Burglary risk and WIDE security availability (odds ratios)
              of ethnic minority groups in comparison to White ethnicity,
              1993–2011/2012 128
Fig. 5.3   Burglary risk and WIDE security availability (odds ratios)
              of single and more than two-adult households in comparison
              to two-adult households. 131
Fig. 5.4   Burglary risk and WIDE security availability (odds ratios)
              of households with children under 16 years old in comparison
              to households without children 131
Fig. 5.5   Burglary risk and WIDE security availability (odds ratios)
              of single adult with children under 16 years old households
              in comparison to other households 132
Fig. 5.6   Burglary risk and WIDE security availability (odds ratios)
              of households in rented from local authorities or private
              landlords accommodation in comparison to households living
              in their own (outright or mortgaged) home 134
Fig. 5.7   Burglary risk and WIDE security availability (odds ratios)
              of non-affluent and affluent households in comparison
              to average income households. 137
Fig. 5.8   Burglary risk and WIDE security availability (odds ratios)
              of no or one-car households in comparison to households
              with two or more cars 139
Fig. 5.9   Burglary risk and WIDE security availability (odds ratios)
              of households in inner-city or urban areas in comparison
              to rural areas. 140
Fig. 6.1   Burglaries per 1000 households in test, control
              and other areas, September 2013–January 2016
              (moving 3-month average). 183
Fig. 7.1   Security features according to housing type (%).
              (Source: French CVS survey Insee-ONDRP-SSMsi,
              2007–2015). Area covered: permanent residences,
              households in Metropolitan France 202
Fig. 7.2   Number of security devices according to housing type.
              (Source: French CVS survey Insee-ONDRP-SSMsi,
              2007–2015). Area covered: permanent residences,
              households in Metropolitan France. Note: The security
              devices are alarms, security doors, digital lock and cameras. . . 203
Fig. 7.3   Combination of security devices according to housing
              type (%). (Source: French CVS survey Insee-ONDRP-SSMsi,
              2007–2015). Area covered: permanent residences,
              households in Metropolitan France 203

Fig. 7.4   Effect of variables on the probability of houses being targeted.
              (Source: French CVS survey Insee-ONDRP-SSMsi,
              2007–2015). Area covered: permanent residences, households
              in Metropolitan France. Note to the reader: only variables
              which are significant at the 0.1 level are displayed
              in the figure 207
Fig. 7.5   Effect of variables on the probability of apartments
              being targeted. (Source: French CVS survey Insee-ONDRP-
              SSMsi, 2007–2015). Area covered: permanent residences,
              households in Metropolitan France 207
Fig. 7.6   Effect of variables on the probability of forced entry
              in houses. (Source: French CVS survey, Insee-ONDRP-
              SSMsi, 2007–2015). Area covered: permanent residences,
              households in Metropolitan France 208
Fig. 7.7   Effect of variables on the probability of forced entry
              in apartments. (Source: French CVS survey, Insee-ONDRP-
              SSMsi, 2007–2015). Area covered: permanent residences,
              households in Metropolitan France 209
Fig. 7.8   Effect of the factors on the probability of thefts in houses.
              (Source: French CVS survey, Insee-ONDRP-SSMsi,
              2007–2015). Area covered: permanent residences,
              households in Metropolitan France 210
Fig. 7.9   Effect of the factors on the probability of thefts in apartments.
              (Source: French CVS survey, Insee-ONDRP-SSMsi,
              2007–2015). Area covered: permanent residences,
              households in Metropolitan France 210
Fig. 7.10   Effect of combinations of security devices on the probability
              of forced entry in houses. (Source: French CVS survey,
              Insee-ONDRP-SSMsi, 2007–2015). Area covered: permanent
              residences, households in Metropolitan France. Note:
              The 'other combinations' field includes the devices
              mentioned in less than 1% of the responses.
              More particularly, the combinations included
              in the 'other' category are those with cameras
              and the combination of digital lock and alarm. 212
Fig. 7.11   Effect of combinations of security devices on the probability
              of forced entry in apartments. (Source: French CVS survey,
              Insee-ONDRP-SSMsi, 2007–2015). Area covered: permanent
              residences, households in Metropolitan France 212
Fig. 7.12   Effect of combinations of security devices on the probability
              of theft for houses. (Source: French CVS survey,
              Insee-ONDRP-SSMsi, 2007–2015). Area covered:
              permanent residences, households in Metropolitan France 213

Fig. 7.13   Effect of combinations of security devices on the probability
             of theft for apartments. (Source: French CVS survey,
             Insee-ONDRP-SSMsi, 2007–2015). Area covered:
             permanent residences, households in Metropolitan France. . . . 214
Fig. 7.14   Effect of repeat victimisation on the probability of forced
             entry to houses. (Source: French CVS survey, Insee-ONDRP-
             SSMsi, 2007–2015). Area covered: permanent residences,
             households in Metropolitan France 216
Fig. 7.15   Effect of repeat victimisation on the probability of forced
             entry to apartments. (Source: French CVS survey, Insee-
             ONDRP-SSMsi, 2007–2015). Area covered: permanent
             residences, households in Metropolitan France 216

Fig. 8.1   Trends in the proportion of dwellings with security devices
             installed or 'no security' in England and Wales, CSEW/BCS
             1992–2011/2012. Note: Adapted from Tseloni et al.
             (2017), p. 6 231
Fig. 8.2   Average protective effects against burglary with entry
             across numbers of security devices in combination, CSEW
             2008/2009–2011/2012. Note: Calculated from
             Tseloni et al. (2014), Table 4 232
Fig. 8.3   Security Protection Factors (SPFs) for selected home security
             against burglary with entry (significant at 5% level unless
             shaded in grey) based on the 2008/2009–2011/2012, CSEW
             (Source: Tseloni and Thompson (2015), p. 34) 233
Fig. 8.4   Trends in most effective security combinations
             and single security, 1992–2011/2012, CSEW/BCS. Note:
             The 1992–1996 CSEW data about lights is assumed
             to correspond to external lights, internal lights or both
             in the post-1996, CSEW sweeps (Source: Tseloni et al.
             (2017), p. 8) 234
Fig. 8.5   Over time Security Protection Factors against burglary with
             entry for security device combinations (pairs or triplets)
             1992–2011/2012 (significant in at least burglary with entry
             or attempted burglary) (p-value <0.05 unless shaded in white).
             Note: For the 1992–1996 CSEW sweeps, external and internal
             lights are confounded hence values for E and I equal L.
             In addition, security chains and dummy alarms were
             not recorded in this period, hence they are missing 234
Fig. 8.6   Over time Security Protection Factors against burglary
             with entry for security device combinations (four or more)
             1992–2011/2012 (capped at 80) (significant in at least burglary
             with entry or attempted burglary) (p-value <0.05 unless shaded
             in white). Note: For the 1992–1996 CSEW sweeps, external and
             internal lights are confounded hence values for E and I equal L.
             In addition, security chains and dummy alarms were not
             recorded in this period, hence they are missing 235

Fig. 8.7   Over time Security Protection Factors against burglary
              with entry for single security devices 1992–2011/2012
              (significant in at least burglary with entry or attempted burglary)
              (p-value <0.05 unless shaded in white). Note: For the
              1992–1996 CSEW sweeps, external and internal lights
              are confounded hence values for E and I equal L. In addition,
              security chains and dummy alarms were not recorded
              in this period, hence they are missing 235
Fig. 8.8   Trends in burglary risk of dwellings with no security relative
              to national risk in England and Wales, CSEW/BCS 1992–1996
              to 2008/2009–2011/2012. Note: Adapted from
              Tseloni et al. (2017), p. 7 236
Fig. 8.9   Trends in burglary mode of entry, CSEW/BCS
              1992–2011/2012 (Source: Tseloni et al. (2017), p. 10) 237
Fig. 8.10   Trends in percentage of dwellings with double glazing
              and burglaries per 1000 dwellings, 1996–2008
              (Source: Farrell et al. (2016), p. 8). 238
Fig. 8.11   Trends in percentage of dwellings rented
              from councils, 1980–2012 (Source: Live tables on dwelling
              stock (including vacants), Table 101, DCLG. Dwelling stock
              by tenure, 1980 to 2012) 239
Fig. 8.12   Marginal effects of burglar alarms on burglary with entry,
              CSEW/BCS 1992–1996 and 2008/2009–2011/2012
              (Source: Tilley et al. (2015a), pp. 11 and 12) 240

Diagrams

Diagram 2.1   Rationale for crime prevention publicity   27

Diagram 4.1   Conceptual drawing of 'deter' and 'thwart' mechanisms   91

Diagram 5.1   The association between burglary risk and security
              availability considering contextual influences 110

Diagram 6.1   Flow chart detailing the steps of the pilot target hardening
              process in the control and test areas. (Source: NCDP 2014) . . . 176
Diagram 6.2   Example of aerial map for identification of the cocoon area
              around a burgled property. (Source: NCDP 2015b). 178

Diagram 7.1   Question sequencing on burglary victimisation.
              (Source: French CVS survey Insee-ONDRP-SSMsi,
              2007–2015). Area covered: permanent residences,
              households in Metropolitan France 199
Diagram 7.2   The process of burglary. (Source: ONDRP) 200
Diagram 7.3   Sequential victimisation rate of burglary.
              (Source: French CVS survey Insee-ONDRP-SSMsi,
              2007–2015). Area covered: permanent residences,
              households in Metropolitan France 201

List of Tables

Table 1.1	International variations in rates of burglary	6
Table 1.2	Variations in risk of burglary, CSEW 2001/2002	10
Table 1.3	Expected and observed prevalence of multiple victimisation (burglary and theft in a dwelling): combined CSEW data, 1982 and 1984	12
Table 2.1	Approaches to tacking burglary	23
Table 2.2	Safer Cities burglary reduction schemes: contexts and measures implemented (Tilley and Webb 1994)	30
Table 2.3	Biting Back responses to burglary	31
Table 2.4	Types of displacement following Barr and Pease (1990)	36
Table 3.1	Description of 16 images	56
Table 3.2	Proportion of burglars referencing the five key principles of SBD	62
Table 3.3	Number of references to the key principles of SBD	62
Table 4.1	Examples of CSEW security device combinations that fit the ONS (2013) 'basic/enhanced' categories	79
Table 4.2	The most common environmental cues considered by burglars in relation to burglary target selection as suggested by previous research (acronym FAVOR)	83
Table 4.3	Expected mechanism of different security devices	88
Table 4.4	What are the odds of burglary with entry and the odds of attempted burglary given a particular security device combination? With that security combination, are the odds of being a burglary with entry victim higher than attempted burglary? (CSEW, 2008/2009–2011/2012)	92
Table 4.5	Descriptive statistics (2008/09–2011/12) for no security, WD and WIDE populations	96

Appendix Table 4.6   Modules of the 2011–2012 CSEW questionnaire
                      and subset of respondents who were asked
                      each module.   98
Appendix Table 4.7   Household security measures over time,
                      as measured by the CSEW (in both the Victim
                      and Non-Victim Forms) 1992–2011/2012. Note:
                      Adapted from Tseloni et al. (2017)   99
Appendix Table 4.8   Alternative deter/thwart risk calculations
                      (where the population selected comprises only
                      victims with particular security combinations)
                      (CSEW, 2008/2009–2011/2012)   101
Appendix Table 4.9   Alternative deter/thwart odds calculations
                      (where the population selected comprises
                      all victims) (CSEW, 2008/2009–2011/2012)   101

        Table 5.1   Census and employed CSEW sample percentages
                      of selected population subgroups   123
        Table 5.2   Sample percentages and burglary risk of selected
                      population groups via bivariate (contingency tables)
                      analysis of the entire CSEW data 2008/2009–
                      2011/2012   126
Appendix Table 5.3   Descriptive statistics of household and area
                      characteristics, burglary risk and WIDE security
                      availability from the WIDE sample of the CSEW
                      aggregate data sets (1994–2011/2012)   148
Appendix Table 5.4   Correlation (standard error of covariance)
                      between effective security availability
                      and burglary risk during the crime drop   152
Appendix Table 5.5   Estimated odds ratios, *Exp(b)*, of fixed effects
                      of household characteristics, area type and region
                      of England and Wales on joint burglary risk and WIDE
                      security availability over time, 1993–2011/2012 –
                      based on bivariate logit regression models
                      of CSEW aggregate data sets (1994–2011/2012)   153
Appendix Table 5.6   Burglary risks in absolute values in the entire
                      and employed (WIDE security focused) CSEW
                      samples over time   159
Appendix Table 5.7   Household Reference Person (HRP) ethnicity
                      as a percentage of respondents' ethnicity,
                      2006/2007 CSEW   161

        Table 6.1   The percentage of repeat and near repeat burglaries
                      within 200 metres of initial burglary during
                      the year prior to the initiative, January to December
                      2013 (Source: NCDP 2014).   174

Table 6.2 Population and household size and burglary rate per 100 households of Lower Super Output Areas participating in the pilot target hardening initiative (Source: NCDP 2014)................. 175

Table 6.3 Summary of the Activity Log for the burglary pilot project 181

Table 6.4a Total number of burglaries per 1000 households before (October 2013–September 2014) and after (January–December 2015) the pilot and percent change in test, control and other areas excluding the pilot period............................... 184

Table 6.4b Total number of burglaries per 1000 households during the calendar years before and after the mid-point of the pilot and percent change in test, control and other areas including the pilot period 184

Appendix Table 6.5 Burglary Pilot (Target Hardening) Process 188

Appendix Table 6.6 Data Capture Form 189

Appendix Table 6.7a Security Survey 190

Appendix Table 6.7b Details of security works carried out................ 190

Table 7.1 List of environmental, housing unit's specific and control factors........................... 204

Appendix Table 7.2 References for all of the variables used............. 220

Table 8.1 Hypotheses to explain the crime drop and their a priori plausibility as indicated by four crucial tests .. 227

Chapter 1
Introduction

Andromachi Tseloni, Rebecca Thompson, and Nick Tilley

Burglary has formed, and continues to form, a major focus for both research and policy in many industrialised societies. This is for two main reasons. First, burglary comprises a high-volume crime, which rose rapidly in decades following the Second World War (Van Dijk et al. 2012). Second, the experience of burglary has a major impact on many of its victims, who have suffered not only losses but also psychological damage (Dinisman and Moroz 2017). There has, therefore, been a strong interest in understanding what leads to vulnerability to burglary and what can be done to reduce its incidence. This book reports recent original research on burglary patterns, the role of security and efforts to implement preventive strategies based on its findings. It also takes stock of previous major initiatives that have been used to address the problem, especially in the UK.

This book is aimed at researchers, postgraduate or honours students, policymakers and practitioners interested in domestic burglary and its prevention. We have also tried to make it accessible to members of the public interested in crime and crime prevention. Whilst the remainder of this book presents new research findings, in this opening chapter, we introduce you to some basic facts about domestic burglary as they emerge from quite a large volume of existing research. We rehearse what decades of research have revealed about major patterns of domestic burglary and the effects of burglary on victims, before giving an overview of the chapters to follow.

A. Tseloni (✉) · R. Thompson
Quantitative and Spatial Criminology, School of Social Sciences, Nottingham Trent University, Nottingham, UK
e-mail: andromachi.tseloni@ntu.ac.uk

N. Tilley
Jill Dando Institute, Department of Security and Crime Science, University College London, London, UK

© Springer Nature Switzerland AG 2018
A. Tseloni et al., *Reducing Burglary*,
https://doi.org/10.1007/978-3-319-99942-5_1

1.1   Domestic Burglary: Definition, Data Sources and Counts

Domestic burglary includes burglaries in dwellings and attached buildings. It is to be distinguished from non-domestic burglary, which occurs in other types of premises including all types of commercial building, such as shops, banks, factories, warehouses and private offices, as well as other non-residential buildings, such as schools, hospitals and universities. Domestic burglary is sometimes referred to as residential burglary or burglary-dwelling. Because this book is specifically about domestic burglary, in the interests of making the text readable where we refer to 'burglary', we mean only domestic burglary. Where we refer to non-domestic burglary, this will be made clear in the text.

In formal terms, a burglary takes place when someone enters any building or part of a building as a trespasser and with intent to steal anything in the building or to inflict grievous bodily harm or to effect unlawful damage to the building or anything in it (Theft Act 1968). A burglary is committed where the attempt is made to commit the crime, even if it is thwarted. Some crime statistics distinguish between attempted burglaries and burglaries with entry, between burglary with loss and burglary where nothing was taken and between burglary of the dwelling itself and burglary of attached buildings.

There are two main sources of burglary statistics: police recorded crime and crime survey data. Police recorded burglaries are incidents like the ones described in the previous paragraph that have been reported by victims (or any members of the public) and recorded as such by the police. In the Crime Survey for England and Wales (CSEW), burglary is determined via the following questions:

> During the last 12 months, … has anyone GOT INTO this house/flat without permission and STOLEN or TRIED TO STEAL anything? (ONS/TNS 2015, p. 41).

> [Apart from anything you have already mentioned] in that time did anyone GET INTO your house/flat without permission and CAUSE DAMAGE? (ONS/TNS 2015, p. 42).

> [Apart from anything you have already mentioned], in that time have you had any evidence that someone has TRIED to get in without permission to STEAL or to CAUSE DAMAGE? (ONS/TNS 2015, p. 42).

> [Apart from anything you have already mentioned], in that time was anything STOLEN out of your house/flat? (ONS/TNS 2015, p. 42).

> And [apart from anything you have already mentioned], in that time was anything (else) that belonged to someone in your household stolen from OUTSIDE the house/flat - from the doorstep, the garden or the garage for example? NOTE: DO NOT COUNT MILK BOTTLE THEFT (ONS/TNS 2015, p. 43).

Crime surveys can be international, such as the International Crime Victims Survey (ICVS) that covers 78 countries (Van Dijk et al. 2007), national or local. Many countries now operate national victimisation surveys. The first was in the USA – the National Crime Victimisation Survey (NCVS) – and goes back to the 1970s (Cantor and Lynch 2000). Other examples of national crime surveys include the CSEW and the Cadre de Vie et Sécurité (CVS) in France which the coming chapters of this book draw upon (Jansson 2007; Flatley 2014). A notable local crime survey, the data from which has delivered a rich set of criminological insights and innovative empirical methodology, is the Seattle multistage hierarchical data set (Miethe 2006; Rountree and Land 1996). These types of surveys capture crimes not reported to the police as well as a rich set of factual information about victims and non-victims, their households and areas of residence as well as their opinions and perceptions in relation to crime and the criminal justice system. Victimisation surveys are now conducted to measure levels of crime against businesses as well as individuals and households. These surveys have become core indicators of crime and are also used to inform policy. Questions go beyond crime experience and cover a range of other issues that are important for understanding crime patterns, attitudes towards the criminal justice system and security measures taken. Therefore they are an invaluable source of information about crime, crime perceptions and crime prevention (Tilley and Tseloni 2016).

Large-scale national victimisation surveys provide the most robust data for measuring variations in rates of crime within a country, provided that the sample size is large enough and appropriately allocated between densely and sparsely populated geographical areas (Flatley 2014). In England and Wales, national victimisation surveys have been undertaken going back to 1981. The survey was originally called the British Crime Survey, or BCS, but is now called the Crime Survey for England and Wales, or CSEW, better to capture its geographic coverage (details are provided in Appendix A). Throughout this book it will be referred to as the CSEW. Initially the CSEW sample sizes were quite small, at around 10,000, and waves were irregular. Surveys covered 1981, 1983, 1987, 1991, 1993, 1995, 1997 and 1999, before becoming annual or continuous from April 2001 to March 2002 onwards. Sample sizes increased in 2005/2006 when they reached more than 45,000, since which time they have fallen slightly, as shown in Fig. 1.1. The CSEW is (at the time of writing) the only measure of crime designated by the Office for Statistics Regulation as 'National Statistics', meaning they are fully compliant with the Code of Practice for Official Statistics. They are therefore a major resource for understanding crime, crime patterns and changes in crime experience in England and Wales. The research we report in this book leans very heavily on it as the most robust data we have on the issue.

In some instances, an offence that could be classified as a burglary is categorised as another type of offence, in particular where the possible alternative classification relates to a more serious offence. For example, if a rape is committed when unlawful entry has been made, the incident will typically be classified as a rape rather than as a burglary. According to the rules of crime surveys' offence classification, an incident is recorded as the most serious offence that took place during its occurrence. As burglary is the most serious household crime, it can only escalate to a crime against the person, such as rape or assault (ONS 2015, pp. 112–118). A small percentage of crimes reported in the CSEW entail both household and personal offences which, for need of a better term, have been termed composite crimes

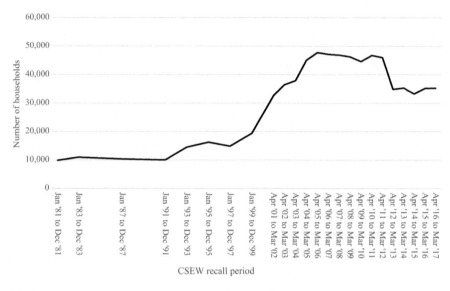

Fig. 1.1 Sample size CSEW 1981–2017

(Tseloni et al. 2010).[1] In police recorded crime, the changed classification can sometimes, however, reflect efforts to downplay the seriousness of the offence in the interests of massaging crime statistics, for example, when what appear to be attempted burglaries are classified as criminal damage or simply suspicious incidents (HMIC 2014).

1.2 The Distribution of Burglary

Crime statistics sometimes distinguish between *prevalence* and *incidence*. Prevalence refers to the number or rate at which potential victims of burglary have experienced one or more burglaries over a given period, normally a year. Incidence refers to the number of burglaries or rate of burglaries committed against members of a given population over a given period, again normally a year. Confusingly, the prevalence and incidence rates are often expressed with different denominators – prevalence as a percentage and incidence as per 1000, 10,000 or 100,000. Furthermore, rates are sometimes expressed in relation to households and sometimes in relation to population. In practice, police recorded crime statistics on burglary generally refer to incidence, and rates are given in relation to the population. In contrast, crime survey data often refer to burglary prevalence in relation to

---

[1] Therefore burglaries which escalated into a more serious offence cannot be identified as such from the publicly available data set; one would need to investigate the incident narrative to decipher how many crimes against the person were committed in the course of a burglary.

households. The Crime and Justice Statistics which are published quarterly by the Office for National Statistics (ONS) provide burglary statistics which refer to both prevalence and incidence rates over 100 households based on the CSEW (ONS 2017). Prevalence rates are always lower than incidence rates. The difference springs from the fact that some victims suffer more than one incident – a phenomenon known as 'repeat victimisation' – an issue that has proved a rich topic for research and an important focus for preventive strategies, as discussed below (Farrell 1992; Pease 1998). Where crime rates are referred to casually in the press, it is generally not clear whether they refer to prevalence or incidence. If the report draws on police data, the reference is almost invariably to incidence.

Rates of burglary per household are strictly preferable to those per population, given that all members of a household are victims where an offence takes place: all have had their private space invaded even if some have not suffered a loss. Crime surveys generally use addresses as the sampling frame, with one eligible member selected as the respondent (with the exception of the NCVS which interviews all household members). When asked specifically about burglary, they are answering on behalf of the whole household. When findings are reported as rates per population (as in the case of police recorded crime), slightly misleading impressions can be given.

At every level, burglary is unevenly distributed (Tseloni and Pease 2005; Van Dijk et al. 2007). Research has identified some widespread patterns in burglary victimisation. Table 1.1 shows the variations in national incidence rates per 100 population for domestic burglary for 2004/2005, as presented by the United Nations Office on Drugs and Crime (UNODC) and found by the ICVS for 2004/2005. The table also shows the rates for reporting domestic burglary to the police, as found in the ICVS.

The UNODC uses official police recorded crime rates. The ICVS is designed to iron out variations in definitions of crimes (including burglary) and to avoid falling foul of national differences in rates of reporting and recording offences. It also uses a standard methodology so that valid comparisons can be made between levels of crimes in different countries (Van Dijk and Tseloni 2012). There have been five main 'waves' or 'sweeps' of the ICVS (Van Dijk et al. 2007; UNIL 2018). Table 1.1 picks out countries that took part in the 2004/2005 ICVS and that are also included in the UNODC data. You should bear in mind that there are wide error margins in ICVS estimates, given that the sample sizes for ICVS surveys in individual countries are quite small – usually around 2000 – and burglary events are quite rare.

Three important points emerge from Table 1.1. First, there were large variations between countries in incidence rates of burglary, however measured. The ICVS rates shown in the table go from 0.8 percent for Sweden to 4.6 percent for England and Wales. The UNODC rates vary from 0.08 percent for Mexico to 0.92 percent for New Zealand. Second, in every case the ICVS rates are much higher than the UNODC rates. As measured by the UNODC, the rate for England and Wales is still relatively high, but not the highest. Likewise, Sweden's rate becomes middling according to the UNODC figures rather than low as found in the ICVS, and Mexico's rate goes from being the lowest according to the UNODC to being second only to

Table 1.1 International variations in rates of burglary

	ICVS incidence/100 population 2004/2005	ICVS reporting rate to the police %	UNODC incidence rates/100 population 2005
Australia	3.1	86	0.91
Austria	1.2	73	0.24
Belgium	2.1	90	0.55
Canada	2.6	74	0.47
England and Wales	4.6	88	0.56
Finland	1.2	68	0.14
Germany	1.1	86	0.13
Greece	2	71	0.10
Hungary	2.5	76	0.18
Italy	2.8	78	0.21
Japan	1.2	63	0.11
Mexico	4.5	3	0.08
Netherlands	1.4	92	0.57
New Zealand	3.9	80	0.92
Northern Ireland	1.6	88	0.42
Norway	1.4	72	0.18
Portugal	1.9	55	0.21
Scotland	2.2	90	0.42
Sweden	0.8	77	0.48
USA	4.1	77	0.47

Sources: UNODC (http://www.unodc.org/documents/data-and-analysis/statistics/crime/CTS12_Burglary.xls) and Van Dijk et al. (2007) for ICVS

England and Wales in the ICVS. Part of the discrepancy in relative rates is explained by the middle column, which shows wide variations in reporting rates as found in the ICVS. Evidently only 3 percent of burglaries are reported in Mexico compared to 92 percent in the Netherlands. Note that England and Wales were found to have one of the highest rates of reporting, at 88 percent. Reporting is normally required for a burglary to be recorded, but it is not sufficient. Discretion is used by police officers in deciding that the offence is actually a burglary, and as already noted they may not always record what the victim takes to be a burglary to be one.

Figure 1.2 shows recent data on variations in prevalence rates of household crime, including burglary, household theft, vehicle crime and criminal damage, by police force area in England and Wales in the year to September 2016, as found in the CSEW. The chart shows substantial variations, from a high of 15 percent in Northamptonshire to a low of 5 percent in Dyfed-Powys. Two points need to be considered here. First, each estimate falls within quite large confidence limits because of the infrequency of burglary and the limited sample size in each police force area. Thus, for Northamptonshire the confidence limits go from 11 to 18 percent (the unweighted base – number of respondents – was 644) and for Dyfed-Powys

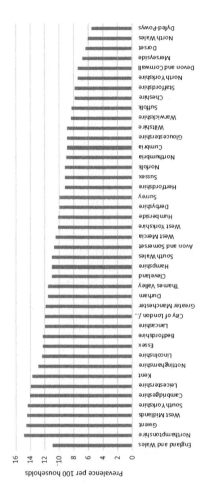

Fig. 1.2 Variations in prevalence rates of household crime, England and Wales, year ending September 2016 (Source: Office for National Statistics CSEW table at: https://www.ons.gov.uk/peoplepopulationandcommunity/crimeandjustice/adhocs/006558csewperceptionandasbdatabypoliceforceareayearendingseptember2016). Accessed 13 June 2018

the confidence limits go from 3 to 8 percent (unweighted base, 663). For England and Wales as a whole, the overall rate is 11 percent with confidence limits of 10–11 percent (unweighted base, 36,629). It is clear that getting estimates for single, relatively rare offences would not be feasible albeit areas can be reliably ordered from lowest to highest crime quartiles from statistical analyses of the CSEW (Lynn and Elliot 2000).

One alternative to the CSEW is to use recorded rates instead of survey estimates, acknowledging that this may fall foul of variations in reporting and recording practices and the fact that they require improvements to become National Crime Statistics. Figure 1.3 shows the range of incidence rates for recorded burglary, in the year to September 2016. They run from a low of 2.3 per 1000 population in Dyfed-Powys to a high of 11.3 per 1000 population in West Yorkshire. Variations in crime rates at levels below the police force level tend to depend on recorded crime. They show huge variations. For example, on the Kirkholt Estate in Rochdale, the annual incidence rate for domestic burglary in 1985 was equivalent to 24.6 percent, based on the numbers recorded by the police in the first 5 months of the year (Forrester et al. 1988, p. 2). This far exceeded the rates elsewhere. It doubles the then incidence rate found in the CSEW at the time. This is especially striking given that recorded crime rates miss out many offences that are not reported to the police and, of those reported, not all are recorded (HMIC 2014).

Figures 1.4, 1.5 and 1.6 show variations in levels of burglary at different geographical levels. Figures 1.5 and 1.6 show such variations within a region and also that even in a very high burglary area, there remain massive differences in rates of recorded burglary within it.

Risks of burglary vary by the demographic characteristics of households as well as by place. Table 1.2 shows what was found in the 2001/2002 BCS (Simmons et al. 2002). This shows that the young are at more risk than the old; single adult households with children are more at risk than those without children; the poorer are at

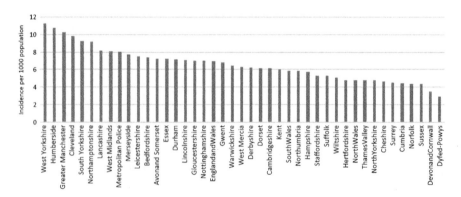

Fig. 1.3 Variations in incidence rates of domestic burglary per 1000 population, England and Wales, year ending September 2016, recorded crime data (Source: Office for National Statistics table at: https://www.ons.gov.uk/peoplepopulationandcommunity/crimeandjustice/datasets/policeforceareadatatables). Accessed 13 June 2018

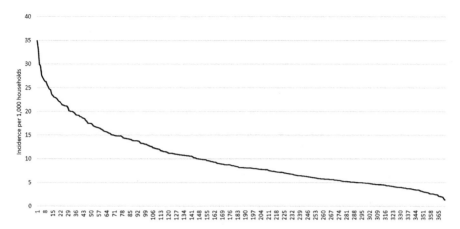

Fig. 1.4 Variations in recorded domestic burglary incidence rates per 1000 households for 371 Crime and Disorder Reduction Partnership areas in 2007–2008. Note: Data drawn or downloaded from https://data.gov.uk/dataset/crime-england-wales-2008-2009@2012-06-27T16:12:50.324579. Accessed 13 June 2018

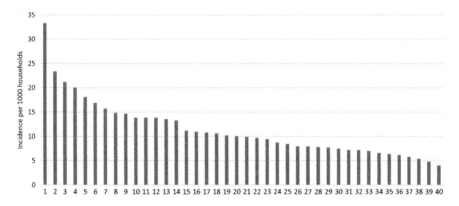

Fig. 1.5 Variations in recorded domestic burglary incidence rates per 1000 households for 40 Crime and Disorder Reduction Partnership areas within the East Midlands region in 2007–2008. Note: Data drawn or downloaded from https://data.gov.uk/dataset/crime-england-wales-2008-2009@2012-06-27T16:12:50.324579. Accessed 13 June 2018

more risk than the richer; renters are at more risk than owner occupiers; the unemployed are more at risk than the employed; those living in flats or maisonettes are more at risk than those living in detached houses; those who go out more are at greater risk than those spending more time at home; those living in inner cities are more at risk than those living in rural areas; those in public (also known as social or council) housing are more at risk than those living in private housing; and those living in areas with high levels of physical disorder are more at risk than those living in areas with low levels.

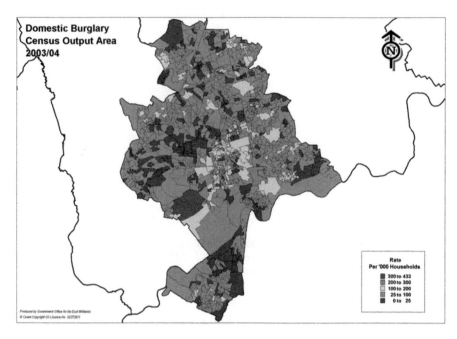

Fig. 1.6 Variations in recorded domestic burglary incidence rates across census output areas within the CDRP with the highest domestic burglary rate in the East Midlands in 2003/2004. Note: Data and analysis by Home Office research team, Government Office for the East Midlands, unpublished

Table 1.2 Variations in risk of burglary, CSEW 2001/2002

| | % Victims once or more | | |
	All burglary	With entry	Attempts
Age of head of household			
16–24	9.1	5.8	3.8
25–44	4.2	2.4	1.9
45–64	2.9	1.6	1.3
65–74	2.0	1.2	0.9
75+	2.4	1.6	0.8
Head of household under 60			
Single adult and child(ren)	9.3	6.0	3.7
Adults and child(ren)	3.5	1.8	1.8
No children	3.7	2.2	1.6
Head of household over 60	2.2	1.3	1.0
Household income[a]			
Less than £5000	4.9	3.4	1.9
£5000 less than £10,000	4.5	2.4	2.3
£10,000 less than £20,000	3.3	1.6	1.8

(continued)

Table 1.2 (continued)

	% Victims once or more		
	All burglary	With entry	Attempts
£20,000 less than £30,000	3.5	2.2	1.4
£30,000 or more	2.8	1.6	1.2
Tenure			
Owner occupiers	2.6	1.5	1.1
Social renters	5.3	2.9	2.6
Private renters	5.7	3.9	2.0
Head of household employment status[b]			
In employment	3.5	2.0	1.6
Unemployed	5.1	4.3	1.3
Economically inactive	5.1	3.0	2.3
Accommodation type			
Houses	3.2	1.9	1.4
Detached	2.5	1.5	1.0
Semi-detached	2.9	1.7	1.3
Terraced	4.0	2.3	1.8
Flats/maisonettes	4.7	2.8	2.1
Hours home left unoccupied on an average weekday			
Never	3.4	2.4	1.1
Less than 3 h	2.9	1.6	1.4
3 but less than 5 h	2.8	1.6	1.3
5 h or more	4.2	2.5	1.9
Area type			
Inner city	5.5	3.0	2.7
Rural	2.0	1.2	0.8
Urban	3.6	2.1	1.6
Council estate[c]	4.7	2.7	2.3
Non-council estate	3.2	1.9	1.3
Level of physical disorder[d]			
High	6.8	4.3	2.8
Low	3.1	1.8	1.4
All households	**3.5**	**2.0**	**1.5**

Source: Simmons et al. (2002, Table 4.03)

[a]The 2001 BCS sweep introduced additional prompts on equivalent monthly as well as annual income. This means that crime risks broken down by household income may not be directly comparable with past sweeps

[b]Based on men aged 16–64 and women aged 16–59

[c]Council areas are those that fall into ACORN types 33, 40–43 and 45–51

[d]Based upon the interviewer's perception of the level of (a) vandalism, graffiti and deliberate damage to property, (b) rubbish and litter and (c) homes in poor condition in the area. For each the interviewer had to code whether it was 'very common', 'fairly common', 'not very common' or 'not at all common'. For both variables 'very' and 'fairly' common were set to 1 and 'not very' and 'not at all' to 0. These variables were then summated for each case. The incivilities scale ranged from 0 to 3. Those with a score of 2 or 3 were classified as being in high-disorder areas

1.3   Repeat Victimisation

We have shown that burglaries are distributed unevenly by area at all levels of geographical resolution, for example, country, region, area or census tract (or output area). They are also unevenly distributed by address. Even if we take as our population addresses that have been burgled, some experience more subsequent burglary than others. This is the phenomenon of 'multiple' or as it is now more commonly termed 'repeat' victimisation (RV) (Pease 1998; Farrell 1992). Sparks et al. (1977) had noticed repeat patterns in an early victimisation survey. Al Reiss (1980) had noticed that members of the same households experienced at least two crime types within 6 months and created a matrix of multiple victimisation from early NC(V)S data.[2]

The significance for burglary in particular and a major impetus for further study of the phenomenon emerged from the Kirkholt Burglary Prevention Project (Forrester et al. 1988, 1990). The very high incidence rate referred to previously was produced in part by the repeat incidents experienced at some addresses. As Forrester et al. (1988, p. 9) say, 'An analysis of the 1996 burglaries on Kirkholt clarified that the chance of a second or subsequent burglary was over four times as high as the chance of a first…'. This finding prompted the authors to look at the screening questions of the 1982 and 1984 sweeps of the CSEW. Table 1.3 reproduces their findings.

A large volume of research has since found that these repeat patterns are found for many types of crime (Chenery et al. 1996) and across countries (Farrell et al. 2005). The findings for burglary in particular are found across places in England and Wales and across other jurisdictions also. Moreover, high crime neighbourhoods experience high levels of crime in part as a result of the very high rates of repeat victimisation within them (Trickett et al. 1992). When crime happens, the chances of a subsequent victimisation increase (Pease and Farrell 2014; Pease and Tseloni 2014).

It is hard to overestimate the significance of repeat victimisation for understanding burglary and for devising strategies to reduce levels of burglary. In some ways, it might seem odd that it remained unrecognised for a long time. Indeed, for a long time, false comfort was given to victims of burglary through the suggestion that having experienced burglary they had, so to speak, had their turn, so they should not

Table 1.3 Expected and observed prevalence of multiple victimisation (burglary and theft in a dwelling): combined CSEW data, 1982 and 1984

	2+	3+	4+	5+
Observed	111	38	24	15
Expected	32	1	0	0

Source: Forrester et al. (1988, p. 9)
Note: Weighted data, unweighted $n = 21{,}232$

---

[2] The NCVS before being redesigned in 1994 was called the National Crime Survey (NCS).

worry about experiencing another. The then unknown fact that one burglary made another more likely was flatly contradicted in this advice. There seem to be two reasons why repeat burglary was long overlooked. One is that the same police officer is unlikely to be sent again to an address of a property that has already been burgled so individual officers will not get a sense of heightened vulnerability. The second is that the police will only come to have a record of repeat incidents when they are reported and recorded.

Repeat victimisation has some interesting patterns. The first is that a second incident is more likely than a first and a third more likely than a second (Pease and Tseloni 2014). The second is that the risks of a repeat incident fade over time. This is sometimes referred to as 'decay'. The reduction of the risk of a repeat falls quite quickly, as shown in Fig. 1.7, using data from Victoria, Australia (Sagovsky and Johnson 2007). This shows that the risk of repeats is much higher than expected immediately after an incident but that this heightened risk decays quickly till around 13 weeks, when it reaches parity or a little below. The third is that in addition to repeat burglaries of the same address being at heightened risk in the short term, the same goes for nearby dwellings which is known as near repeats. The closer in time and space a dwelling is to one that has suffered a burglary, the greater the increased risk that it will experience a burglary. As time and space increase, the heightened risk of nearby dwellings decays (Bowers and Johnson 2005). These features of repeat and near-repeat burglaries have been drawn on in initiatives aiming to reduce burglary, for example, by targeting measures where and when they are most needed.

Measuring repeat victimisation poses a few challenges. For recorded crime one has already been mentioned: the dependency between incidents that are reported and recorded and the observation that rates of both fall far short of 100 percent.

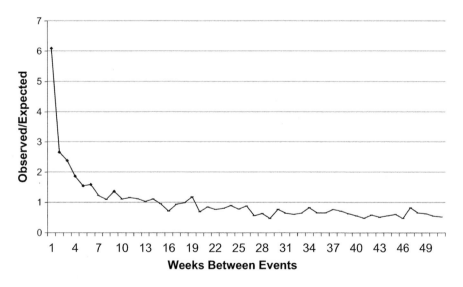

Fig. 1.7 The ratio of observed to expected repeat burglaries in Victoria, Australia (Source: Sagovsky and Johnson 2007, pp.1–26)

A further problem with recorded crime measurements of repeat incidents relates to recording practices. Inconsistencies in spelling and address format mean that repeats are liable to be missed, although the growth in use of gazetteers has ameliorated this particular problem. There is also a 'time window' problem that arises in particular when victimisation survey data are used (Pease and Farrell 2014). The time window problem refers to the variable time available for repeats over the 1-year recall period normally used. If a burglary occurred on the first day of the year in question, there are 364 days for one or more subsequent incidents, but if a burglary occurs on the last day of the recall year, there are no more days for further incidents. Therefore, if the same household is burgled once again in the following 12 months, it will be recorded as single victim, whereas there have been other incidents but outside the recall period. It should also be noted that victimisation surveys depend on accurate recall from respondents. Whilst well-designed surveys minimise the risk of flawed recall, the problem cannot be altogether avoided (Schneider 1981).

1.4 Burglary Trends

Levels of burglary rose in England and Wales between the first sweep of the CSEW until 1993, since which time rates have been falling steadily, as shown in Fig. 1.8. Similar falls in burglary rates in England and Wales have been found in many other countries also. They were part of a trend towards reducing crime rates for a range of offences (Tseloni et al. 2010). The falls were most marked for vehicle theft and burglary, which had accounted for a high proportion of all crime in the early 1990s, so much so that burglary and theft of and from vehicles accounted for 37 percent of all CSEW crime in 1993 but only 22 percent in 2013/2014.

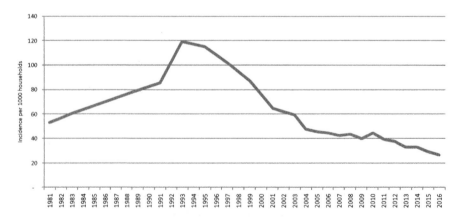

Fig. 1.8 Domestic burglary incidence rate per 1000 households trend in England and Wales 1981–2016, CSEW. Note: The figures up to 2000 refer to calendar years and those after that to financial years (see also Fig. 1.1) (Source: ONS 2017)

1.5 The Impact of Burglary

Burglary leads to financial costs, of course. For the victim, these relate to the cash and goods stolen, repairs to damage caused by the burglar and in some cases earnings where the burglary has led to time away from work. Financial costs are also borne by the criminal justice system, as offences are investigated, cases tried and penalties imposed on those found guilty. In addition to these financial costs, there are psychological harms, as victims may feel that their homes have been defiled and their privacy invaded. Further costs relate to insurance and security measures. Brand and Price (2000) estimated the overall costs of a domestic burglary, including both tangible and intangible items. They give the average cost for each burglary in a dwelling as £330 (security) plus £100 (insurance administration) plus £830 (property stolen and damaged), plus £550 (emotional and physical damage to victims), plus £40 (lost output), plus £4 (victim services), plus £490 (criminal justice, including the police). This gives a total per incident of £2300 at 2000 prices.

It might seem callous to put a monetary value on non-monetary harms and to lump together costs to victims, police and the private sector. Economists do this, however, to develop some common and inclusive unit of account to gauge the overall costs to society of harms, so that they can work out whether preventive efforts are worthwhile overall and to decide how to prioritise expenditure (Manning et al. 2016). Reliance on only the most obviously given a monetary value would ignore what matters most for many victims in terms of harms felt. Another non-monetary (often overlooked) cost of burglary relates to the impact on the environment; recent research has identified the carbon cost of burglary to be equal to 2750 miles of driving a car or to be precise 1154 kg of carbon dioxide emissions (Skudder et al. 2018).

1.6 Responding to Burglary

Research on burglary, its patterns and methods of prevention has made it increasingly clear that the police alone are poorly placed to prevent domestic burglary, albeit that they have crucial roles to play. In 2002/2003 it was found that across England and Wales, only 8 percent of burglaries were detected other than through an admission made by the offender in connection with other offences (so-called 'TICs' or offences taken into consideration) (Tilley and Burrows 2005, p. 3). It looked unlikely that the criminal justice system was going to provide sufficient deterrence to offenders whose chances of being caught were quite slim. As this book reveals, research has also shown that preventive interventions that do not rely on offender detention have been effective in reducing burglary.

As a high-volume, high-harm crime, burglary has attracted substantial attention by policymakers and practitioners, as later chapters in this book show. In the UK the police have been interested in the prevention of burglary as well as its detection. Voluntary groups, most notably Neighbourhood Watch and Victim Support, have been concerned with both prevention and with providing support to those who have experienced burglary. Government departments, notably the Home Office but others, have developed policies and programmes aimed at preventing burglary, as discussed in later chapters. Much of this has drawn on research of the kind outlined in this chapter.

1.7   Outline of the Remainder of the Book

Chapter 2 provides a brief history of national burglary security initiatives in England and Wales since the 1980s. It is written by Professors Gloria Laycock and Nick Tilley, the former having established and headed the Police Research Group in the UK Home Office. It highlights the importance of major projects that were found to have reduced domestic burglary as well as lessons learned from approaches that have produced disappointing results.

Chapter 3 focuses on domestic burglary offenders and their preferred targets. It draws on the literature on the modus operandi used by burglars, the attributes of dwellings and their immediate surroundings that make them attractive to burglars as reported by burglars. It also draws on Professor Armitage's wealth of experience in relation to the development and implementation of Secured by Design in England and Wales. It proposes that housing and residential neighbourhood layout, planning and engineering shape opportunities for domestic burglary. As a result, we should seek to eradicate poor design having considered offenders' choices of what to burgle and how to commit their offences.

Chapter 4 is co-authored with Professors Farrell and Pease, proponents of repeat victimisation theory and experts in crime prevention and the crime drop. The chapter is concerned with the kinds of security device used to try to prevent burglary and their effectiveness. It presents results from the first ever study to estimate the effectiveness of different anti-burglary security devices. The findings have been obtained from detailed analyses of the CSEW, which have explored the impact of security device combinations on the burglary risks faced by householders. It compares security device effectiveness against burglary with entry and attempted burglary separately and offers a new perspective in relation to the distinctive mechanisms performed by different devices.

Chapter 5 explores variations in burglary risk amongst different population socio-economic groups and the relationship of this to the security devices in place. It presents results from the first study to estimate the relationship between burglary and security amongst different population groups and areas. It uses data from the CSEW to estimate the risk that different households face in relation to burglary as well as the likelihood of having installed the most effective combinations of security devices. It identifies specific subgroups whose relatively high risks endure and the role of continuing security weaknesses in maintaining their vulnerability. This has implications for how we respond to victims. It also has implications in terms of housing policy and offering security upgrades to those most in need.

A practical example of security upgrades that had encouraging results in terms of burglary reductions follows on. Chapter 6 describes a demonstration project implemented in an English city, which has drawn on the research findings relating to burglary and security devices presented earlier in this book. It is co-authored by Dr. James Hunter who has a wealth of expertise in the geography of crime and health inequalities and Professor Andromachi Tseloni. The chapter outlines preliminary results from that project and is an example of implementing research evidence in practice.

Chapter 7 sets the findings of the research in England and Wales in the context of findings for other countries. It is written by Amandine Sourd and Dr. Vincent Delbecque from the National Observatory of Crime and Criminal Justice in France which produces and analyses the French national crime survey data: Cadre de Vie et Sécurité. The chapter suggests we should view burglary as a process of distinct but sequential events to be studied. As such different security devices as well as neighbourhood, property and household characteristics are found to influence burglary at different stages of completion.

Chapter 8 focuses on explanations for the unexpected and widespread drop in burglary that has occurred in England and Wales but also in many other countries. It briefly outlines and critically evaluates major explanations of the international crime drop, focusing in detail on their relevance to falls in domestic burglary. It goes on to suggest that household security improvements have played a major part in directly producing the crime drop, drawing extensively on analyses of successive sweeps of the CSEW.

Chapter 9 is concerned with research and how it can inform policy and practice. It is co-authored by Dr. Rebecca Thompson and Kate Algate, former Chief Executive of the Neighbourhood and Home Watch Network (England and Wales). The chapter presents their collective personal reflections on working together as part of an 18-month project involving a range of organisations from across the public and third sectors. They discuss some of the lessons learnt from trying to conduct impactful research.

Chapter 10 draws together the main themes discussed in the preceding chapters on burglary patterns and trends, which security devices work and how and burglary prevention lessons. It offers potential avenues for burglary prevention that householders, landlords and an array of public, voluntary and private sector organisations and businesses can take up. The conclusions of this book end unsurprisingly with suggestions for future research.

References

Bowers, K., & Johnson, S. (2005). Domestic burglary repeats and space-time clusters. *European Journal of Criminology, 2*(1), 67–92.

Brand, S., & Price, R. (2000). *The economic and social costs of crime*. London: Home Office.

Cantor, D., & Lynch, J. P. (2000). Self-report surveys as measures of crime and criminal justice. In *Criminal justice 2000; measurement and analysis of crime and justice* (Vol. 4, pp. 86–138). Washington, D.C.: United States Department of Justice.

Chenery, S., Ellingworth, D., Tseloni, A., & Pease, K. (1996). Crimes which repeat: Undigested evidence from the British Crime Survey 1992. *International Journal of Risk, Security and Crime Prevention, 1*, 207–216.

Dinisman, T., & Moroz, A. (2017). *Understanding victims of crime*. London: Victim Support.

Farrell, G. (1992). Multiple victimisation: Its extent and significance. *International Review of Victimology, 2*, 85–102.

Farrell, G., Tseloni, A., & Pease, K. (2005). Repeat victimization in the ICVS and NCVS. *Crime Prevention and Community Safety: An International Journal, 7*, 7–18.

Flatley, J. (2014). British crime survey. In G. Bruinsma & D. Weisburd (Editors in Chief). *Encyclopedia of Criminology and Criminal Justice (ECCJ)* (pp. 194–203). New York: Springer.

Forrester, D., Chatterton, M., Pease, K., & Brown, R. (1988). *The Kirkholt burglary prevention project, Rochdale* (Crime Prevention Unit Paper 13). London: Home Office.

Forrester, D., Frenz, S., O'Connell, M., & Pease, K. (1990). *The Kirkholt burglary prevention project: Phase II* (Crime Prevention Unit Paper 23). London: Home Office.

Her Majesty's Inspectorate of Constabulary (HMIC). (2014). *Crime-recording: Making the victim count. The final report of an inspection of crime data integrity in police forces in England and Wales*. London: HMIC.

Jansson, K. (2007). *British crime survey – Measuring crime for 25 years*. London: Home Office.

Lynn, P., & Elliot, D. (2000). *The British crime survey: A review of methodology*. London: National Centre for Social Research.

Manning, M., Johnson, S., Tilley, N., Wong, G., & Vorsina, M. (2016). *Economic analysis and efficiency in policing, criminal justice and crime reduction: What works?* London: Palgrave Macmillan.

Miethe, T. D. (2006). *Testing theories of criminality and victimisation in Seattle, 1960–1990* (ICPSR 9741). https://www.icpsr.umich.edu/icpsrweb/ICPSR/studies/9741. Accessed 30 May 2018.

Office for National Statistics (ONS). (2015, October). *User guide to crime statistics for England and Wales*. London: ONS.

Office for National Statistics (ONS). (2017). *Crime and justice: Figures on crime levels and trends for England and Wales based primarily on two sets of statistics: the Crime Survey for England and Wales (CSEW) and police recorded crime data.* https://www.ons.gov.uk/peoplepopulation-andcommunity/crimeandjustice. Accessed 29 May 2018.

Office for National Statistics/TNS UK Limited. (2015). *2017–18 Crime Survey for England and Wales questionnaire (from April 2017).* https://www.ons.gov.uk/file?uri=/peoplepopulation-andcommunity/crimeandjustice/methodologies/crimeandjusticemethodology/201718csewque stionnaire.pdf. Accessed 30 May 2018.

Pease, K. (1998). *Repeat victimisation: Taking stock* (Crime Detection and Prevention Series Paper No. 90). London: Home Office.

Pease, K., & Farrell, G. (2014). Multiple victims and super-targets. In G. Bruinsma & D. Weisburd (Editors in Chief), *Encyclopaedia of Criminology and Criminal Justice (ECCJ)* (pp. 3184–3190). New York: Springer.

Pease, K., & Tseloni, A. (2014). *Using modelling to predict and prevent victimisation, Springer-Brief criminology series*. New York: Springer.

Reiss, A. J. (1980). Victim proneness in repeat victimization by type of crime. In S. Fienberg & A. J. Reiss (Eds.), *Indicators of crime and criminal justice quantitative studies* (pp. 41–53). Washington, D.C.: Department of Justice.

Rountree, P. W., & Land, K. C. (1996). Burglary victimisation, perceptions of crime risk, and routine activities: A multilevel analysis across Seattle neighbourhoods and census tracts. *Journal of Research in Crime and Delinquency, 33*, 147–180.

Sagovsky, A., & Johnson, S. D. (2007). When does repeat burglary victimisation occur? *Australian & New Zealand Journal of Criminology, 40*(1), 1–26.

Schneider, A. (1981). Methodological problems in victim surveys and their implications for research in victimology. *The Journal of Criminal Law and Criminology, 72*, 818–838.

Simmons, J., et al. (2002). *Crime in England and Wales 2001/2002* (Home Office Statistical Bulletin 07/02). London: Home Office.

Skudder, H., Brunton-Smith, I., Tseloni, A., McInnes, A., Cole, J., Thompson, R., & Druckman, A. (2018). Can burglary prevention be low-carbon and effective? Investigating the environmental performance of burglary prevention measures. *Security Journal, 31*, 111–138.

Sparks, R., Genn, H., & Dodd, D. (1977). *Surveying victims.* Chichester: Wiley.

Tilley, N., & Burrows, J. (2005). *An overview of attrition patterns* (OLR 45/05). London: Home Office.

Tilley, N., & Tseloni, A. (2016). Choosing and using statistical sources in Criminology – What can the Crime Survey for England and Wales tell us? *Legal Information Management, 16*(2), 78–90. https://doi.org/10.1017/S1472669616000219.

Trickett, A., Osborn, D. R., Seymour, J., & Pease, K. (1992). What is different about high crime areas? *The British Journal of Criminology, 32*(1), 81–89.

Tseloni, A., & Pease, K. (2005). Population inequality: The case of repeat victimisation. *International Review of Victimology, 12*, 75–90.

Tseloni, A., Mailley, J., Farrell, G., & Tilley, N. (2010). The cross-national crime and repeat victimization trend for main crime categories: Multilevel modelling of the International Crime Victims Survey. *European Journal of Criminology, 7*(5), 375–394.

University of Lausanne (UNIL) (2018). *Key publications.* https://wp.unil.ch/icvs/key-publications/keypublications. Accessed 13 June 2018.

Van Dijk, J., & Tseloni, A. (2012). Global overview: International trends in victimisation and recorded crime. In J. van Dijk, A. Tseloni, & G. Farrell (Eds.), *The international crime drop: New directions in research* (pp. 11–36). Hampshire: Palgrave Macmillan.

Van Dijk, J., van Kesteren, J., & Smit, P. (2007). *Criminal victimisation in international perspective: Key findings from the 2004-2005 ICVS and EU ICS* (Onderzoek en beleid 257). Den Haag: Boom Juridische Uitgevers.

Van Dijk, J., Tseloni, A., & Farrell, G. (2012). Introduction. In J. van Dijk, A. Tseloni, & G. Farrell (Eds.), *The international crime drop: New directions in research* (pp. 1–8). Hampshire: Palgrave Macmillan.

Chapter 2
A Short History of the England and Wales National Burglary Security Initiatives

Gloria Laycock and Nick Tilley

Abbreviations

ALO	Architectural Liaison Officer
BCS	British Crime Survey
CDA	Crime and Disorder Act 1998
CPDA	Crime Prevention Design Advisor
CPTED	Crime Prevention through Environmental Design
CPU	Crime Prevention Unit
CRP	Crime Reduction Programme
CSEW	Crime Survey for England and Wales
DOCO	Designing Out Crime Officer
PEP	Priority Estates Project
RBI	Reducing Burglary Initiative
SCP	Situational crime prevention
SDP	Strategic Development Projects

2.1 Introduction

Domestic burglary (hereafter burglary) is a long-standing challenge to governments, police agencies and communities. Rates of burglary are typically measured through police statistics, but in 1981 the UK Government began funding the British Crime Survey (now the Crime Survey for England and Wales, CSEW) as a more accurate measure of offending, reflecting as it does the experience of victims, much of which is not reported to the police. Figure 2.1 below shows the rise and fall of burglary from 1981 to 2017 based on both police recorded crime and crime survey data. The rise in both police recorded crime and the results from the crime surveys were

G. Laycock (✉) · N. Tilley
Jill Dando Institute, Department of Security and Crime Science, University College London, London, UK
e-mail: g.laycock@ucl.ac.uk

© Springer Nature Switzerland AG 2018
A. Tseloni et al., *Reducing Burglary*,
https://doi.org/10.1007/978-3-319-99942-5_2

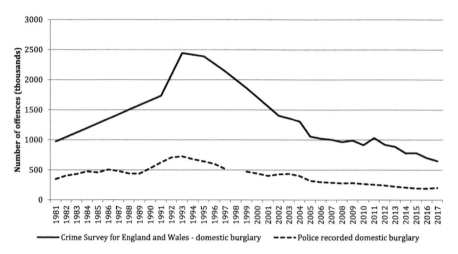

Fig. 2.1 Trends in Crime Survey for England and Wales and police recorded burglary, year ending December 1981 to year ending March 2017. Notes: new Home Office counting rules were introduced in April 1988; the National Crime Recording Standard was introduced in April 2002; police recorded crimes up till 1997 refer to the calendar year and from 1999 to the years ended 31 March; crime surveys were not conducted for 1982, 1984–1986, 1988–1990, 1992, 1994, 1996, 1998 and 2000–2001; and the figures for these are interpolated (Source: Crime Survey for England and Wales, Office for National Statistics and Police Recorded Crime)

particularly noticeable from 1981 to 1993. Victimisation surveys were also finding high levels of fear of burglary, especially amongst victims where psychological impacts were high (Hough 1984). This concentrated central government attention: something had to be done. In this chapter, we will be discussing some of the UK Government initiatives related to crime in general and to burglary in particular, which may have influenced these rates. Of course, it is not easy to demonstrate a causal link between what is done at national level and changes in rates of crime in a jurisdiction, but it is possible to discuss the plausibility of such relationships as is touched on in the remainder of this chapter.

The next section of this chapter outlines differing ways to prevent burglary and explains the approaches that came to be adopted in England and Wales from the 1980s. The chapter then describes ten significant programmes and initiatives, which included burglary prevention as a main concern, that have been implemented since the 1980s. Because many of these involved situational crime prevention (SCP), which has often been dismissed on the grounds that crime is merely displaced, we go on to review what is known about this from empirical research – basically (a) that assumptions that displacement will necessarily occur are contradicted by the evidence and (b) that situational measures often produce a 'diffusion of benefits' by which we mean preventive effects beyond their operational range. The chapter goes on to summarise the main lessons that can be taken from burglary reduction work since the 1980s in England and Wales, before a brief conclusion, which stresses that what emerges most clearly is the finding that SCP comprises a highly effective method for reducing burglary.

2.2 Background

Tackling burglary can be approached in several ways; in broad terms it can be prevented, disrupted, detected or prosecuted, and the sentences to which offenders are subject can be modified. Table 2.1 below lists possible activities under each of these headings, all of which, to varying degrees, have been influenced by national crime policies over the years.

Some of these activities were relatively neglected in the 1980s, and others were judged to be failing. For example, detection rates for burglary are typically low and have been for many years, although the practice of allowing offenders to have possibly undetected offences taken into consideration at the time of sentencing for a known offence artificially inflated the burglary clear-up figures for some forces (Tilley and Burrows 2005; Burrows et al. 2005). Now that this practice is no longer allowed, there is a significant apparent reduction in detection figures. The rates in the 12 months up to March 2014 (Home Office data, as reported by Her Majesty's Inspectorate of Constabulary and Fire and Rescue Services, at https://www.justice-inspectorates.gov.uk/hmicfrs/crime-and-policing-comparator/, accessed September 17th 2017), for example, ranged from a low of 8 percent in Wiltshire (a small rural force with a relatively high percentage of second home owners) to a high of 45 percent in the City of London (which has a very low resident population in a small geographical area). The figure for England and Wales over that period was 12 percent, but it should be remembered that this figure is derived by dividing the number of detected offences by the number recorded by the police not the (unknown) number that were committed. From the perspective of the offender, the odds of getting away with the offence are higher than these official clear-up figures would suggest. Offenders themselves seemed to have some appreciation of the low chances of detection. Bennett and Wright found that only 12 percent of the 128 burglars they interviewed 'believed that there was a chance of getting caught for any particular offence' (Bennett and Wright 1983, p. 186). They either thought there was no chance (34 percent) or refused to think about the chances (46 percent).

Table 2.1 Approaches to tacking burglary

Generic approach	Possible activities
Prevention	Community crime prevention, early intervention, situational crime prevention, high-profile policing and publicity campaigns, e.g. 'lock it or lose it', focused deterrence on known burglars
Disruption	Attention to stolen goods markets, local publicity announcing initiatives, stings, stakeouts, letters, etc. to known burglars, close monitoring post prison release
Detection	Rapid response to calls for service, door-to-door enquiries, hotspot policing, effective use of intelligence, surveillance
Prosecution and sentencing	Sentencing policy, restorative justice, offender treatment processes

There are obvious implications from the low detection rates for the efficacy of prosecution and sentencing as deterrents to burglary. As Beccaria reminds us, it is the probability of capture that is more salient than the potential sentence: most offenders do not expect to get caught for their offence. To make matters worse, the review of the effect of offender treatment by Lipton et al. was published in 1975, and an earlier article based on its supposed conclusions by Martinson (1974) led to the depressing (and overstated) conclusion that nothing worked (Sarre 1999). By the 1980s the time was ripe for a rethink on how to control crime, and especially how to reduce burglary and car crime, which in the UK were numerically high and also of high public concern, as measured by the fear of crime figures emerging from the crime surveys.

On the back of all this, the Home Office began to take an increasing interest in the potential of primary crime prevention (Brantingham and Faust 1976) as a crime control measure. Primary prevention refers to the prevention of crime events as against more traditional secondary prevention, which focuses on those at risk of becoming criminal or tertiary prevention, which focuses on preventing continuing criminal activity amongst those already involved in it. In particular, SCP (Clarke and Mayhew 1980), as it came to be known, was developed as a major plank of central government policy and was potentially associated with a number of significant changes in the crime figures as discussed elsewhere in this volume (Chaps. 6 and 8), although from the start there were concerns that the use of situational measures, such as security devices, might simply lead to displacement of some kind.

By the early 1980s, the UK Home Office had launched the British Crime Survey. It had also established the Crime Prevention Unit (CPU) in the Police Department. In addition to exercising its responsibility for preventive policing, the Unit oversaw nationwide publicity campaigns mainly directed at burglary and car crime, which were intended to encourage the public to take greater care and to protect themselves against potential victimisation. It also incorporated a resident research capacity, which initiated a research series of publication aimed at practitioners (Crime Prevention Unit Papers).

Perhaps most significantly the Home Office issued a guidance note in 1984 (Circular 8/84) to all police forces and local authorities in consultation with eight other government departments emphasising the need for agencies to work together to reduce crime. In doing so they were acknowledging that the police could not be expected to take sole responsibility for crime reduction. This multi-agency approach was substantially reinforced in 1990, with Circular 44/90 and then in 1992, when the incoming Labour Government drafted legislation requiring local services to work together. The Crime and Disorder Act (1998 CDA) was probably one of the most powerful signals that crime prevention needed to be given priority locally, and the requirement within this legislation that local plans be based on the analysis of crime figures reflected the influence of research on the developing agenda. The CDA required the police and local authorities to draft a local crime strategy reflecting local crime patterns and to agree this with the communities they served. This spawned action across a wide area and exposed the lack of skills available effectively to analyse crime data and interpret results.

Many of the initiatives encouraged by the CPU were familiar, but they came with a requirement to evaluate their effect and demonstrate reductions. Some of the more significant evaluations of initiatives funded by central government and others are described below.

2.3   Programmes

2.3.1   *Neighbourhood Watch*

Neighbourhood Watch: The first Neighbourhood Watch scheme in England and Wales was established in 1982 in Mollington in Cheshire following earlier initiatives in the USA. The movement grew very quickly, and there were 42,000 schemes by 1988 covering 2.5 million households (Husain 1988). This expansion resulted in part from the popular appeal of a movement that seemed to promise protection from offenders, especially burglars, a close working relationship with the police and the establishment of community control within neighbourhoods. Moreover, police enthusiasm was bolstered by the use of Neighbourhood Watch counts in an area as a performance indicator. Schemes were rather more easily established in relatively stable, relatively well-off low crime neighbourhoods than in relatively poor, relatively unstable high crime ones. The initial proposal for a local Neighbourhood Watch scheme generally came from the community rather than the police, thus favouring areas where there was already a sense of community. Neighbourhood watches typically included four main components: property marking, household surveys to identify ways in which security might be improved, mutual support and surveillance amongst neighbours and two-way communications with local police where suspicious activities were reported to the police and the police reported emerging local threats to Neighbourhood Watch members. From the police point of view, Neighbourhood Watch was seen to provide them with 'eyes and ears' in local areas. An early evaluation by Trevor Bennett (1990) produced disappointing findings. Before and after crime rates in two areas in London where Neighbourhood Watch schemes were introduced were compared with matched comparison areas with no Neighbourhood Watch. Neighbourhood Watch appeared to have no positive effects on crime levels. Recent research drawing on the CSEW suggests that Neighbourhood Watch schemes do now tend to spring up in *high crime* neighbourhoods, even where these do not provide otherwise favourable social circumstances (Brunton-Smith and Bullock 2018). Neighbourhood Watch continued to thrive, and a national Neighbourhood Watch organisation, which still functions, was established in 2007.

2.3.2 Property Marking

As mentioned above Neighbourhood Watch comes as a 'package' and includes encouragement to mark significant and potentially vulnerable items with an identifying mark using, for example, an ink which is invisible unless viewed under ultraviolet light. In the UK, choosing an identifying mark was easy as there is a national system of post codes, which together with the house number uniquely locates the property: a typical post code might look like AB1 2CD. A CPU evaluation of property marking was carried out by Laycock (1985) and showed a significant reduction in domestic burglary for those households in the scheme compared to those who were not, but it also raised questions as to causality. The most popular target of domestic burglary is cash – and cash cannot be property marked – which led to speculation about the mechanism through which property marking might have worked in the experimental area. Further analysis suggested that the observed reductions were more likely due to the intense local publicity associated with the launch and evaluation of the scheme (Laycock 1992).

2.3.3 Publicity Schemes

In the 1980s, there was substantial investment in national crime prevention publicity schemes. Evaluations of national publicity were also beginning to suggest effectiveness as a crime prevention measure, albeit over a long time period. For example, car owners were encouraged to 'lock it or lose it' when leaving vehicles. An early evaluation of such schemes was carried out by Riley and Mayhew (1980) who suggested four 'aims' of government-funded publicity schemes. These, and their potential effects, are set out in Diagram 2.1 below (taken from Laycock and Tilley 1995, p. 548).

As can be seen, there are two routes which might plausibly result in reductions in crime, one encouraging potential victims to protect themselves and one discouraging potential offenders from offending. The first relies on SCP and the second on deterrence. Riley and Mayhew concluded that changes in behaviour did not follow publicity campaigns, but importantly they did acknowledge that some campaigns led to increased knowledge and a change in attitude. This led them to conclude that short-term evaluations of the kind that they had carried out may be incapable of demonstrating longer-term effects. In relation to car crime, for example, Webb and Laycock (1992) suggested that drivers were increasingly locking their vehicles. Data from several studies over a 20-year period showed that the number of cars left with unlocked doors or trunks declined from 22 percent in 1971 to 4 percent in 1992. Of course, this trend will have been hugely reinforced by the increasing availability of central locking systems but at the time of these research studies central locking was not as common as it is today. We might conclude from this that although a single publicity campaign may have a small impact on behaviour in the context of

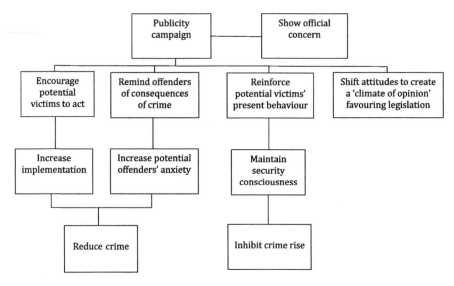

Diagram 2.1   Rationale for crime prevention publicity

crime prevention, persistent and relevant campaigns at a time of rising crime and associated public concern may have a longer-term effect.

In contrast, publicity at the local level has been shown to have an immediate but short-lived effect (Johnson and Bowers 2003). This was the case in the evaluation of property marking mentioned above, but Johnson and Bowers were also able to demonstrate 'anticipatory benefits' at the local level when publicity campaigns in the local newspaper were shown to lead to declines in domestic burglary even before an initiative was launched.

2.3.4   The Kirkholt Burglary Prevention Project

The Kirkholt Burglary Prevention Project has been the single most influential individual crime prevention initiative in the UK. Our concern here is with its immediate effect in reducing burglary, although its impact has been much broader. 'Kirkholt' is the name of the council estate in Rochdale where the eponymous project was run in the late 1980s. The estate included 2280 dwellings and had a very high rate of domestic burglary (around 25 burglaries per 100 households per annum at the start of the project). The brief for the action research team, including those from Manchester University and Greater Manchester Police, was simple: reduce burglary and tell us how you did it. The project followed a problem-solving approach. It began with extensive data collection and analysis. Recorded crime, interviews with local burglars, interviews with victims and their neighbours comprised the main sources, complemented with other information from local agencies and the census

(Forrester et al. 1988, 1990). Key findings related to the very high rates of repeat victimisation (echoing findings from the British Crime Survey), the frequency of fuel prepayment meters as targets of theft (49 percent of burglaries) and the low distances travelled by prolific local burglars to commit their crimes. Findings were presented to a problem-solving group made up of representatives of the North West Electricity Board, Rochdale Victims Support Scheme, Rochdale Education Authority, Rochdale Borough Housing Department, Greater Manchester Police, Greater Manchester Probation Service, the Association of British Insurers and the Home Office Crime Prevention Unit. A range of responses were put in place including upgrading security, removal of prepayment meters and the establishment of cocoon-type neighbourhood watches around dwellings that had suffered a burglary (basically half a dozen properties either side of the burgled household). The most important decision was to focus efforts on victims. Based on analysis of prior burglary patterns, they were an obvious high-risk group through which crime prevention could be drip fed. The results were impressive. Over a 1-year period – from January to September 1986 to January to September 1987 – the total number of burglaries fell from 316 to 147, when the level in the rest of the police area increased slightly. The falls in burglary level were not matched by changes in other crime types. Moreover, the rates of repeat burglary dropped more than would be expected simply as a result of the overall fall in the number of burglaries. The Kirkholt project had major implications for policy and practice. In general terms, it showed what could be achieved by focusing on repeat incidents. This is especially significant given findings that higher than expected rates of repeat incidents are consistently found across countries and across crime types. The Kirkholt project directly inspired efforts at its replication and wider application in the UK.

2.3.5 Safer Cities

The reports of findings from Kirkholt were published at about the time the Safer Cities programme was starting. Safer Cities was a Home Office programme that ran from 1988 to 1995 in local authorities in England and Wales. Each of the first 20 cities included in the programme had a small staff and £250,000 per annum to spend on projects. Against the backdrop of concern about burglary, many of these cities made efforts to emulate Kirkholt's achievements on high crime estates, which at the time were mostly owned and run by local councils. In the event the uses made of the Kirkholt experience were diverse (Tilley 1993). The resources put into the projects varied widely in terms not only of cash but also of agency involvement; the effort put into research on local patterns was included in some but not other projects; repeat victimisation was a focus in some but not others; fuel prepayment meters were removed in some places but not others; cocoon home watch was included in some places but not others, and problem-solving animated some but not other projects. Moreover, initial burglary rates differed widely from one another and from Kirkholt. In other words, the replicated components varied widely as did the initial conditions in which the initiative developed. Unsurprisingly, in view of this, the outcomes also varied widely. The explanation for this diversity lies not in incompetence on the part of those attempting the replications but rather in uncertainty at the time over the key components that need to be reproduced. These have become clearer in ensuing work, building explicitly on the Kirkholt work in follow-up action projects.

The Safer Cities interest in domestic burglary prevention led to a range of projects using diverse methods, only a fraction of which drew their inspiration from Kirkholt (Tilley and Webb 1994). Safer Cities burglary reduction projects involved the use of diverse interventions that can be broadly classified as varieties of target hardening only and target hardening plus varieties of community mobilisation, as shown in Table 2.2. All were put in place in relatively poor neighbourhoods. What is evident is (a) the broad use of SCP measures and (b) diverse forms of targeting, for example, victims, those at risk, the vulnerable, tenants and those living in high crime neighbourhoods.

One project adopted a similar problem-solving approach to Kirkholt but devised a somewhat different strategy. This was run in Wolverhampton on the Lunt, a sink estate of 776 dwellings (72 percent council owned), where analysis suggested that the poor conditions and high crime vulnerability on the estate had led stronger members of the community to move elsewhere leaving an increasingly vulnerable population alongside local offenders. This produced a spiral of decline, with high levels of crime and incivility, and a transient population whose members were unable to defend themselves against local predators. The decline had occurred in spite of earlier heavy investments in housing improvements through Estate Action. The strategy adopted on the Lunt was to target harden many of the dwellings, including the addition of rear fences to 60 of them, crack down on prolific offenders, improve the appearance of the estate with clean-ups and physical improvements and try to build collective efficacy using various community-building exercises. In effect, it was hoped that this suite of measures would reverse the spiral of decline and in doing so also reduce domestic burglary. Over the 3-year period of the initiative (1988–1991), burglary fell by 43 percent, while national rates were rising (see Tilley and Webb 1994, p. 54).

A large-scale study across some 300 Safer Cities schemes in 11 of the 20 cities that had focused on domestic burglary compared trends with 8 other cities, drawing on both recorded crime and crime survey data (Ekblom et al. 1996). Altogether 400 high crime neighbourhoods were included in the study across Safer Cities and their comparators. Findings concluded that overall the Safer Cities schemes had reduced rates of burglary and had done so cost-effectively. The cost of each prevented burglary in high crime areas was around £300 and in lower crime areas around £900, whereas the cost per burglary to the state and the victim was around £1100. Overall, Ekblom et al. estimated that some 56,000 burglaries had been prevented. Effects grew with the intensity of the interventions. Physical security seemed to work independently, but community orientated work was ineffective on its own. Mixed strategies worked best. This accords with the success found specifically on the Lunt estate in Wolverhampton.

Table 2.2 Safer Cities burglary reduction schemes: contexts and measures implemented (Tilley and Webb 1994)

	Birmingham	Bradford	Hull	Nottingham Meadows	Nottingham St Ann's	Rochdale Back 'O' The Moss	Rochdale Belfield	Rochdale Wardle-worth	Sunderland	Tower Hamlets	Wolver-hampton
Context											
Recorded burglary rate before	21%	9%	5%	4.2%	6.3%	11.4%[a]	12%[a]	13.9%[a]	6.8%	6.8%[a]	12%
No of households	175	835	2403	3936	9311	700	668	1100	865	10,004	776
Tenure pattern	Local authority	Local authority	Local authority	Mixed	Mixed	Local authority	Local authority	Private	Housing association	Local authority	Local authority
Measures adopted											
Target hardening victims				✓	✓					✓	
Target hardening vulnerable				✓						✓	
Target hardening hot spots					✓	✓	✓	✓			
Target hardening area	✓	✓	✓						✓		✓
Property marking	✓	✓	✓	✓	✓	✓	✓	✓			*[b]
Publicity (continuous)	✓				✓	✓	✓	✓			
Advice/support to victims					✓	✓	✓	✓			✓
Neighbourhood Watch type groups	✓				✓	✓	✓	*			*

[a]Extrapolated to a yearly rate for 6 months prior to scheme implementation in Rochdale-Belfield and Back 'O' The Moss and 3 months for Rochdale Wardleworth and 3 months for Tower Hamlets

[b]Measures marked '*' were planned but were little implemented

2.3.6 Huddersfield and the 'Olympic' Model

'Biting Back' was a project that operated in Huddersfield in the 1990s (Anderson et al. 1995; Chenery et al. 1997). It ran at much the same time as the Safer Cities burglary prevention initiatives but had an explicit focus on repeat victimisation. It was also designed to operate in a whole policing area rather than in a particular high crime community. The 'Olympic' model refers to tiered levels of intervention depending on how often a target had been victimised, going from 'bronze' through 'silver' to 'gold'. Table 2.3 shows what this strategy looked like in practice. The bronze response to the first incident focused on reducing vulnerability by rapid repairs, victim support, increasing security, and the establishment of a cocoon home watch. If a further incident occurred, a silver response would be added, including greater police attention and some proactive efforts at detection including the temporary installation of silent alarms that would sound in a control room and call for a fast response. If victimisation persisted, the response was ratcheted up to gold level which included increased use of measures that would help detect offenders. The Olympic model aimed to direct more intensive police efforts at detection where they promised most dividends both in terms of catching offenders and in dealing with those most likely to be prolific. Where one off offences occurred, or if potential repeat offenders were more easily deterred, the measures put in place called on far fewer resources.

The Olympic model was based on research on the ways in which repeat victimisation is produced. There have been two hypotheses about the source of repeats (Pease 1998). One is that repeats result from variations in the attractiveness of dwellings that could be chosen as targets by any passing offender – this is sometimes referred to as an explanation in terms of 'population heterogeneiry', 'risk heterogeneity' or 'flag'. A second is that an offender having targeted a property would return having learned better how to commit the crime and what there was to

Table 2.3 Biting Back responses to burglary

Bronze	Silver	Gold
Victim letter, UV marker pen and crime prevention advice	Visit from CPO; Search warrant	Visit from CPO; priority AFR
Discount vouchers on security equipment	Installation of Tunstall Telecom monitored alarm	Installation of high-tech equipment, i.e. covert cameras and alarms
Informants check	Police watch visits (twice weekly)	Police watch visits (daily)
Early check on known outlets; Targeting of offenders	Security equipment loan	Index solutions
Loan of temporary equipment (alarms, timer switches, dummy alarms)		Tracker (installed in moveable household items, e.g. boiler)
Cocoon watch; rapid repairs; security uprating		

steal, or they would tell other offenders, or they would leave the property in a more vulnerable state than it had been beforehand – these types of explanation are sometimes described in terms of 'state dependency', 'event dependency' or 'boost'. There is no inconsistency between these two explanations. It is possible that both have merit; both may operate individually or in conjunction with one another (Tseloni and Pease 2003). In the event, the time course of repeat victimisation (the tendency of repeat incidents to occur in quick succession), together with offender reports of their decisions to return to the same target, suggests that event dependency/boost typically plays a big part in repeat burglary patterns and that those who repeat tend to be the more prolific offenders. For this reason, switching attention towards proactive efforts at detection following repeat incidents made sense.

The Huddersfield project was important in developing a *strategy* for dealing with repeat burglaries by increasing the intensity of response and switching attention from prevention to detection with successive incidents. Recorded repeats fell (from an average of 0.23 prior burglaries per burglary in the year preceding the start of the initiative to 0.14 when the project had been in place), and the overall number of recorded domestic burglaries dropped by 30 percent.

2.3.7 Crime Reduction Programme

The Crime Reduction Programme (CRP) was originally envisaged as a 10-year £250 million initiative in England and Wales to implement evidence-based crime reduction and at the same time improve the evidence base on what works. The money was released following a review of what was at the time known from research, which included findings from both the Kirkholt project and Safer Cities (Goldblatt and Lewis 1998). In the event the programme only ran for 3 years from 1998 and has been widely deemed a failure due to the design, development and implementation problems at both national and local levels (Homel et al. 2004). One thread of the CRP comprised the Reducing Burglary Initiative (RBI). Twenty-five million pounds covering some two million households was spent on 240 projects, 63 of which were Phase 1 'Strategic Development Projects' (SDPs) intended to pilot innovative responses that would extend the evidence base of what works and what is cost-effective in burglary reduction. A condition for funding the SDPs was that they be independently evaluated for impact and cost-effectiveness. Projects were run in relatively high crime areas of 3000 to 5000 households – to begin with they had to experience at least twice the national burglary rate to be eligible for funding, although this was later reduced to 1.5 times the national rate. Funding was to be allocated on a competitive basis, where bidders had to show that there was a problem and that the proposed measures were well tailored to its analysis.

The RBI, however, allowed insufficient time for the development of well thought through innovative ways of reducing local burglary problems in light of the particular issues there. Moreover, the evaluations were tricky and findings controversial (Hope 2004). Nevertheless, taking 55 of the 63 SDPs, Kodz and Pease (2003) found that overall levels of burglary were reduced by 20 percent (7 percent compared to changes in reference areas, defined in terms of the wider areas in which the project was lodged), in the 21 months after the launch of the RBI. In a few cases, the falls were dramatic although in 15 projects relative burglary rates increased. These findings, however, suggest overall that efforts at burglary prevention in relatively high burglary areas can have positive effects even when not well worked through. The kinds of intervention included in projects showing steep declines in burglary included publicity campaigns, target hardening the homes of victims of burglary and those deemed vulnerable, home watch, alley gates (see below for more on these), diversionary activities, police crackdowns and drug arrest referrals although in all cases mixed packages were used, which meant that working out the contribution of individual components and knowing what was dispensable was not possible.

2.3.8 Design Against Crime

The notions that properties vary in vulnerability to burglary due to their security and that the design of housing developments influences burglary risk have a long history. Crime Prevention through Environmental Design (CPTED) was envisaged by C Ray Jeffery in 1977. Oscar Newman published his book on defensible space in 1972. Crime prevention has been the stated core mission of the police in England and Wales since 1829. Police Crime Prevention Officers were first mooted for all police services in 1965. More recently Architectural Liaison Officers (ALOs), Designing Out Crime Officers (DOCOs) and Crime Prevention Design Advisors (CPDAs) have become specialists within the police. Design against crime addresses much more than burglary, but includes it. 'Secured by Design' is a British police initiative set up in 1989 to inform and promote design-based crime prevention (http://www.securedbydesign.com). It sets security standards for accreditation of doors and door locks, windows and window locks, intruder alarms and fences. Chapter 3 deals in detail with Secured by Design and its effects on burglary (see also Armitage (2013)).

2.3.9 Alley Gating

Traditional working class homes in the UK are often terraced and 'back to back' with a wide alley running between the rear of the properties. This allows access to service such as rubbish collection or, in the old days particularly, the delivery of coal. Unfortunately, these alleys have also been used as access points for burglars and as a convenient place to dump rubbish and other unwanted items. As a means of preventing burglary and other more minor nuisance, some local authorities adopted the practice of installing gates at each end of the alleyway. The effect of these 'alley gates' on burglary was evaluated by Bowers et al. (2004) and showed a 37 percent reduction in burglary relative to a suitable comparison area. They were also able to demonstrate that there was a diffusion of benefits to surrounding areas and that the initiative was cost beneficial with a saving of £1.86 for every pound spent.

2.3.10   Estate Action, Single Regeneration Budget and Priority Estates

Some areas of local authority housing experienced multiple problems. Crime problems were mixed in with poor housing, transient populations, poor educational performance and high levels of unemployment. The estates were very unattractive. As Foster and Hope (1993, p. 6) note, 'By the mid-1970s, the (then) Department of the Environment had become concerned about a growing number of unpopular or difficult-to-let estates, sometimes ones that had been completed only relatively recently'. The physical and social fabrics were both poor and produced spirals of decline of the kind as described earlier in the Lunt Estate (see Sect. 2.3.5). Various government initiatives attempted to address these problems. Crime was included but was not the exclusive or even major focus. Some of the estates were laid out on Radburn principles (Shaftoe and Read 2005). These were criss-crossed with paths to the fronts of houses, natural surveillance was poor, and the physical security of buildings was often weak. In retrospect, this design has been widely seen as a disaster. There were efforts to remodel estates and repair the physical fabric. Likewise, there were efforts to empower communities to manage and take control of their estates. Concerns with security were swept into broad efforts to improve the lot of residents. There was substantial expenditure. Foster and Hope (1993) focused on Priority Estates Projects (PEP), using two case studies, one in London the other in Hull. Foster and Hope note that crime prevention was not a specific task of the model, but nevertheless there were reasons to believe that it might produce that outcome. They describe four means by which this might happen: creating better dwelling security, halting the spiral of deterioration and thereby reducing signs of disorder, investing in the estate to give residents a more positive view of it and hence a greater stake in their community and increasing informal community control through increased surveillance and supervision. Thus, improvements in physical security were included in an initiative that did more than that and also had broader aims than the prevention of crime.

Although details vary in important ways by funding stream, the point we wish to make here is that government initiatives including upgrades in security relevant to burglary were occurring separately from programmes whose primary or exclusive focus was on the prevention of crime in general or of burglary in particular. Moreover, in relation to this, Foster and Hope (p. 11) state that 'research evidence tends to suggest that security and design changes alone may have less impact than when they act in conjunction with other aspects of social life, which will help to promulgate greater community control over the residential environment'. One of the main findings from Foster and Hope's study related to the challenges in successfully implementing PEP, in terms both of the way the measures were applied and of the nature of the estates themselves. In terms of the application of the measures, Foster and Hope wryly remark that in London, 'many of the elements of the PEP model were actually *better* implemented on the control estate…than they were on the experimental estate, despite the attentions of PEP' (p. 84). In terms of the nature of the estates, increasing population turnover and heterogeneity made the stimulation of community controls especially challenging. Having said that, comparing changes in burglary prevalence and incidence on the targeted estate with changes on the control estate in Hull showed the former to have performed better in statistically significant terms at the 0.05 level (the effect was largely a function of substantial increases in burglary on the control estate). It is clear that crime in general and burglary in particular were part of a wider range of concerns within broader government initiatives targeting impoverished communities.

2.3.11   *Improved Street Lighting*

Street lighting has what academics call 'face validity'; it 'makes sense' to assume that the brighter the lighting the less crime. Research on lighting and crime appeared to support this proposition. For example, Painter and Farrington (1997, 1999a, b) carried out a series of studies on the effect of street lighting improvements in targeted areas of towns in the UK. These showed significant reductions in crime and fear of crime, a diffusion of benefits and cost-effectiveness. These results led to pressure on central government from the lighting industry to press local authorities for greater investment in improved lighting as a crime reduction measure. In response, the Home Office funded a project by Atkins et al. (1991) who studied the effects on crime of an upgrade in street lighting across the whole of the Borough of Wandsworth in South London, where the improvement programme was needed for a variety of reasons including the possible reduction of crime. Atkins et al. were, in effect, testing what might happen if the requests to promote street lighting across the board were to be encouraged as requested by the industry. Unsurprisingly the results did not show a significant reduction in night-time crime (which was the focus of the evaluation). A subsequent discussion of this study, and a careful analysis of the reported results by Pease (1999), suggests that although Atkins et al. reported that the lighting improvements did not reduce crime across the Borough, this was partly because they failed to recognise 'that crime is highly localised spatially and that an overall upgrading is a scattergun approach to a series of localised problems'. But this was precisely what Atkins et al. were asked to test. They were asked to do so in response to the requests by the lighting industry to invest millions of pounds of public money in untargeted street lighting improvements.

We can conclude therefore that as Welsh and Farrington (2008) say in their systematic review, '… improved street lighting significantly reduces crime, …' (2008,

p. 3). However, we need to add that the evidence for this tends to come from focused studies and that, as Atkins et al. show, untargeted street lighting improvements across wide areas do not show similar effects (see also Steinbach et al. (2015)).

2.4 The Vexed Question of Displacement

Much of the research described in the chapter so far has been about SCP, reflecting the Home Office interests from around the mid-1970s. One of the 'Achilles heels' of SCP was displacement: the notion that suppressing crime in one area would simply lead offenders to do something else, somewhere else or at some other time. Following Repetto (1976), Barr and Pease (1990) outline six ways in which crimes might be displaced following an intervention as shown in Table 2.4.

Although displacement is seen as a negative consequence of crime prevention, it can be interpreted as a positive in a number of respects. First if there is a switch to another crime type, for example, that new offence may be less serious than that which was prevented. Reductions in burglary might result in increases in attempted burglary. Geographic displacement might also occur following a burglary initiative, but its extent may still allow for a net reduction. This is summarised by Guerette and Bowers (2009) who say that '..."benign" displacement could occur when the displacement is of lower volume, results in less harm, or is less severe' (2009, p. 1335). They also describe the diffusion of benefits which may follow the introduction of a crime prevention scheme. This was first described by Clarke and Weisburd (1994) and is not uncommonly found following a crime prevention scheme when the area surrounding the experimental area also enjoys a reduction in offending.

The conclusion drawn by Guerette and Bowers, from their substantial review of the relevant research, is that there is continued support for the view that crime does not simply displace following a crime prevention initiative; indeed displacement is the exception rather than the rule. Insofar as it does happen, it is often, as they term it, 'benign', and a diffusion of benefits is sometimes more likely.

Table 2.4 Types of displacement following Barr and Pease (1990)

Displacement type	Description
Time	Crimes move to a different time
Geography	There is a shift from one place to another
Target	A different victim is targeted with the same crime
Crime type	Offenders switch to a different crime
Method	The modus operandi changes to reflect the intervention
Offender	New offenders replace old offenders who may have been arrested or deterred

2.5 Lessons Learned

In the previous section, we described a range of initiatives, many of which were research based, that were introduced in England and Wales from around the early 1980s. The chapter is about 'national burglary security initiatives', and although many of the projects that have been described were local in application, the vast majority were conceived, funded and managed by central government – in this sense they were national. In this final section, we look at the lessons that we feel can be drawn from this experience and briefly consider the prospects for the influence of research on policy and practice.

2.5.1 What Worked?

One of the most important lessons from the experience recounted above is that government-funded research can and did affect both policy and practice. This was achieved through a systematic process of research and development over a prolonged period, facilitated by the structural arrangements for linking research, policy and practice in England and Wales (Laycock and Clarke 2001). That said there were a couple of key contributions which singly shifted the discourse and perspectives on crime control. The first was the introduction of the British Crime Survey (BCS) in 1981, which quantified the sometimes large gap between police reported/recorded crime and crime experienced by victims, which may not have been reported or recorded, provided opportunities to assess the public's view of the police, enabled the more accurate monitoring of changes in crime rates over time, illustrated the extent of repeat victimisation and provided a measure of public fear of crime.

Another example of an influential initiative was the Kirkholt project, as noted above. This was important in four senses:

(a) Although not the first piece of work to highlight repeat victimisation (see Farrell (1992) for a comprehensive review), the Kirkholt project kick-started a programme of applied work looking at the extent to which repeat victimisation occurs and might act as the focus of prevention (Laycock and Clarke 2001). This programme has demonstrated the wide range of offences which are marked by repeat victimisation. Clearly the quintessential example is domestic violence, but there are other crimes ranging from criminal damage to attempted murder.

(b) It adopted a problem-solving approach, where (i) the specific problem was identified, (ii) analyses were undertaken and (iii) a strategy developed out of the analysis and then (iv) the outcomes were evaluated systematically. This approach had been advocated by Herman Goldstein for policing more generally (Goldstein 1979, 1990) and by Paul Ekblom for crime prevention specifically (Ekblom 1988). Kirkholt provided an outstanding case study for an approach to dealing with crime problems, which is now being adopted widely both for reducing burglary and for dealing with other crime problems.

(c) It involved a wide range of security-focused interventions. Although the specific focus on repeat victimisation was a crucial element of the initiative, there were diverse specific measures that related in part to resources that were available and in part to the specific needs of the housing on the Kirkholt estate. There are particular measures that may be effective on their own (e.g. alley gating), but it is becoming clearer that crime prevention measures tend to produce their effects interactively. In relation to domestic burglary (and car theft), combinations of measures have been found to tend to be more effective than would be expected from the sum of their individual effects (Farrell et al. 2014; Tseloni et al. 2017).

(d) It led to further research and development aimed at better understanding the ways in which burglars operate and how they can be prevented from doing so. Research has shown that following the burglary of a particular home, the next burglary can be predicted within an area small enough for the police plausibly to patrol over a relatively short time frame. This is possible because it is not only the originally burgled home that is at heightened risk following a burglary but also their immediate neighbours. In trying to understand this process, it is possible to draw an analogy with the way in which animals forage for food – they 'hit' an area to graze but then move on before predators attack them. Researchers speculated that similar behaviour might be exhibited by burglars who burgle a relatively small area before moving on (see Johnson and Bowers (2004), Johnson et al. (2008, 2009) and Johnson (2014)). The approach is now known as 'predictive policing' and has itself spawned a series of research studies across several jurisdictions. In retrospect, it may be that the cocoon home watch introduced in Kirkholt alerted those living close to victims of the risk of burglary and thereby motivated precautionary behaviour.

Other research is more diffuse in its effects, eventually affecting ways in which we now think about crime and how to prevent it. The work on SCP was carried out over several years, but its effect has been of major significance in crime reduction. Clarke published a seminal article in 1983 entitled *Situational Crime Prevention: Its Theoretical Basis and Practical Scope* in which he argued that the overemphasis on the characteristics of the criminal rather than the situational characteristics of the crime was poorly serving crime control efforts. Quite apart from the fact that it had proven extremely difficult to 'treat' offenders and reduce offending rates, researchers had for many years been stressing the major role played by the environment in determining behaviour. Ross and Nisbett (1991), for example, described the fundamental attribution error, which is to say that we interpret the behaviour of others as a function of their personality but attribute our own behaviour to the situation in which we find ourselves. Applying this to crime we can see why there is such a concentration on the characteristics of the offender at the expense of the situation. Just remind yourself of something you have done of which you might be ashamed or embarrassed – was it because of the situation?! Clarke referred to this in his acceptance speech as cowinner (with Pat Mayhew, a long-time collaborator), of the 2015 Stockholm Prize for Criminology, and indeed the subsequent publication (Clarke 2016) was called *Criminology and the Fundamental Attribution Error*.

2.5.2 What Didn't Work

The problem-solving approach takes time, skill and care. It also requires attention to detail. Evaluation is technically complex and needs to be bespoke to what was done. The Crime Reduction Programme and the Reducing Burglary Initiative within it were too hurried in terms of the development of proposals, assessment of proposals, funding decisions and evaluations of their effectiveness. It also lacked the personnel competent to deliver it. In retrospect, much money was wasted. Learning important lessons has followed only from specific well-targeted initiatives with evaluations focused on expected outcome patterns as occurred in the Kirkholt and Huddersfield initiatives and the alley gating project in Merseyside. Encouraging the adoption of problem-solving in policing was maintained throughout the period, and at times it appeared to be going well. For example, in 1999 the Tilley Award was launched, following the earlier Goldstein model in the USA which began in 1993. These schemes are intended to encourage and celebrate problem-solving projects with a national award. The Tilley Award has not been run from the Home Office since 2012 (although it is to be relaunched in 2019), and the Federal Government in the USA stopped supporting the Goldstein Award in 2013, although it continues with funding from other sources.

With the fall in police budgets, the pressure on delivery and the move to evidence-based policing, there is renewed interest in problem-solving, but it remains subject to political whims and fancies both at national level and within police agencies where it is still heavily dependent upon the interests of particular senior officers. We have not yet learned how best to embed the ideas and methodologies from research into service delivery at local level.

2.5.3 The Importance of Context

Experience over the past 40 years confirms the importance of the perhaps obvious observation that burglary prevention measures are always introduced in specific contexts. These relate to neighbourhood characteristics, underlying crime rates, street layout, criminal justice practices and other programmes that may complement or cut across the measures aiming to reduce burglary, housing attributes, levels of wealth, quality of implementation, etc. Where a single measure is introduced, its potential to reduce crime depends on other conditions that are in place. This has been found to be the case, for example, in relation to burglar alarms, property marking, lighting improvements, Neighbourhood Watch (itself a blend of measures) and alley gating. Furthermore, burglary prevention initiatives rarely include a single measure. Worse still, conditions are apt to vary not only by place but also by time. All this means that we are unlikely to come across magic bullets that 'work' unconditionally. Specific measures generally produce their effects (or not) according to context – a vital element in the 'what works' conversation. The general basis for this across social programmes of all types, alongside its significance for evaluation, is described in Pawson and Tilley (1997).

The ubiquity of contextual contingency has posed a challenge for those who would like to produce catalogues of what works and what doesn't do so at the level of the individual intervention. Those attempting this are reduced to looking across disparate evaluations to try to find a net effect, where any apparent net effect will simply reflect the distribution of contexts in which measures have so far been tried. The College of Policing has developed a crime reduction 'toolkit' for national use that includes summaries of reviews of research that cover a host of crime prevention measures, including some such as street lighting upgrades, alley gating and Neighbourhood Watch that are highly relevant to domestic burglary. The College of Policing toolkit tries to summarise research findings speaking to the effects found as well as the context needed for the relevant preventive mechanisms to be activated, issues of implementation and the economic costs of the intervention and the returns that have been achieved to date or can be expected (see http://whatworks.college.police.uk/toolkit/Pages/Toolkit.aspx).

2.5.4   Strategy

What national initiatives have brought out are broad frameworks and strategies that can be applied to burglary. There are three of these and they are linked to one another. The first relates to repeat victimisation, to which we have already referred. Those victimised are at greater risk of a subsequent crime, and this heightened risk fades quite quickly. This ubiquitous pattern provides a basis for targeting preventive efforts both in the interests of effectiveness – it gets the preventive grease to the squeak – and in the interests of distributive justice where it counters the tendency of crimes to concentrate on the few who therefore suffer disproportionately. The second relates to problem-solving to which we have again referred above. In determining what is possible and what is promising in the face of any situation where burglary is a problem, an analysis of the conditions in which a specific problem arises may suggest particular interventions, for example, where burglars access properties in a high crime neighbourhood by using rear alleys that would suggest that gating access to them might be a promising intervention. Third, the successes of SCP suggest that attention to near causes (without having to tackle so-called 'root causes' of criminality) can be sufficient to prevent many offences without leading to displacement. A whole raft of interventions that increase perceived risks to burglars, perceived effort to commit burglary or reduce expected rewards from burglary is available.

Where one specific measure (again such as alley gating) may not be relevant in one high burglary area, it is likely that there are other possibilities.

More promising than efforts to find specific magic bullets are broad strategies that speak to the mechanisms that can be activated to reduce burglary and the kinds of condition needed for the activation. In general, the less restrictive the contextual conditions, the better.

2.6 Conclusions

Readers will recall that the impetus for the investment in SCP grew from the pessimistic conclusion of the 1970s that 'nothing worked'. As we now know, that was an overstated conclusion, and indeed more recent systematic reviews are suggesting that lots of things work sometimes. In the most recent and probably most substantial review of this type, Weisburd et al. (2017) conclude that across seven broad areas of crime prevention and rehabilitation,[1] and drawing on 118 systematic reviews, there is 'persuasive evidence of the effectiveness of programs, policies and practices...'. Interestingly SCP was the least studied area of the 7 that they reviewed with the number of relevant systematic reviews ranging from 33 for developmental prevention covering 1580 primary studies to 7 for situational prevention which drew on 153 studies. And yet based on what has been observed in relation to the crime drop, situational prevention would appear to have been the most effective. There is, therefore, a case to be made that more investment in primary research, studying the effects of situational measures on crime, might be a priority.

References

Anderson, D., Chenery, S., & Pease, K. (1995). *Biting back: Tackling repeat burglary and car crime* (pp. 1–57). London: Home Office Police Research Group.

Armitage, R. (2013). *Crime prevention through housing design*. Basingstoke: Palgrave Macmillan.

Atkins, S., Husain, S., & Storey, A. (1991). *The influence of street lighting on crime and fear of crime* (Crime Prevention Unit Paper No. 28). London: Home Office.

Barr, R., & Pease, K. (1990). Crime placement, displacement and deflection. In M. H. Tonry & N. Morris (Eds.), *Crime and justice: A review of research* (Vol. 12, pp. 277–318). Chicago: University of Chicago Press.

Bennett, T. (1990). *Evaluating neighbourhood watch*. Aldershot: Gower.

Bennett, T., & Wright, R. (1983). Constraints to burglary: The offender's perspective. In R. Clarke & T. Hope (Eds.), *Coping with burglary* (pp. 181–200). Dordrecht: Kluwer.

Bowers, K. J., Johnson, S. D., & Hirschfield, A. F. G. (2004). Closing off opportunities for crime: An evaluation of alley-gating. *European Journal on Criminal Policy and Research, 8*(4), 285–308.

---

[1] These were developmental prevention; community prevention; situational prevention; policing, sentencing and deterrence; correctional interventions; drug treatment interventions.

Brantingham, P. J., & Faust, F. L. (1976). A conceptual model of crime prevention. *Crime & Delinquency, 22*, 284–296.

Brunton-Smith, I., & Bullock, K. (2018). Patterns and drivers of co-production in Neighbourhood Watch in England and Wales: From neo-liberalism to new localism. *British Journal of Criminology,* https://doi.org/10.1093/bjc/azy012.

Burrows, J., Hopkins, M. Robinson, A. Speed, M., Tilley, N. (2005). Understanding the attrition process in volume crime investigation (Home Office Research Study 295). London: Home Office.

Chenery, S., Holt, J., & Pease, K. (1997). *Biting back II: Reducing repeat victimisation in Huddersfield* (Vol. 2). London: Home Office Police Research Group.

Clarke, R. V. (1983). Situational crime prevention: Its theoretical basis and practical scope. In M. Tonry & N. Morris (Eds.), *Crime and justice: A review of research* (Vol. 4, pp. 225–256). Chicago/London: The University of Chicago Press.

Clarke, R. V. (2016, May/June). Criminology and the fundamental attribution error. *The Criminologist, 41*(3).

Clarke, R., & Mayhew, P. (1980). *Designing out crime*. London: HMSO.

Clarke, R. V., & Weisburd, D. (1994). Diffusion of crime control benefits: Observations on the reverse of displacement. In R. V. Clarke (Ed.), *Crime prevention studies* (Vol. 2, pp. 165–182). Monsey, New York: Criminal Justice Press.

Ekblom, P. (1988). *Getting the best out of crime analysis* (No. 10). London: Home office.

Ekblom, P., Law, H., Sutton, M. with the assistance of Crisp, P., Wiggins, R. (1996). Safer cities and domestic burglary (Home Office Research Study 164). London: Home Office.

Farrell, G. (1992). Multiple victimisation: Its extent and significance. *International Review of Victimology, 2*, 89–111.

Farrell, G., Tilley, N., & Tseloni, A. (2014). Why the crime drop? *Crime and Justice, 43*, 421–490.

Forrester, D., Chatterton, M., Pease, K., & Brown, R. (1988). *The Kirkholt burglary prevention project, Rochdale* (Crime Prevention Unit Paper 13). London: Home Office.

Forrester, D., Frenz, S., O'Connell, M., & Pease, K. (1990). *The Kirkholt burglary prevention project: Phase II* (Crime Prevention Unit Paper 23). London: Home Office.

Foster, J., & T. Hope. (1993) Housing, community and crime: the impact of the priority estates project home office research study No. 131. London: HMSO (ISBN 0–11–341078-6).

Goldblatt, P., & Lewis, C. (1998). *Reducing offending: An assessment of research evidence on ways of dealing with offending behaviour* (Home Office Research Study 187). London: Home Office.

Goldstein, H. (1979). Improving policing: A problem-oriented approach. *Crime & Delinquency, 25*(2), 236–258.

Goldstein, H. (1990). *Problem-oriented policing*. New York: McGraw-Hill.

Guerette, R. T., & Bowers, K. J. (2009). Assessing the extent of crime displacement and diffusion of benefits: A review of situational crime prevention evaluations. *Criminology, 47*, 1331–1368. https://doi.org/10.1111/j.1745-9125.2009.00177.x.

Homel, P., Nutley, S., Webb, B., & Tilley, N. (2004). *Investing to deliver: Reviewing the implementation of the UK crime reduction programme* (Home Office Research Study 281). London: Home Office.

Hope, T. (2004). Pretend it works: Evidence and governance in the evaluation of the Reducing Burglary Initiative. *Criminal Justice, 4*(3), 287–308.

Hough, M. (1984). Residential burglary: A profile from the British crime survey. In R. Clarke & T. Hope (Eds.), *Coping with burglary* (pp. 15–28). Dordrecht: Kluwer.

Husain, S. (1988). *Neighbourhood watch in England and Wales: A locational analysis* (Crime Prevention Unit Paper 12). London: Home Office.

Jeffery, C. R. (1977). *Crime prevention through environmental design*. Beverly Hills: Sage Publications.

Johnson, S. J. (2014). How do offenders choose where to offend? Perspectives from animal foraging. *Legal and Criminological Psychology, 19*, 193. https://doi.org/10.1111/lcrp.12061.

Johnson, S. D., & Bowers, K. J. (2003). *The role of publicity in crime prevention: Findings from the Reducing Burglary Initiative* (Home Office Research Study 272). London: Home Office.

Johnson, S. D., & Bowers, K. J. (2004). The burglary as clue to the future: The beginnings of prospective hot-spotting. *European Journal of Criminology, 105*, 237–255.

Johnson, S. D., Bowers, K. J., Birks, D., & Pease, K. (2008). Predictive mapping of crime by ProMap: Accuracy, units of analysis and the environmental backcloth. In D. Weisburd, W. Bernasco, & G. Bruinsma (Eds.), *Putting crime in its place: Units of analysis in spatial crime research* (pp. 171–198). New York: Springer.

Johnson, S. D., Summers, L., & Pease, K. (2009). Offenders as forager: A direct test of the boost account of victimization. *Journal of Quantitative Criminology, 25*, 181–200.

Kodz, J., & Pease, K. (2003). *Reducing Burglary Initiative: Early findings on burglary reduction* (Home Office Research Findings 204). London: Home Office.

Laycock, G. K. (1985). *Property marking: A deterrent to domestic burglary?* (Crime Prevention Unit Paper 3). London: Home Office.

Laycock, G. K. (1992). Operation identification or the power of publicity? In R. V. Clarke (Ed.), *Crime prevention: Successful case studies* (pp. 230–238). New York: Harrow and Heston.

Laycock, G., & Clarke, R. V. (2001). Crime prevention policy and government research: A comparison of the United States and United Kingdom. *International Journal of Comparative Sociology, XLII*(Spring), 235–256.

Laycock, G. K., & Tilley, N. (1995). Implementing crime prevention programs. In M. Tonry & D. Farrington (Eds.), *Building a safer society: Crime and Justice, a review of research* (Vol. 19, pp. 535–584). University of Chicago Press.

Lipton, D. S., Martinson, R., & Wilks, J. (1975). *The effectiveness of correctional treatment: A survey of treatment valuation studies*. New York: Praeger Press.

Martinson, R. (1974). What works? – Questions and answers about prison reform. *The Public Interest, 35*, 22–54.

Newman, O. (1972). *Defensible space*. New York: Macmillan.

Painter, K. A., & Farrington, D. P. (1997). The crime reducing effect of improved street lighting: The Dudley project. In R. V. Clarke (Ed.), *Situational crime prevention: Successful case studies* (2nd ed., pp. 209–226). Guilderland: Harrow and Heston.

Painter, K. A., & Farrington, D. P. (1999a). Street lighting and crime: Diffusion of benefits in the stoke- on-Trent project. In K. A. Painter & N. Tilley (Eds.), *Surveillance of public space: CCTV, street lighting and crime prevention, Crime prevention studies* (Vol. 10, pp. 77–122). Monsey: Criminal Justice Press.

Painter, K. A., & Farrington, D. P. (1999b). Improved street lighting: Crime reducing effects and cost- benefit analyses. *Security Journal, 12*, 17–32.

Pawson, R., & Tilley, N. (1997). *Realistic evaluation*. London: Sage.

Pease, K. (1998). *Repeat victimisation: Taking stock* (Crime Detection and Prevention Series Paper 90). London: Home Office.

Pease, K. (1999). A review of street lighting evaluations: Crime reduction effects. In K. A. Painter & N. Tilley (Eds.), *Surveillance of public space: CCTV, street lighting and crime prevention, Crime prevention studies* (Vol. 10, pp. 47–76). Monsey: Criminal Justice Press.

Reppetto, T. A. (1976). Crime prevention and the displacement phenomenon. *Crime & Delinquency, 22*(2), 166–177.

Riley, D., & Mayhew, P. (1980). *Crime prevention publicity: An assessment* (Home Office Research Study No. 64). London: HMSO.

Ross, L., & Nisbett, R. E. (1991). *The person and the situation: Perspectives of social psychology*. New York: McGraw-Hill.

Sarre, R. (1999, December 9–10). Beyond 'what works?' A 25 year jubilee retrospective of Robert Martinson. Paper presented at the History of Crime, Policing and Punishment Conference convened by the Australian Institute of Criminology in conjunction with Charles Sturt University and held in Canberra. http://www.aic.gov.au/media_library/conferences/hcpp/sarre.pdf. Accessed 19 July 2017.

Shaftoe, H., & Read, T. (2005). Planning out crime: The appliance of science or an act of faith? In N. Tilley (Ed.), *Handbook of crime prevention and community safety* (pp. 245–265). Cullompton: Devon.

Steinbach, R., Perkins, C., Tompson, L., Johnson, S., Armstrong, B., Green, J., Grundy, C., Wilkinson, P., & Edwards, P. (2015). The effect of reduced street lighting on road casualties and crime in England and Wales: Controlled interrupted time series analysis. *Journal of Epidemiology and Community Health, 69*(11), 1118–1124.

Tilley, N. (1993). *After Kirkholt: Theory, method and results of replication evaluations.* Home Office Police Department.

Tilley, N., & Burrows, J. (2005). *An overview of attrition patterns* (RDS Online Report OLR 45/05). London: Home Office.

Tilley, N., & Webb, J. (1994). *Burglary reduction: Findings from Safer Cities schemes* (Vol. 51). London: Home Office Police Research Group.

Tseloni, A., & Pease, K. (2003). Repeat personal victimization: "Boosts" or "flags"? *British Journal of Criminology, 43*(1), 196–212.

Tseloni, A., Farrell, G., Thompson, R., Evans, E., & Tilley, N. (2017). Domestic burglary drop and the security hypothesis. *Crime Science: An Interdisciplinary Journal, 6*(3). https://doi.org/10.1186/s40163-017-0064-2.

Webb, B., & Laycock, G. K. (1992). *Tackling car crime* (Crime Prevention Unit Paper 32). London: Home Office.

Welsh, B., & Farrington, D. F. (2008). *Effects of improved street lighting on crime. Campbell Collaboration Systematic Review.* Norway: Campbell Collaboration.

Chapter 3
Domestic Burglary: Burglar Responses to Target Attractiveness

Rachel Armitage

Abbreviations

ACPO	Association of Chief Police Officers
ALO	Architectural Liaison Officer
CPDA	Crime Prevention Design Advisor
CPTED	Crime Prevention through Environmental Design
DCLG	Department for Communities and Local Government
DOCO	Designing Out Crime Officer
ODPM	Office of the Deputy Prime Minister
SBD	Secured by Design

3.1 Introduction

The reduction of burglary can take many forms and involve many different individuals and organisations. Whatever form that intervention takes, a vital factor in achieving success is targeting those at most risk of burglary victimisation. Whilst we know that many factors play a part in increasing burglary risk, household-level characteristics have been shown to be the most significant predictor of burglary risk: '…household victimisation first and foremost relates to households' profile and lifestyle' (Tseloni 2006, p. 228). Just as households vary in their victimisation risks, so do places. Indeed, this has been proposed as a law – Weisburd's law of crime concentration '…for a defined measure of crime at a specific microgeographic unit, the concentration of crime will fall within a narrow range of bandwidths of percentages for a defined cumulative proportion of crime' (Weisburd 2015, p. 138). Focusing crime prevention interventions at the micro- or household level is vital and allows key agencies, where possible, to modify those factors or, where factors cannot be altered, to flag heightened risk and intervene to assist prevention. Many household

R. Armitage (✉)
University of Huddersfield, Huddersfield, UK
e-mail: R.A.Armitage@hud.ac.uk

© Springer Nature Switzerland AG 2018
A. Tseloni et al., *Reducing Burglary*,
https://doi.org/10.1007/978-3-319-99942-5_3

factors cannot be changed. A person's age, marital stage or income cannot be influenced by external agencies (should we even want them to), yet other factors can be – those specific to the design, build and management of individual homes and the areas that surround those homes. It is these factors that form the basis of this chapter – to establish the extent to which residential housing design impacts upon burglary and how we can use offender accounts to improve those elements of design. This chapter focuses predominantly on the Secured by Design (SBD) award scheme – the planning policy and guidance that facilitate the delivery of the scheme and its effectiveness as a crime reduction measure. The chapter concludes by exploring burglar perceptions of housing design and the extent to which those interpretations of risk are aligned with existing studies that focus upon police-recorded crime data.

3.2   Reducing Burglary Through Secured by Design

3.2.1   Place-Based Crime Reduction

Place-based approaches to crime prevention are not new. Situational crime prevention (Clarke 1992) focused the preventer on the crime event and the context in which that crime took (or might take) place. Interventions range from property marking or tagging products to closing off streets through measures such as alley gating, and evaluations have consistently demonstrated the effectiveness of removing opportunities for crime at the target location, in the case of burglary, that target being the house (see, e.g. Bowers and Guerette 2014; Sidebottom et al. 2017a, b). Enhancing physical security or target-hardening properties is an effective situational crime prevention intervention and research examining offender's perceptions (Cromwell and Olson 1991; Armitage 2017a), and self-reported or police-recorded crime (Budd 1999; Pease and Gill 2011; Vollaard and Ours 2011; Tseloni et al. 2014) has confirmed this association. We know that those least able to afford often costly security devices are those with the highest risk of burglary victimisation (and repeat victimisation) with the household characteristics 'lone parent', 'private' and 'social renting' experiencing the highest levels of victimisation (Tseloni 2006; Tseloni et al. 2010). Significantly, research has shown that whilst, 'Generally speaking, protection increases with the number of security devices' (Tseloni et al. 2014, p. 658), more security devices (thus more expenditure) do not necessarily equate to more burglary protection. Spending what might be limited resources on numerous security products isn't necessary, '…burglary protection does not consistently increase with the number of devices that make up each configuration' (Tseloni et al. 2014, p. 659). It is the specific combination of security devices that drives the reduction of burglary, with the combination of WIDE (window locks, internal lights, door locks and external lights) recommended as a cost-effective grouping of security devices.

3.2.2   *Crime Prevention through Environmental Design (CPTED)*

A place-based crime prevention with a broader focus upon the wider environment and its impact on crime is Crime Prevention through Environmental Design (CPTED). Defined by Armitage (2013, p.23) as: 'The design, manipulation and management of the built environment to reduce crime and the fear of crime and to enhance sustainability through the process and application of measures at the micro (individual building/structure) and macro (neighbourhood) level', CPTED incorporates a series of components or principles – namely, defensible space, movement control, surveillance, physical security and management and maintenance – as a means of preventing and controlling crime. CPTED emerged largely from the work of Oscar Newman in the 1970s, although the term 'Crime Prevention through Environmental Design' itself was taken from Jeffery's 1971 book of the same title.

3.2.3   *Secured by Design (SBD): Development, Management and Implementation*

Recognising the impact of enhanced and targeted physical security measures, in addition to the benefits of wider design and layout-based modifications, in 1989 the award scheme Secured by Design (SBD) was created. SBD was conceived by a small group of police officers in the south-east of England who were becoming increasingly aware of both rising burglary rates and poor design and build quality in residential housing (Brooke 2013). The scheme is based upon a series of specific technical standards and broader design principles that are contained within design guides that apply to a variety of property types including residential housing, schools, railway stations, commercial premises and hospitals. The technical standards are continually evolving and responding to emerging weaknesses or patterns of *modus operandi* (e.g. the response to evidence specific to mole gripping of euro profile locks in the north of England). The broader principles of limiting through movement, maximising natural surveillance and defensible space and ensuring that programmes are in place to manage and maintain an area are less fluid, although the scheme's guide does state that: 'The Police service continually re-evaluates the effectiveness of Secured by Design and responds to emerging crime trends and independent research findings, in conjunction with industry partners, as and when it is considered necessary and to protect the public from crime' (Secured by Design 2016, p. 4).

Owned and managed by Police Crime Prevention Initiatives, renamed from ACPO Crime Prevention Initiatives following the dissolution of ACPO in 2015, the SBD award scheme is delivered by Designing Out Crime Officers (DOCOs), intermittently referred to as Architectural Liaison Officers (ALOs) or Crime Prevention Design Advisors (CPDAs) who work with architects, planners, building control and developers to ensure that developments meet the specific criteria before awarding a scheme SBD status. Unlike the Dutch equivalent of SBD – Police Label Secured Housing – once a development is awarded SBD, following a final 'sign off' by the relevant DOCO, there are no post-award assessments to ensure compliance, and the title of SBD cannot be revoked if a development falls into disrepair or standards are not maintained. Whilst national figures for SBD builds are not systematically collected, estimates suggest that approximately 43 percent of all new-build properties (between 2013 and 2016) achieved SBD status (J. Cole, personal communication, June 9, 2017).

3.2.4 Consideration for Crime Prevention Within the Planning System

Whilst there is little doubt that the growth of the SBD scheme was linked to evidence of impact (see Armitage 2013 for a summary) and the enthusiasm and passion of those managing and implementing the scheme, one of the key drivers within the UK has been the inclusion of SBD as a requirement, or recommendation, within planning and housing policy, guidance and regulation. Before discussing the current status of SBD within such policy and regulations – the position in 2017, it is important, albeit briefly, to summarise the drivers that *have* existed to incentivise and therefore encourage the growth of the SBD scheme, particularly the period between 1998 (the Crime and Disorder Act) and 2011 (the introduction of the Localism Act). Whilst these policies are no longer in place, their relevance to the historical development of SBD remains acutely relevant.

In 1998, the Crime and Disorder Act was introduced in England and Wales. Though broad reaching, one of the most relevant sections of the Act (for the subject of crime prevention within the planning system) was Section 17, which required a number of key agencies including the police service, fire service and local authorities to: 'Without prejudice to any other obligation imposed upon it...exercise its functions with due regard to...the need to do all it reasonably can to prevent crime and disorder in its area' (Great Britain 1998). Whilst this brought to the fore the possibility that legal action could be taken against organisations that did not do all that they reasonably could to prevent crime within their service delivery (Moss and Pease 1999), its significance lay more in the message it portrayed – that crime control is not the sole responsibility of the police and other agencies must integrate and embed crime prevention principles within their day-to-day activities.

Prior to 1998, crime prevention had little prominence within the planning system. The circular *5/94: Planning Out Crime* (Department of the Environment 1994) made reference to the importance of consulting DOCOs (referred to as ALO/CPDAs at the time), and it referenced SBD as a good practice scheme for designing out crime. However, its 11 pages gave very little in the way of detailed guidance regarding *how* to maximise consideration for security within housing design. The era post-1998 saw much more prominence given to urban planning, not just for crime reduction but for wellbeing, health and sustainability. The Urban Policy White

Paper *Our Towns and Cities: The Future* (ODPM 2000) took forward many of the recommendations that had been raised by Lord Rogers' Urban Task Force (Department of Environment, Transport and the Regions 1999) with a focus upon revitalising towns and cities, improving social and environmental sustainability, improving transport networks and reducing development on greenfield sites, making specific reference to the importance of designing out crime as well as recommending a review and update of circular *5/94*: '...good design of buildings and the way buildings and public spaces are laid out can help prevent crime' (ODPM 2000).

Circular *5/94* was officially replaced by *Safer Places: The Planning System and Crime Prevention* (ODPM/Home Office 2004) with the publication of *Planning Policy Statement 1: Delivering Sustainable Development* (DCLG 2005) that set out the government's national policies on land-use planning in England. It guided planning authorities to 'Deliver safe, healthy and attractive places to live' (DCLG 2005, p. 7); to 'Promote urban and rural regeneration to improve the wellbeing of communities, improve facilities, promote high quality and safe development' (DCLG 2005, p. 11); and to 'Promote communities which are inclusive, healthy, safe and crime-free' (DCLG 2005, p. 11).

Crime prevention-specific planning guidance was updated in 2004 with the publication of *Safer Places: The Planning System and Crime Prevention* (ODPM/Home Office 2004). At 108 pages in length, as opposed to its predecessor's 11, it included much more detail regarding how to create safe spaces and introduced the 7 principles – access and movement, surveillance, structure, ownership, physical protection, activity and management and maintenance – much the same as the principles of both CPTED and SBD.

In 2004 the *Planning and Compulsory Purchase Act (2004)* introduced the requirement to produce Design and Access Statements when submitting applications for outline and full planning permission. The DCLG Circular 01/2006: *Guidance on Changes to the Development Control System* (DCLG 2006) outlined what was required within a Design and Access Statement, and paragraph 87 specifically stated that Design and Access Statements must demonstrate how crime prevention measures and in particular the principles outlined in Safer Places have been addressed.

> Design and Access Statements for outline and detailed applications should therefore demonstrate how crime prevention measures have been considered in the design of the proposal and how the design reflects the attributes of safe, sustainable places set out in Safer Places – the Planning System and Crime Prevention (ODPM/Home Office 2004). (DCLG 2006, p. 15)

Alongside these policy and guidance documents, *Planning Policy Statement 3: Housing* (DCLG 2011), originally published in 2006, highlighted the importance that planning authorities should place upon the creation of safe developments.

> Matters to consider when assessing design quality include the extent to which the proposed development...Is easily accessible and well-connected to public transport and community facilities and services, and is well laid out so that all the space is used efficiently, is safe, accessible and user-friendly. (DCLG 2011, p. 8)

Alongside these, the introduction of several key policy documents, specific to sustainability and social housing, saw a major shift in the priority afforded to crime prevention within the planning system. Whilst not *requiring* developments to be built to the SBD standard, these policies incentivised the attainment of SBD, allowing developers both financial and marketing benefits from that achievement. In 2007, the *Code for Sustainable Homes* (DCLG 2008) became a voluntary standard for which developers could apply to market their properties as being built to minimise the environmental impact of that development. The Code measured the sustainability of a home against nine categories: energy and CO_2 emissions, water, materials, surface water run-off, waste, pollution, ecology, health and wellbeing and management, the latter category awarding two credits for compliance with part two (physical security) of SBD. As well as the incentive of two credits, for a property to be defined as a minimum of one star rating for sustainability, a development had to meet this security criterion.

Another key area of policy incentivising the attainment of SBD status was that published by the Housing Corporation and then Homes and Communities Agency – the organisation that regulates social housing providers across the UK (social housing providers being Housing Associations, Registered Social Landlords – not local authority housing). The then Housing Corporation's *Design and Quality Standards* (2007) and English Partnerships' (2007) *Quality Standards* were two such policies.

The Housing Corporation's *Design and Quality Standards*, which replaced the original *Scheme Development Standards*, set out the *requirements* and *recommendations* for all new homes that required Social Housing Grants. Two elements of those policies were important in explaining the rise in consideration for crime prevention within social housing. To receive Social Housing Grants, social housing providers had to meet a minimum level of criteria set out within the Code for Sustainable Homes, and in meeting that criteria, they had to build homes that met, at least, the physical security elements of SBD (physical security being locks, doors, windows and not the design and layout elements). Thus, in order to obtain Social Housing Grants, providers must build to a minimum Part 2 of SBD (physical security). Within the Recommendations Annex, Enhanced standards were defined as requiring full SBD certification and designing homes in accordance with the advice obtained from the local police ALO/CPDA (now DOCO). The policy stated that developments which met Recommended standards would: '...subsequently find reflection in the Corporation's assessment of affordable housing providers through the Value for Grant Comparator tool' (Housing Corporation 2007, p. 2), and: 'Some enhanced aspects will be reflected in the Grant Index during the bid assessment process and improve the value for money ranking' (Housing Corporation 2007, p. 3). Thus, in building homes to SBD standards, social housing providers would receive additional funding from the Homes and Communities Agency. This varied little from the Housing Corporation's previous policy – *Scheme Development Standards* (Housing Corporation 2003) – that included SBD as a *Recommended* criterion but highlighted how Registered Social Landlords that built schemes to a standard over and above the Essential criteria (to incorporate Recommended items) would receive Enhanced Quality Assessments. These were reflected in compliance audit results and in turn influenced the level of funding from the Housing Corporation.

Between 2008 and 2011, the Homes and Communities Agency also imposed English Partnerships' *Quality Standards* for developments on land where they retained an interest. This included developments on land that was entirely or partly owned by the Homes and Communities Agency or public land regeneration programmes that they managed. Where these standards were applied, all development on that land had to be designed and developed in line with the principles of SBD.

In 2011 things began to change (and not for the better). The UK had entered a period of austerity, government cuts were affecting all public services, and the government at the time were pushing for deregulation within the building industry. The Localism Act (2011) saw the end of the 44 Planning Policy Statements to be replaced by the National Planning Policy Framework (just 55 pages). The Code for Sustainable Homes was abolished in 2015, and the Department for Communities and Local Government's internal order to the Homes and Communities Agency to refrain from requesting SBD or offering financial incentives to build to SBD also removed these incentives. The Taylor Review of housing standards (DCLG 2012a) also saw the abolition of *Safer Places* in 2014.

There remain many positives. The National Planning Policy Framework (DCLG 2012b) and its associated National Planning Guidance were introduced, and these refer clearly to the importance of crime prevention within SBD – the National Planning Guidance going as far as to encourage pre-application discussions with the CPDA (now DOCO). In October 2015, building regulations were updated for England and Wales to include Approved Document Q that requires some level of physical security – although not at the level required by SBD (it should be noted that this already existed in Scottish building regulations). Local authorities must also develop their own individual Local Plans, and these must be in line with the National Planning Policy Framework – remembering that this states that planning must: 'Create safe and accessible environments where crime and disorder, and the fear of crime, do not undermine quality of life or community cohesion' (DCLG 2012b, p. 15). Whilst there have been cuts to police budgets seeing the number of DOCOs fall from 347 in 2009 to just 125 in November 2014 (Armitage 2017a, b), each local authority area retains a minimum of one DOCO whose role is (amongst other things) to influence that Local Plan to ensure that crime prevention is given adequate consideration.

3.2.5   Evaluating the Effectiveness of Secured by Design (SBD)

As was referred to earlier in the chapter, SBD is based upon the principles of CPTED, and for that reason any measure of its effectiveness should consider not only evaluations of the SBD scheme itself but also empirical evidence regarding the effectiveness of surveillance, physical security, through movement, defensible space and implementing systems to manage and maintain an area. Whilst SBD has been in place in England and Wales since 1989, a recent review of all research conducted to establish the impact of SBD on crime levels found just 7 from a total of 331 related articles that reported quantitative data on the effectiveness of SBD (Sidebottom et al. 2017a, b). It should be noted that many of these studies are dated; publication dates for the seven include 1994, 1999 (for two), 2004, 2011, 2012 and 2016. As data would be collected prior to publication, it is likely that the sample housing for these studies could be as old as 1991. Keeping in mind that SBD is an evolving standard for which technical specifications improve on an annual basis, these studies are unlikely to report the current impact of SBD on burglary levels.

The methodological standard of the seven studies also varied considerably with the most prominent weakness being the selection of a control or comparison sample. Four of the seven studies reported the impact of SBD in new build developments (as compared to non-SBD), the remaining three focused upon developments refurbished to the SBD standard – comparing crime levels pre- and post-refurbishments. Of the four studies that focus upon new build developments, Sidebottom et al. (2017a, b) report a tentative finding (the findings are presented from the draft systematic review) that the SBD groups are 54 percent less likely to experience a burglary. All four studies favour the treatment. However, for two, the difference is not statistically significant.

For the studies that evaluated refurbished developments, the results were all positive in direction. Pascoe (1999) reported reductions in victimisation surveys across ten refurbished developments with 39 percent of respondents reporting three or more burglaries pre-SBD as compared to 18.4 percent post-refurbishment (to the SBD standard). Armitage (2000) reported that crime fell by 55 percent on developments refurbished to the SBD standard. Teedon et al. (2009) reported 61 percent reductions in housebreaking within the SBD sample, as compared to 17 percent in the comparison area. Jones et al. (2016) reported significant reductions in burglary in both the treatment and control groups, albeit the reductions were greater in the SBD sample.

The author is only aware of two evaluations that have measured the impact of SBD on repeat, as opposed to single, victimisation. Armitage (2000) found that although the concentration rate for total crime was higher within the non-SBD sample, levels of repeat burglary were higher within the SBD sample. Residents appeared to be protected against total crime and burglary; however, once they had experienced a burglary offence, they were more likely than their non-SBD counterparts to experience a subsequent burglary offence. Armitage and Monchuk (2011) confirmed this finding with 36 percent of crimes against the SBD sample representing a repeat offence, as compared to 27 percent of the crimes against the non-SBD sample. However, closer scrutiny of their data suggests that once the offence-type assault was excluded, whilst the repeat victimisation levels remained the same for the non-SBD sample, the proportion of offences that represented a repeat within the SBD sample reduced to just 10 percent. The extent to which SBD schemes aim to or can indeed impact upon personal crime types appears questionable. It is clear from these findings that repeat assault offences are higher within the SBD sample – a finding that warrants (although not in this chapter) further investigation.

3.2.6   The Principles of Secured by Design (SBD) and Their Individual Impact on Crime

Whilst a scheme would not achieve SBD status unless it met all standards included within the guidance, it is useful briefly to reflect on the evidence relating to those principles of defensible space, through movement, surveillance, management and maintenance and physical security (the latter having been discussed earlier within this section).

Defensible space or territoriality (Newman 1973) refers to the extent to which the physical design of a neighbourhood can increase or inhibit an individual's sense of control over the space in which they reside. This is often achieved through the demarcation of private, semiprivate, semipublic and public space (using design features such as a change of road colour and texture or the narrowing of the road) to ensure that it is clear who should and who should not be within a given area. Brown and Altman (1983) found that compared with non-burgled houses, properties that had experienced a burglary had fewer symbolic (as well as actual) barriers. In their study of housing in the Netherlands, Montoya et al. (2016) found that houses with a front garden had less than half the burglary risk of those without. Defensible space should, if working effectively, create territorial responses amongst the owners or managers of a space. This might take the form of challenging a stranger, calling the police or simply making their presence known as a means of assuring the stranger that they are being observed. Brown and Bentley (1993) interviewed offenders, asking them to judge (from pictures) which properties would be more vulnerable to burglary. The results revealed that properties showing signs of territorial behaviour (such as the installation of a gateway at the front of the property or a sign on the gate/door marking the area as private) were perceived by offenders to be less vulnerable to burglary.

Research studies have utilised a variety of methods to establish the extent to which crime risk varies according to levels of through movement within a housing development. These include analysis of police-recorded crime, interviews with convicted burglars and, more recently, agent-based modelling to simulate street networks and crime risk. Limiting through movement within the design and layout of developments is based upon the premise that increased connectivity provides enhanced opportunities for offenders to attach a target to their awareness space (Brantingham and Brantingham 1993) – to notice it as a suitable target. Limiting through movement also works on the principle that less connectivity means that offenders have fewer opportunities to access, move through and exit a development before and after a burglary event. However, what Armitage (2004, p.48) and later Birks and Davies (2017) refer to as the 'encounter versus enclosure' debate is not so clear-cut. Whilst less movement opportunities reduce the likelihood that an offender will become aware of a target (and should they become aware, there will be less access and escape routes), reducing connections also reduces the opportunities for legitimate users to move through a space, users that could act as guardians – what Jacobs (1961) refers to as eyes on the street. The presence of these users has the *potential* capability (note the caution here) of acting to deter abusers. Armitage (2006, p. 82) refers to this as: 'The main area of conflict within the field of designing out crime...' suggesting that the debate has: 'dominated much of the discussion surrounding SBD'.

The majority of research confirms that enhanced movement within a housing development increases the risk of burglary (Wiles and Costello 2000; Armitage et al. 2010; Johnson and Bowers 2010; Johnson and Bowers 2014; Davies and Johnson 2014). Whilst enhanced movement increases the *opportunity* for guardianship, this opportunity does not necessarily equate to actual policing or challenging of strangers or potential abusers (Reynald and Elffers 2009; Reynald 2010). In their paper on street networks and crime risk, Birks and Davies (2017) conduct a series of simulated experiments to randomly reduce permeability by removing connections between property nodes using agent-based simulation. They found a curvelinear relationship between permeability and crime, suggesting that moderate reductions in connectivity resulted in increased levels of victimisation but that at an inflexion point of 30 percent road closures and further reductions in permeability led to overall reductions in crime. This confirms the argument (Armitage 2013) that the encounter debate has merit but that the existence of footpaths has to meet a genuine need so that those footpaths will be utilised and provide adequate real (as opposed to hypothesised) guardianship. In 2002, Taylor argued that: 'Neighbourhood permeability is ... one of the community level design features most reliably linked to crime rates, and the connections operate consistently in the same direction across studies: more permeability, more crime' (Taylor 2002, p. 419). Armitage (2017b, p. 273) urged caution, asserting that this statement was: 'overgeneralised'. Birks and Davies (2017) reason that: 'Ultimately, the absence of a simple relationship between permeability and offending may be the most significant outcome of our analysis' (p. 43).

Research suggests that surveillance plays a major part in offenders' decision-making processes. When selecting targets for burglary, offenders prefer to avoid confrontation and, where possible, select properties which are unoccupied (Reppetto 1974; Brown and Altman 1983; Cromwell and Olson 1991; Brown and Bentley 1993; Nee and Meenaghan 2006). Physically assessing levels of surveillance opportunities within residential developments confirms that households with higher levels of surveillance from neighbouring properties and passers-by experience significantly lower levels of burglary (Armitage 2006; Armitage et al. 2010; Winchester and Jackson 1982; Van der Voordt and Van Wegen 1990). Recognising the complexities surrounding surveillance, with potential opportunities not necessarily equating to actual surveillance, Reynald (2009) conducted a study that measured the relationship between guardianship intensity and surveillance opportunities on a sample of 814 residential properties in The Hague. Reynald measured guardianship intensity using a four-stage model which moves from stage one, invisible guardian stage (no evidence that the property is occupied), to stage two, available guardian stage (evidence that the property is occupied), to stage three, the capable guardian stage (fieldworkers are observed by residents), and to stage four, intervening guardian stage (fieldworkers are challenged by residents). Surveillance opportunities were measured by observing the extent to which the view of a property's windows was obstructed by physical features such as trees and walls. The results revealed a positive statistically significant correlation between surveillance opportunities and guardianship intensity (0.45), suggesting that guardianship intensity increases as opportunities for surveillance increase. When assessing the relationship between crime and guardianship intensity, the results were positive and statistically significant. The analysis revealed that crime decreases consistently at each stage of the four-stage model. Crime drops significantly between the invisible and available guardian stages, decreasing even more at the capable guardian stage and slightly more at the intervening stage.

Looking at the links between levels of management and maintenance and subsequent crime experiences, research generally confirms that the presence of low-level disorder influences offenders' perception of risk – a lack of care or concern for an area being associated with less risk of challenge or apprehension (Taylor and Gottfredson 1987; Cozens et al. 2001).

3.3 Accounting for Burglar Perceptions

3.3.1 Methodology

The research presented within this chapter was conducted between 2014 and 2016 in West Yorkshire, England. The sample included 22 incarcerated adult males convicted of burglary offences and identified by the Integrated Offender Management Team (based at one prison) to be prolific. Offending levels within the sample ranged from one burglary a day (Participant 19) to five to ten burglaries a day (Participant 16). The offenders took part in the research voluntarily, and recruitment took place post-sentencing to avoid involvement for bargaining purposes.

Interviews took place within the prison with offenders shown a series of 16 images of housing and asked to describe: 'From what you can see from the photo, can you describe what would *attract* you to this property when selecting a target for burglary'. And 'From what you can see from the photo, can you describe what would *deter* you (put you off) from selecting this property as a target for burglary'. Participants were encouraged to talk openly about their thoughts, to describe their thinking and to keep in mind that there were no right or wrong responses.

The 16 images are described in Table 3.1 and illustrated in Fig. 3.1. They included a series of socially and privately owned housing as well as a mix of housing types and styles.

Interviews were transcribed, and thematic analysis was used to identify patterns or themes in responses. Content analysis was used to count the regularity with which those themes were discussed and to assess the levels of consistency between offender accounts.

Table 3.1 Description of 16 images

Image	Description of image
1	Row of red brick terraced houses and rear gardens. The fences between gardens are low. Children's play equipment is present in some gardens. Wheelie bins are present, some with overflowing rubbish
2	Rear view of two semi-detached properties. One property is bounded by a fence, one by a hedge. One property has an open downstairs window. The upstairs curtains are closed on one property
3	Gable end of a corner plot on a cul-de-sac. The property is red brick with a burglar alarm. The side door is wood. Cars are parked in the drive
4	Front view of a property bounded by a high hedge and gate. The house is covered in overgrown ivy. There is no evidence of a burglar alarm
5	Front view of a row of terraced properties. A lamp post with Neighbourhood Watch sign clearly displayed
6	Rear view of a row of terraced properties – one with an extension. The windows and door are UPVC. The garden has a patio table and chairs. There is a ladder on the floor
7	Rear view of a row of terraced back to back properties. The rear of the properties is accessed via alleyways. The garden shows a lack of maintenance; there are old toys and a bike in the garden
8	Rear view of a row of terraced properties. Some windows have net curtains. There is an old sofa and rolled-up carpet in the garden. Access to the rear of the properties is via an alleyway
9	Close up of a UPVC door within a recess. Wheelie bins are present next to the front door
10	Front view of a row of semi-detached bungalows. Front gardens are bounded by low fencing
11	Footpath bounded by houses. The footpath is well lit with well-tended gardens acting as a buffer between property and footpath
12	Gated development with a locked metal fence. The properties have cars parked in allocated spaces and windows facing the parking area
13	Small cul-de-sac of semi-detached properties. The entrance to the estate is narrowed with a change in road colour and texture. At the entrance, the word 'private' is painted on the floor in large white letters
14	Footpath running at the rear of properties. The footpath is bounded by a wall and high (approx. 8 metre) solid fencing
15	Rear view of a row of large detached properties with rear conservatories and sash wooden windows
16	Front view of a row of semi-detached bungalows set within a communal space. The communal space is well tended with garden furniture and plants. The properties are bounded by a low (waist height) metal fence

3.3.2   Limitations of the Research

It should be highlighted that whilst the accounts of active offenders can provide valuable information (Copes and Hochstetler 2014), there are risks associated with such accounts and these should be made clear. The first and most perceptible risk is false narratives from participants. There is a possibility that offenders will approach the responses with an element of bravado – *I'm not deterred by anything*, thus

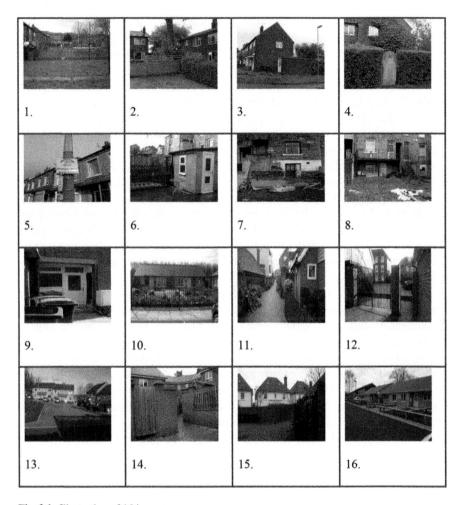

Fig. 3.1 Illustration of 16 images

underplaying the deterrent effect of certain design features. Conversely, offenders may downplay their boldness – *No, I wouldn't burgle them*, thus risking overestimating the deterrent effect of certain design features. There is also a risk of downplaying their offending or elements of that offending that might be conceived or morally wrong (such as targeting the elderly) – what Shaw and Pease (2000) refer to as *false morality*. Spinning aside, there is a genuine risk of narrator inaccuracy (Kearns and Fincham 2005; Elffers 2010; van Gelder et al. 2017) that may be due to the passing of time, drug use or simply that aspects of the story are lost or forgotten.

The second limitation relates to the focus upon prolific offenders. Previous research suggests that a large proportion of burglaries committed by prolific offenders are repeats (Everson 2000). One risk here relates to the potential to overemphasise the importance of design features that influence repeat offending. Another is that prolific or experienced burglars may have very different decision-making processes to those less experienced. For example, Homel et al. (2014) found that the number of cues used to decide on a target decreased with burglary expertise.

The third limitation relates to the drug-use patterns of the sample of offenders. Of the 22 offenders, 17 described themselves as drug users – not just taking drugs but committing burglaries whilst under the influence of drugs. The combination was predominantly heroin and cocaine; however, some described taking mephedrone, alcohol, cannabis, crack, amphetamine and MDMA. There is a risk that focusing upon the responses of drug-using offenders will produce findings that minimise the deterrent effect of specific security features. For example, 'nothing deterred me', 'I would keep going until I got in', and 'I felt invincible'. These security features could have prevented a complete burglary (we are relying on the offender accounts regarding effectiveness) and could have deterred offenders not under the influence of drugs. In addition, the offender responses given whilst in prison and no longer under the influence of drugs may differ from the responses they would have given at the time of the offence(s).

An additional risk relates to the sample being selected from those burglars who have been detected and sentenced. To what extent does this sample represent unsuccessful, overconfident offenders – those making poor decisions regarding suitable targets? Finally, the use of images of residential housing as opposed to physically or virtually placing an offender at a specific residential location. Participants were able to see the property itself but were not able to assess the wider area. Nee et al. (2015) reported on a small study that utilised both virtual reality and a mock property in assessing ex-burglar and student behaviour in conducting simulated burglaries. Although the sample was small ($n = 7$), there was little difference between virtual reality – thus not being in a physical property – and search behaviour in a mock target property.

3.4   Burglar Accounts of Target Attractiveness: Research Findings

The analysis of offender accounts reveals some interesting patterns regarding the features that make a suitable or unsuitable target and how these align with the principles and standards of the SBD scheme.

3.4.1   What Makes a Suitable Target?

First, what makes a suitable target? The images which received the highest proportion of 'yes' responses were image 4, with 91 percent (20) of participants stating that they would burgle this property; image 6, with 86 percent (19) of participants stating that they would burgle this property; and image 15, with 86 percent of participants stating that they would burgle this property. The visual features described by the 22 offenders as influencing their decision-making were (1) a lack of surveillance, (2) poor quality physical security and (3) signs of wealth. Only the first two could be modified by the agencies responsible for designing, implementing and managing the SBD scheme. A brief summary of these responses confirms that offenders prefer properties with limited surveillance, where the entry and exit from the property are concealed and where others are unlikely to notice them whilst inside the property.

Image 4 shows the front of a property bounded by a high hedge and gate. The house is covered in overgrown ivy, and there is no evidence of a burglar alarm. Offenders explained their reasoning as specifically relating to an ability to commit the burglary with less risk of observation.

Image 6 shows a rear view of a property that has a small extension. The extension has one UPVC door and window and a flat roof. The main justification from the 19 participants who stated that they would burgle this property focused upon the poor quality of the physical security evident within the image. Offenders stated that it was clear that the door lock would be easy to mole grip, that the door hinges were on the outside of the door frame (so could be removed) and that the beading around the glass within the door could easily be jemmied (forced with a screwdriver, crowbar or similar).

Image 15 shows the rear view of a row of large detached properties with rear conservatories and sash wooden windows. Whilst participants were clearly attracted to these properties (and not deterred or discouraged by any design features), this attraction appeared largely to be related to the perception of wealth – that the rewards would be greater than any risks. This interpretation was based upon the size and style of the properties (detached, painted white, sash windows) and the extent to which the properties were well managed and maintained (tidy gardens), thus representing residents that cared about the goods that they owned and had money to spend on those goods.

3.4.2 What Makes an Unsuitable Target?

What features of housing design deemed a target unsuitable for burglary? The three images with the lowest number of 'yes' responses were image 16, with just 1 participant (5 percent) stating that he/she would burgle this property; image 10, with 2 participants (9 percent) stating that they would burgle this property; and image 1, with 6 participants (27 percent) stating that they would burgle this property. The visual features described by offenders as deterring them from burgling a property included:

1. The perception (wide doors, bungalows) that the properties were sheltered housing; thus this would be morally wrong.

2. The perception (red brick, rubbish in the garden, overflowing bin) that the properties were council owned; thus this would be morally wrong.
3. Low, transparent front boundary fences enhancing natural surveillance.

Only one of these factors is within the direct influence of those designing and managing housing, including award schemes such as SBD.

Image 16 shows a row of semi-detached bungalows set within a communal space. The communal space is well tended with garden furniture and plants. The properties are bounded by a low (waist height) metal fence. When asked why they would not burgle this property, the vast majority of participants explained their decision based on moral grounds – stating that they perceived these properties to be sheltered housing for the elderly. Participants who gave this response appeared extremely confident from the visual cues that these were sheltered houses.

> This is blatantly old people's housing and I would not burgle there. A lot of people would and if I look back, a lot of people I mixed with would and they would even target old people. I would never as my mum brought me up to have pride and morals. (Participant Four)

Participant Six expressed the same view, adding that should you get caught, you would be singled out in prison as *scum*.

> It's old peoples' homes so no. When you go to prison people would know you had targeted an old person's house and you would be scum. I could get in easily but this is a moral judgement. (Participant Six)

Participant Nine expressed the same moral judgement but added reference to the risk that the shock of burglary could kill an old person.

> Sometimes a job can go wrong with an old person and it can kill them, everything is just too risky. I might have looked at these when I was much younger as a kid, but I've got morals now and it's just wrong doing a job on an old person' house. (Participant Nine)

There were non-moral justifications for avoiding these properties; these related to the low, transparent fencing in the front gardens and the lack of demarcation between front gardens. This was perceived as aiding surveillance from neighbouring properties.

> There's no barriers between fronts of the properties either, so it'd put me off if next door were in. (Participant Ten)

Very similar to image 16, image 10 shows a row of semi-detached bungalows with low, transparent wooden fencing marking the boundary of the front gardens. The houses have in-curtilage car parking, and the entrance to the driveway/garden path has a change in road colour and texture. Only two participants (9 percent) stated that they would burgle the properties. As with image 16, the main justification for avoiding these properties was moral.

> I wouldn't touch this. Bungalows are associated with pensioners and old people…I've got a heart. My nana lives in a bungalow and she means a lot to me, so there's no way. (Participant Eight)

Participant Seven referred to the risk of being treated as a *nonce* in prison, should you be found to have targeted old people.

No, it's old peoples' homes. In prison you know if people have targeted old people and these people should be on a wing with the nonces. (Participant Seven)

The design features that appeared to distinguish these properties as old people's houses/sheltered housing included the width of the doors and the extent to which the gardens were managed and maintained.

The doors are quite wide so I'd think that the people living in these houses were wheelchair users and would leave. (Participant Ten)

Image 1 shows a row of red brick terraced houses and rear gardens. The fences between gardens are low. Children's play equipment is present in some gardens. Wheelie bins are present, some with overflowing rubbish. Only six participants (27 percent) stated that they would burgle this property. The rationale for avoiding this property related largely to these properties being perceived as council housing – thus this would be morally wrong, and the risk would not outweigh the benefits due to a perceived lack of attractive products within the house.

You don't sh*t in your own back yard. I don't like burgling in poor areas when they don't have much. (Participant Four)

It looks like a sh*t hole to be honest, grass isn't cut well,   the fence is old and the trampoline is going rusty. Even the bins are overflowing. It just doesn't look worthwhile, it looks like you'd have to do 10 houses there for it to be worthwhile. (Participant Ten)

Whilst the main justification for avoiding these properties was moral, several offenders did comment on the gardens being *too open* with the risk of observation from neighbouring properties too great.

It's an open space at the back of the houses which is a good thing and a bad thing. You can be seen, but you can also see too. (Participant Nine)

Offender Perceptions of the Principles of Secured by Design (SBD)
Looking more specifically at the responses according to the key principles and standards of the SBD scheme, it was clear that offender perceptions of target attractiveness largely confirmed those principles as key deterrent features. All participants referred to *surveillance* and *physical security* as key deterrents in their target selection. Eighty-two percent ($n = 18$) referred to *movement control*, 77 percent ($n = 16$) to *management and maintenance* and 36 percent ($n = 8$) to the concept of *defensible space* (Table 3.2).

Surveillance was referenced most regularly, 133 times; *physical security*, 103 times; *management and maintenance*, 40 times; and *movement control*, 39 times, and the concept of *defensible space* was just referred to on 11 occasions (Table 3.3).

3.4.3   *Surveillance*

Based on the principle of surveillance, SBD explicitly requires developments to be designed to maximise natural or informal surveillance. This includes ensuring that dwelling frontages are free from obstructions and that sightlines between properties are clear.

Table 3.2 Proportion of burglars referencing the five key principles of SBD

SBD principle	Proportion of participants who referred to each principle ($n = 22$)
Surveillance	100%
Physical security	100%
Movement control	82%
Management and maintenance	77%
Defensible space	36%

Table 3.3 Number of references to the key principles of SBD

SBD principle	Number of references to the principle
Surveillance	133
Physical security	103
Management and maintenance	40
Movement control	39
Defensible space	11

> For the majority of housing developments, it will be desirable for dwelling frontages to be open to view, so walls, fences and hedges will need to be kept low. (Secured by Design 2016, p. 18)

> Planting should not impede the opportunity for natural surveillance. (Secured by Design 2016, p. 23)

Surveillance is also maximised through the positioning of properties within a development. Dwellings should face onto the street; entrance doors should be at the front of the property (as opposed to the side); and windows should be positioned to gain maximum surveillance from 'active rooms' such as the living room and kitchen.

> Dwellings should be positioned facing each other to allow neighbours to easily view their surrounding (Secured by Design 2016, p. 21).

Remembering that offenders were not asked to comment on the level of 'surveillance', 'visibility' and 'likelihood of observation' per se, they were simply asked to describe their reaction (in terms of levels of attractiveness) to a series of images that subtly included different levels of natural surveillance. Offenders confirmed that many elements of the risk of being surveilled would deter them from selecting a target.

Image 4 (high fence and gate with overgrown shrubbery) received the greatest number of 'yes' responses when asked whether offenders would target the property – 20 of the 22 (91 percent). Participants regularly expressed the view that the lack of surveillance – resulting from the high fence/gate and shrubbery – would limit the opportunity for neighbours and passers-by to see them entering and exiting the property, thus enhancing their confidence in committing crimes at that location.

> This is a burglar's dream. There are high trees at the back, the hedge is high so blocks the view from the road, the gate is high so no-one can see you. (Participant Six)

That's well tempting as a burglar. It's got a six-foot high hedge around it - no one would see nothing whilst you're in there. You could spend as much time as you needed in it. (Participant Nineteen)

Yeah the high hedge would stop you being seen. Burglar's dream that one. (Participant Eleven)

Again confirming the SBD guidance, offenders regularly referred to the risks of having large windows at the front of a property, allowing guardians to see them committing the burglary: 'The front windows are nice and big too, so it'd mean that I could be seen easier if I was inside' (Participant Ten).

Offenders were deterred by housing developments where surrounding properties faced onto the street, allowing neighbours to observe them entering and exiting a development and property and creating a perception of enhanced community cohesion and guardianship: 'I'd keep away. Would want nothing to do with that. They could be gawping out the windows – you only need one of them on that street' (Participant Sixteen).

Whilst the offenders largely confirmed the guidance set out within SBD, there were elements of the findings that cast doubt on some of the basic principles of SBD's approach to informal surveillance. SBD New Homes states that: 'Vulnerable areas, such as exposed side and rear gardens, need more robust defensive barriers by using walls or fencing to a minimum height of 1.8 m' (p. 18). Offenders countered this advice, suggesting that solid fences act to restrict surveillance, thus acting as an attractor: 'I like solid fences like these as no-one can see you. Once you are over these fences you are safe – in a comfort zone' (Participant Twelve). Reinforcing this finding, others confirmed that open boundaries, low or transparent fences would deter them from selecting a target for burglary.

Open fences would put me off. (Participant Twelve)

I'd feel more exposed if the walls and fences were lower. (Participant Fifteen)

Visible displays indicating the presence of Neighbourhood Watch did not deter this sample of offenders. Whilst they generally agreed with the concept of Neighbourhood Watch encouraging community cohesion, guardianship and ultimately enhanced surveillance, they did not believe that this translated from the principle to practical risk: 'It's like the signs are up there but there is no action' (Participant Five).

3.4.4 Movement Control

The impact of through movement on levels of crime is a much-debated topic. Research largely confirms that greater opportunities for offenders to access, pass through and egress a development do enhance levels of crime risk, although the pattern is not necessarily linear and there are additional factors that must be accounted for, given that the presence of guardians/users of that space can act as a deterrent for potential offenders. SBD guidelines clearly align with the majority of criminological research, suggesting that road layout will impact on offender decision-making and that lower levels of connectivity are desirable.

> There are advantages in some road layout patterns over others especially where the pattern frustrates the searching behaviour of the criminal and his need to escape. (Secured by Design 2016, p. 14)

The guidance accepts that through movement has benefits in terms of enhancing walkability and addressing concerns regarding sustainability. Through movement is therefore discouraged, whilst acknowledging that, where unavoidable, it must be designed to minimise risk – footpaths should be wide and well-lit, avoiding hiding spaces, and should not run at the rear of properties.

> Whilst it is accepted that through routes will be included within development layouts, the designer must ensure that the security of the development is not compromised by excessive permeability. (Secured by Design 2016, p. 14)

The cul-de-sac is a favoured design layout but only where that cul-de-sac is '… short in length and not linked by footpaths' (Secured by Design 2016, p. 14), in line with research demonstrating that leaky (linked) culs-de-sac experience higher levels of crime (Armitage et al. 2010; Johnson and Bowers 2010) and that true, sinuous culs-de sac experience the lowest levels of crime (Armitage et al. 2010).

The 22 offenders were shown a series of images that portrayed through movement (or a lack of it), specifically images 11 and 14. Offenders regularly commented on the concept of through movement (or the lack of it) with the benefits summarised as (1) through movement enables them to 'root' for a suitable target; (2) through movement allows them to enter a development, commit an offence and exit the development without retracing their steps, thus reducing risk of observation; (3) through movement allows them to evade the police, footpaths benefiting those with enhanced knowledge of the area and those on foot; and (4) through movement provides them with a legitimate reason to be in an area.

Offenders spoke of how the presence of footpaths within housing developments facilitated the selection of suitable targets, allowing them to 'root' within a development and to conduct this searching legitimately because a footpath is a public space.

> I would first walk up and down the footpath and have a look at what I could see in the houses. The houses are on a public footpath, no one would give me a second glance if I walked up and down. Even if a tramp walked up and down they wouldn't look out of place. It's a footpath, no-one can question you. (Participant Six)

Here the offender is confirming the importance of footpaths in facilitating their searching behaviour, allowing them to search a development to establish which properties are the most suitable in terms of access, occupancy and gains. Footpaths also permit the offender to conduct this behaviour without fear of challenge; they could be legitimate users of that public space going about their daily activities. Crucially, residents cannot claim ownership of, or territorial responses over, that space.

Offenders also confirmed the benefits of footpaths as a means of accessing, moving through and exiting a development – allowing them to plan their route to and from a target: 'The appeal of a footpath is that you know how you are getting in and how you escape' (Participant Three).

Participant Nine confirmed the benefits of footpaths ('ginnels') as a means of reducing the risks of accessing and egressing a development. They also expanded upon this to discuss how footpaths reduce their need to scan for risk, with just two possible directions – up and down the footpath – as opposed to a full 360-degree inspection.

> Pretty perfect that, straight in and out and you're covered. If someone sees me there, I'd be done doing what I need to and off before anyone got there. I've only got to look up and down the ginnel too, not all around me. That means I've got more time to concentrate getting in the actual houses. The ginnel takes a lot of work out of the job for me. (Participant Nine)

These sentiments were expressed by the majority of offenders. A small selection of these responses is highlighted below:

> It's appealing as you can go either way, you're not stuck and you're in and out quickly. I like footpaths, you can just go up and down them – bad idea putting them into housing estates. (Participant Nineteen)

> Those ginnels and footpaths are more or less escape routes. (Participant Fourteen)

> Burglars like paths – you are close to the houses but it's not like a street – you are covered. All burglars like alleyways. (Participant Twelve)

Offenders also described how footpaths provide an advantage for those with knowledge of the area and those on foot, leaving the police with a distinct disadvantage in relation to the offender: 'Burglars like footpaths, it makes it easy as the police can't get there easily' (Participant Seventeen). Participant Nine reiterating that benefit of enhanced knowledge as well as ease of access/exit for pedestrians: 'Having ginnels on an estate is great, cos you know the area better than the police you'll easily lose them...you know the routes'.

Offenders also described the risks associated with cul-de-sac developments in terms of physical limitation in accessing and egressing a development but also in terms of the enhanced risks of being observed by residents and legitimate users of that space. True culs-de-sac have one point of entry/egress for both vehicles and pedestrians. You exit the development the way you entered, increasing the likelihood of being observed.

> If it's a cul-de-sac then it's usually one way in and one way out. You would be stupid to do a cul-de-sac. (Participant Eight)

Offenders spoke about the 'bluff' or pretence involved in committing an act of domestic burglary. You might pretend to be lost, create a fake friend/address that you are visiting or claim that you have lost your dog (whilst carrying a lead or dog biscuits). The vast majority of the 22 offenders described how committing a burglary on a true cul-de-sac enhances the risk involved in conducting this *bluff*. Exiting the development the way that you entered (after being challenged) enhances the likelihood that you will be observed and/or challenged.

> I wouldn't target houses on a cul-de-sac because you feel trapped and it's difficult if someone challenges you. They might say what are you doing and you say you are lost and then you have to walk back out the way you came in as they are looking at you. (Participant Five)

Image 14 did elicit some concern regarding the presence of footpaths within a development. This image displayed a very narrow pathway running at the rear of properties with high walls and fences delineating the public from private space. This, in the view of SBD, would be a poorly designed development. Three offenders expressed the view that the image portrayed an environment in which they would feel trapped and unable to escape, should the police attend the scene.

> This is good and bad. It's closed off so if you are trapped you are f***ed. (Participant One)

> I don't like this footpath, I would feel cornered and I can't see a way out. (Participant Two)

> I'm not comfortable with that footpath as it looks like a dead end and I want to know how I am getting out. (Participant Three)

3.4.5   Defensible Space

The creation of defensible space includes clearly demarcating public, semipublic, semiprivate and private space and ensuring clarity regarding the ownership of that space, thus enabling residents to make territorial responses in protecting what is defined as their space. SBD guidance clearly encourages the creation of defensible space, stating that: 'Where it is desirable to limit access/use to residents and their legitimate visitors, features such as rumble strips, change of road surface (by colour or texture), pillars, brick piers or narrowing of the carriageway may be used'. The guidance goes on to confirm that: 'This helps to define the defensible space, psychologically giving the impression that the area beyond is private' (Secured by Design 2016, p. 14).

The concept of defensible space was demonstrated in several of the images shown to the offenders, in particular image 13 that showed a small cul-de-sac with all properties facing onto the street, a narrowing of the road entrance and a change in road colour and texture at the entrance to the development. The word 'private' was also painted in large white capital letters on the road at the entrance to the estate. Several offenders were deterred by what they described as 'like walking into their own little community' (Participant Fourteen) where residents know each other and know who is a stranger within the area. Participant Eighteen summarised this perfectly: 'Everyone that lives there will be focused on the entrance and what goes on. They'll all know each other and keep an eye out for each other - give the key to the coal man that sort of thing'.

Participant Five was deterred by the sense of community represented within image 13, confirming that he would feel conspicuous and that this would be enough to prevent him from selecting this as a target for burglary.

> These people have a bee in their bonnet....This is a private road for private people. I would feel awkward here. It's all about the bluff and I couldn't pull it off here. (Participant Five)

Whilst the images were not selected to lead offenders to comment on specific components or features of SBD, this was the nature of response expected, in particular from image 13. Rather surprisingly, this image of what can only be described as a textbook example of defensible space produced a somewhat different response. Of the 15 offenders to make comment on the presence of the word 'private' written on the road, 11 interpreted this as an attractor – that 'private' equated to privately owned housing (i.e. not social housing), and thus the residents would have more money or more to steal.

> Private road suggests that it's not council housing so they won't be on benefits. (Participant Three)

> Private road means they've got money, they're middle to high class people – working people and I'd be attracted straight away. I'd think private road they've got coin. (Participant Nine)

> Private road just means they've all bought their houses. (Participant Nine)

Even without the reference to 'private road', offenders largely interpreted that change in road colour and texture and the narrowing of the entrance to the estate as a sign of exclusivity: 'The private road sign and the change in road colour and texture give me the impression that it is an exclusive area – they have more money and that would attract not deter me' (Participant Three). Others simply misinterpreted this as a place that they were not permitted to park: 'The private road just means you can't park there' (Participant Six), or a traffic-calming measure: 'The change in road colour and texture and the private sign wouldn't put me off, I wouldn't even notice it. I just thought it was a speed bump' (Participant Thirteen).

3.4.6 Physical Security

As was highlighted above, the SBD award requires a development to meet both principles related to the design and layout of developments and also specific physical security standards. These include doorsets, windows, lighting, garages, intruder alarms and fencing/gates (list not exhaustive). Physical security standards are reviewed (and tested) regularly; thus, the specifics of these standards are likely to change on an annual basis.

The 22 offenders were shown a series of images that included cues related to physical security. These included doorsets, door locks, windows, gates, alarms and security grilles. As with each of the previous elements of SBD, offenders were not directed to comment on these features; they were simply asked to comment on what they saw in each image.

Physical security was clearly an important factor in influencing burglar decision-making. All offenders referred to the concept with 103 references in total. Offenders made clear that the quality of doorsets and door locks is crucial to the appeal of a property and that they are able to make a clear and rapid judgement on that quality.

> The hinges are on the outside of that door for God's sake. You only need to pop three pins out of them. It's a three-minute job. It's a cheap arse door that one. (Participant Sixteen)

A large proportion of the sample made reference to the ease with which they could mole grip or snap the locks, with offenders regularly referencing these modus operandi as their primary means of entry.

> That would be easy. The lock on that door looks like an old one. The newer ones have thicker handles around them and are harder to get through. I'd use mole grips and a screwdriver. Gold casing around the handle will come off straight away and it'd take me around a minute to get in. (Participant Eight)

> It's quicker mole gripping than using a key sometimes – I reckon I can do most of them in 30 seconds. (Participant Ten)

> I would mole grip the lock, but take the casing away so my grips can't be traced. I would take the casing away or bury it in the garden. I hide the mole grips under the car – just clamp them on! (Participant Twelve)

A large proportion of the sample expressed the view that they were far advanced in what Pease (2001, p. 27) referred to as the 'arms race', almost mocking manufacturers, local authorities and developers for not addressing this clear weakness in housing design.

> If manufacturers know that we can easily mole grip a lock, why don't they change that lock to make it harder to break in? It's common sense. (Participant Seven)

> It's a council door, got a key for those. (Participant Seventeen)

With the exception of one brand of burglar alarm (ADT), the offenders were not deterred by intruder alarms. Justifications for this response were that (1) neighbours and passers-by do not respond/react to an activated alarm; (2) they would simply disable the alarm prior to, or soon after, activation; and (3) they were aware of technical limitations of specific alarms and their installation.

In relation to the first rationale, several participants spoke of alarms being ignored (with the exception of ADT monitored alarms).

> If I smashed the window and the alarm went off I might scuttle away and then come back ten minutes later to see if anyone had bothered dealing with it. From personal experience eight out of ten won't bother doing anything about it. (Participant Ten)

> I have an opinion on alarms. Out of every ten alarms, seven of them won't go off. (Participant Sixteen)

Offenders spoke of spraying expanding polyurethane foam into the external alarm box the night before the burglary – the alarm would be activated but the sound would not be heard: 'I would buy foam sealant from DIY stores. Some took 24 hours to set, some quicker. I'd seal them up during the night and go back and do the houses the day after. The alarms still go off but you can't hear them' (Participant Nine). Another method of deactivating the alarm involved taking the internal alarm box off the wall once inside the property, with several participants stating that for the majority of brands, this would deactivate the alarm: 'Good alarms like *** don't stop when you pull them off the wall. The cheap ones do' (Participant Thirteen).

Even though none of the images showed a specific brand of alarm, of the 12 offenders to reference intruder alarms, 8 mentioned the effectiveness of ADT alarms. It is not clear where the strength of this perception comes from. It was as if the sample believed all ADT alarms to be police monitored (which they are not) and all none ADT alarms to not be police monitored – which is not the case. However, this was a definite pattern to emerge from the 22 interviews.

An ADT alarm would deter me. (Participant One)

I wouldn't go near ADT alarms as they are linked to the police. (Participant Two)

Only ADT alarms would bother me. If it had an ADT alarm I wouldn't give it a second glance. (Participant Six)

Other security measures referred to by the sample included security grilles on windows and security gates. Responses to these two measures were mixed with several being deterred by the presence of security gates: 'It looks secure. The gates are a good idea. I'd like to live somewhere like that to be honest' (Participant Nineteen), and 'Not sure to be honest. The gates are closed for a start and I wouldn't want to draw attention to myself by climbing over. It's a simple thing to shut the gates – not as inviting' (Participant Eight).

Others felt that the gates and grilles conveyed the impression that the properties contained something worth taking. With the security gates in what appeared to be a gated community, this, in the view of offenders, gave an air of exclusivity. For image 7, many offenders interpreted the security grilles as a marker of a cannabis farm. In both cases, these security measures acted as an attractor.

The security grille makes me think there's something worth taking. (Participant Twenty-One)

I would see it as a challenge. (Participant Five)

The bars on the downstairs door make me suspicious. I wonder what they are trying to hide. I would sniff the vent to see if they were growing weed and if they were I would break in. (Participant Five)

It looks like a cannabis farm to me - it would definitely attract some interest from burglars. (Participant Eight)

3.4.7 *Management and Maintenance*

As was alluded to above, SBD is continually reviewed and revised in line with research findings and the monitoring of performance. Not only do specific technical specifications change but so do key elements of the principles of design and layout. Earlier versions of SBD did specify that developments must have a programmed management system in place to maintain an area (this was still in place in 2004); updates have toned this requirement down somewhat – partly as a recognition that with the exception of socially managed housing, there is little that SBD can do to influence management once a development is built and resided within. SBD New Homes (2016) refers to the management of communal areas: 'Adequate mechanisms and resources must be in place to ensure its satisfactory future management' (p. 18) and in relation to planting and vegetation: 'Future maintenance requirements are adequately considered at the design stage and management programmes are put in place to ensure that the maintenance will be properly carried out' (p. 23).

Offenders gave very different responses relating to the concept of maintenance. For those who were attracted to properties with poor management and maintenance, these perceptions related to (1) an abundance of rubbish means the residents will have lots of things to steal. 'They've got four bins. Why do they need so many? It tells me they spend a lot of money on food, so they've got money. Rubbish comes from one thing – buying stuff!' (Participant Nine). (2) A lack of care for your property equates to less care regarding security: 'They're sloppy, which means they might have left their keys in the door, or might have left the door open' (Participant Sixteen), and conversely, for a well-maintained property, residents are more likely to care about security and thus be more vigilant regarding strangers: 'It's manicured so someone takes time to look after it and they're probably looking out for people like me coming along' (Participant Sixteen).

SBD suggests that developments should be well-maintained. A large proportion of the offenders were not only deterred by poor maintenance; they were actually attracted to well-maintained properties, equating a lack of care for external areas with a lack of money (thus there would be insufficient goods to make the effort worthy of the reward) or with a lack of care for goods in general – again suggesting that the risk would not outweigh any reward.

> I wouldn't touch this place. They are scruffy b*stards they aren't going to have owt. Look at the state of their garden, they are like tramps. They haven't even got proper curtains, that's a f*cking bedsheet! (Participant Thirteen)

> Look at the state of them! What are they going to have? (Participant Twenty)

> It's a shit hole, they've got nothing I want. (Participant Sixteen)

Several offenders went on to specify that well-maintained properties would be an attractive target for burglary because the owners are likely to have more money and thus more goods to steal.

Neat and tidy means they look after their house and it'll be like that inside too. (Participant Eight)

I would actively look for places like this. If they have a neat garden you know they have something to steal. You know they look after themselves and the house. (Participant Three)

I'd definitely go there, look how nice and clean it is. (Participant Fifteen)

That's an easy hit. If they take care of the garden they've got something worth stealing. (Participant Seventeen)

3.5   What Can Secured by Design Learn from Burglar Accounts?

Before concluding, we should remind ourselves of the progress made within the UK in embedding crime prevention within the planning system. Advancement has not been straightforward, and there have been setbacks along the way, but the current status of planning regulation, policy and guidance provides an adequate level of regard to designing places to prevent crime. It is 'adequate' because it could be better and it has been better. There is much opportunity to strengthen policy, and there is no better way to do that than to conduct research to improve our knowledge base. It is hoped that this research can form part of that 'lobbying' of government. Research has consistently demonstrated that SBD housing provides enhanced protection against burglary and that this is specific not only to the superior physical security measures – such as window and doorsets and locks – but also to the design and layout of developments.

In-depth interviews with 22 prolific burglars have confirmed that both physical security and design and layout play a key role in influencing decision-making. Offenders are attracted to properties with poor levels of natural surveillance, clear movement opportunities to allow access and escape and poor physical security. They are deterred by properties with robust physical security, abundant surveillance opportunities, a strong sense of community and limited routes for entry and escape. The environmental cues most consistently and regularly referred to by the sample of burglars were surveillance and physical security, with defensible space and management and maintenance less prominent as deterrent factors. Whilst offenders were guided to comment on the physical features of housing design, there were influences referenced that did not relate to housing or development design. The most unattractive targets are being described as such due to moral reasoning – they are resided in by the elderly, disabled or those relying on the welfare state. The extent to which these justifications are accurate is important, but not the focus of this chapter. There are many more relevant findings that strengthen the evidence base regarding the impact of SBD upon burglary but also provide food for thought regarding improvement and enhancement of the scheme; after all SBD does clearly state that:

'The Police service continually re-evaluates the effectiveness of Secured by Design and responds to emerging crime trends and independent research findings, in conjunction with industry partners, as and when it is considered necessary and to protect the public from crime' (Secured by Design 2016, p. 4).

To reiterate, offenders confirmed that they would avoid properties with open frontages, large windows, properties on true culs-de-sac, those with robust physical security and where properties faced onto the street. They were clearly able to distinguish between door and window locks (and frames) that could and could not be easily breached. They were attracted to properties that were concealed from view, those with poor physical security and those on developments that were connected by footpaths – which provided opportunities to root for suitable targets and to access and egress prior to and post-offence. With the exception of one brand, they were not deterred by burglar alarms, and in many instances, security measures such as gating a development, security grilles on windows or other excessive visual security cues appeared to convey wealth and something worth protecting.

Less straightforward findings are related to the concepts of defensible space and management and maintenance. Images of the textbook implementation of defensible space principles – the narrowing of the entrance to a development, a change in road colour and texture and a 'private' sign at the entrance to an estate – received surprising responses. For the vast majority of offenders, 'private' was interpreted as 'privately owned' housing (as opposed to social housing) and thus indicated both wealth and a moral justification for selection. In terms of management and maintenance, there was no straightforward pattern of responses, but the established viewpoint that a lack of concern for the appearance and upkeep of a property equates to a perception that the property is an attractive target due to the residents' lack of concern was not confirmed. The majority of offenders were attracted to well-kept properties – if a resident cares for their garden and outbuildings, they will care for their internal products – thus plentiful pickings. In many cases poor maintenance simply turned the offender off, with responses that the risk would not be worth the reward. This is a difficult finding to respond to as it would be a brave policymaker that would publish guidance to encourage residents to leave their properties unkempt! But it does warrant further consideration.

Research of this kind is time-consuming – it took 18 months to recruit and interview 22 offenders; however, it is vital in reviewing the SBD guidance and standards. Analysing police-recorded and self-reported crime provides a strong indication of risk; however, triangulating that with offender accounts adds detail, depth and specificity. After all, as Nee (2003) explains, these individuals are those making day-to-day decisions regarding target selection, and we should allow: '… the expert in the chosen field, the residential burglar, to lead the course of the enquiry, yielding a rich and increasingly focused understanding of the subject' (Nee 2003, p. 37).

References

Armitage, R. (2000). *An evaluation of secured by design housing within West Yorkshire – Briefing Note 7/00*. London: Home Office.

Armitage, R. (2004). *Secured By design – An investigation of its history, Development and future role in crime reduction*. Unpublished PhD Thesis, University of Huddersfield, Huddersfield.

Armitage, R. (2006). Sustainability versus safety: Confusion, conflict and contradiction in designing out crime. In G. Farrell, K. Bowers, S. Johnson, & M. Townsley (Eds.), *Imagination for crime prevention: Essays in Honour of Ken Pease, Crime prevention studies* (Vol. 21, pp. 81–110). Monsey, New York: Criminal Justice Press and Willan Publishing.

Armitage, R. (2013). *Crime prevention through housing design: Policy and practice, Crime Prevention and Security Management Book Series*. Basingstoke: Palgrave Macmillan.

Armitage, R. (2017a). Burglars' take on crime prevention through environmental design. *Security Journal, 31*, 285. https://doi.org/10.1057/s41284-017-0101-6.

Armitage, R. (2017b). Crime prevention through environmental design. In R. Wortley & M. Townsley (Eds.), *Environmental criminology and crime analysis* (pp. 259–285). New York: Routledge.

Armitage, R., & Monchuk, L. (2011). Sustaining the crime reduction impact of secured by design: 1999 to 2009. *Security Journal, 24*(4), 320–343.

Armitage, R., Monchuk, L., & Rogerson, M. (2010). It looks good, but what is it like to live there? Assessing the impact of award winning design on crime. *Special Volume of European Journal on Criminal Policy and Research, 17*(1), 29–54.

Birks, D., & Davies, T. (2017). Street network structure and crime risk: An agent-based investigation of the encounter and enclosure hypothesis. *Criminology, 55*(4), 900–937.

Bowers, K. J., & Guerette, R.T. (2014). Effectiveness of situational crime prevention. In G. Bruinsma & D. Weisburd (Editors in Chief) *Encyclopaedia of Criminology and Criminal Justice* (pp. 1318–1329). New York: Springer.

Brantingham, P. J., & Brantingham, P. L. (1993). Environment, routine and situation: Toward a pattern theory of crime. *Advances in Criminological Theory, 5*, 259–294.

Brooke, M. (2013). Secured by design – The story so far. *Safer Communities, 12*(4), 154–162.

Brown, B. B., & Altman, I. (1983). Territoriality, defensible space and residential burglary: An environmental analysis. *Journal of Environmental Psychology, 3*, 203–220.

Brown, B., & Bentley, D. (1993). Residential burglars judge risk: The role of territoriality. *Journal of Environmental Psychology, 13*, 51–61.

Budd, T. (1999). *Burglary of domestic dwellings. Findings from the British Crime Survey*. Issue 4/1999. London: Home Office.

Clarke, R. V. (1992). Introduction. In R. V. Clarke (Ed.), *Situational crime prevention – Successful case studies* (pp. 3–36). New York: Harrow and Heston.

Copes, H., & Hochstetler, A. (2014). Consenting to talk: Why inmates participate in prison research. In P. Cromwell & M. Birzer (Eds.), *Their own words: Criminals on crime* (pp. 19–33). New York: Oxford University Press.

Cozens, P., Hillier, D., & Prescott, G. (2001). Defensible space: Burglars and police evaluate urban residential design. *Security Journal., 14*, 43–62.

Cromwell, P. F., & Olson, J. N. (1991). *Breaking and entering: An ethnographic analysis of burglary*. Newbury Park: Sage.

Davies, T., & Johnson, S. D. (2014). Examining the relationship between road structure and burglary risk via quantitative network analysis. *Journal of Quantitative Criminology, 31*, 481. https://doi.org/10.1007/s10940-014-9235-4.

Department for Communities and Local Government. (2005). Planning policy statement 1: Delivering sustainable development. http://www.communities.gov.uk/publications/planningandbuilding/planningpolicystatement1. Accessed 21 Mar 2012.

Department for Communities and Local Government. (2006). *Circular 01/2006: Guidance on changes to the development control system*. London: DCLG.

Department for Communities and Local Government. (2008). *The code for sustainable homes: Setting the standard in sustainability in new homes*. London: DCLG.

Department for Communities and Local Government (2011). Planning policy statement 3: Housing. http://www.communities.gov.uk/publications/planningandbuilding/pps3housing. Accessed 21 Mar 2012.

Department for Communities and Local Government. (2012a). *External review of government planning practice guidance*. London: DCLG.

Department for Communities and Local Government. (2012b). *National Planning Policy Framework*. London: DCLG.

Department of Environment, Transport and the Regions. (1999). *Towards an urban renaissance – Final report of the urban task force*. London: DETR.

Department of the Environment. (1994). *Planning out crime: Circular 5/94*. London: DoE.

Elffers, H. (2010). Misinformation, misunderstanding and misleading as validity threats to accounts of offending. In W. Bernasco (Ed.), *Offenders on offending: Learning about crime from criminals* (pp. 13–22). Cullompton: Willan.

Everson, S. (2000). *Repeat offenders and repeat victims: Mutual attraction or misfortune?* Unpublished PhD thesis, University of Huddersfield.

Great Britain. (1998). *Crime and Disorder Act 1998*. Chapter 37. London: HMSO.

Homel, R., Macintyre, S., & Wortley, R. (2014). How burglars decide on targets: A computer-based scenario approach. In B. LeClerc & R. Wortley (Eds.), *Cognition and crime: Offender decision making and script analyses* (pp. 26–47). Oxford: Routledge.

Housing Corporation. (2003). *Scheme development standards* (5th ed.). London: Housing Corporation.

Housing Corporation. (2007). *Design and quality standards*. London: Housing Corporation.

Jacobs, J. (1961). *The death and life of great American cities*. New York: Random House.

Jeffery, C. R. (1971). *Crime prevention through environmental design*. Beverly Hills: Sage.

Johnson, S., & Bowers, K. J. (2010). Permeability and burglary risk: Are Cul-de-Sacs safer? *Quantitative Journal of Criminology, 26*(1), 89–111.

Johnson, S. D., & Bowers, K. J. (2014). How guardianship dynamics may vary across the street network: A case study of residential burglary. In *Liber amicorum voor Henk Elffers*. NSCR: Amsterdam.

Jones, A., Valero-Silva, N., & Lucas, D. (2016). *The effects of 'secure warm modern' homes in Nottingham: Decent homes impact study*. Nottingham: Nottingham City Homes.

Kearns, J. N., & Fincham, F. D. (2005). Victim and perpetrator accounts of interpersonal transgressions: Self-serving or relationship-serving biases? *Personality and Social Psychology Bulletin, 31*(3), 321–333.

Montoya, L., Junger, M., & Ongena, Y. (2016). The relation between residential property and its surroundings and day- and night-time residential burglary. *Environment and Behavior, 09*(2014), 516–549. https://doi.org/10.1177/0013916514551047.

Moss, K., & Pease, K. (1999). Crime and Disorder Act 1998: Section 17 a wolf in sheep's clothing? *Crime Prevention and Community Safety: An International Journal, 1*(4), 15–19.

Nee, C. (2003). Research on burglary at the end of the millennium: A grounded approach to understanding crime. *Security Journal, 16*(3), 37–44.

Nee, C., & Meenaghan, A. (2006). Expert decision making in burglars. *British Journal of Criminology, 46*, 935–949.

Nee, C., White, M., Woolford, K., Pascu, T., Barker, L., & Wainwright, L. (2015). New methods for examining expertise in burglars in natural and simulated environments: preliminary findings. *Psychology, Crime & Law, 21*(5), 507–513. https://doi.org/10.1080/1068316X.2014.989849.

Newman, O. (1973). *Defensible space: People and design in the Violent City*. London: Architectural Press.

Office of the Deputy Prime Minister. (2000). *Our towns and cities: The future – Delivering an urban renaissance*. London: ODPM.

Office of the Deputy Prime Minister and Home Office. (2004). *Safer places – The planning system and crime prevention*. London: HMSO.

Pascoe, T. (1999). *Evaluation of secured by design in public sector housing – Final report.* Watford: BRE.

Pease, K. (2001). *Cracking crime through design.* London: Design Council.

Pease, K., & Gill, M. (2011). *Direct and indirect costs and benefits of home security and place design.* Leicester: Perpetuity Research and Consultancy International Ltd..

Reppetto, T. A. (1974). *Residential crime.* Cambridge: Ballinger.

Reynald, D. (2009). Guardianship in action: Developing a new tool for measurement. *Crime Prevention and Community Safety, 11*, 1–20.

Reynald, D. (2010). Guardians on guardianship: Factors affecting the willingness to supervise, the ability to detect potential offenders, and the willingness to intervene. *Journal of Research in Crime and Delinquency, 47*(3), 358–390.

Reynald, D. M., & Elffers, H. (2009). The future of Newman's defensible space theory linking defensible space and the routine activities of place. *European Journal of Criminology, 6*(1), 25–46.

Secured by Design. (2016). *Secured by Design New Homes – 2014.* London: Secured by Design.

Shaw, M., & Pease, K. (2000). *Research on repeat victimisation in Scotland: Final report.* Edinburgh: Scottish Executive Central Research Unit.

Sidebottom, A. L., Tilley, N., Johnson, S., Bowers, K., Tompson, L., Thornton, A., & Bullock, K. (2017a). Gating alleys to reduce crime: A meta-analysis and realist synthesis. *Justice Quarterly, 35*, 55. https://doi.org/10.1080/07418825.2017.1293135.

Sidebottom, A.L., Armitage, R., Tompson, L. (2017b, March). *Reducing crime through secured by design: A systematic review.* Secured by Design National Training Event 2017. Northampton.

Taylor, R. (2002). Crime prevention through environmental design (CPTED): Yes, no, maybe, unknowable and all of the above. In R. B. Bechtel & A. Churchman (Eds.), *Handbook of environmental psychology* (pp. 413–426). New York: Wiley.

Taylor, R., & Gottfredson, S. D. (1987). Environmental design, crime and prevention: An examination of community dynamics. *Crime and Justice: An Annual Review of the Research, 8*, 387–416.

Teedon, P., Reid, T., Griffiths, P., Lindsay, K., Glen, S., McFayden, A., & Cruz, P. (2009). *Secured by design impact evaluation final report.* Glasgow: Glasgow Caledonian University.

Tseloni, A. (2006). Multilevel modeling of the number of property crimes: Household and area effects. *Journal of the Royal Statistical Society: Series A (Statistics in Society), 169*(2), 205–233.

Tseloni, A., Ntzoufras, I., Nicolaou, A., & Pease, K. (2010). Concentration of personal and household crimes in England and Wales. *European Journal of Applied Mathematics, 21*(45), 325–348.

Tseloni, A., Thompson, R., Grove, L. E., Tilley, N., & Farrell, G. (2014). The effectiveness of burglary security devices. *Security Journal* (advance online publication 30 June 2014; https://doi.org/10.1057/sj.2014.30).

Van Der Voordt, T. J. M., & Van Wegen, H. B. R. (1990). Testing building plans for public safety: Usefulness of the delft checklist. *Netherlands Journal of Housing and Environmental Research, 5*(2), 129–154.

van Gelder, J.-L., Nee, C., Otte, M., van Sintemaartensdijk, I., Demetriou, A., & van Prooijen, J.-W. (2017). Virtual burglary: Exploring the potential of virtual reality to study burglary in action. *Journal of Research in Crime and Delinquency, 54*(1), 29–62. https://doi.org/10.1177/0022427816663997.

Vollaard, B., & Ours, J. C. V. (2011). Does regulation of built-in security reduce crime? Evidence from a natural experiment. *The Economic Journal., 121*(May), 485–504.

Weisburd, D. (2015). The law of crime concentration and the criminology of place. *Criminology, 53*(2), 133–157.

Wiles, P., & Costello, A. (2000). *The 'road to nowhere': The evidence for travelling criminals* (Home Office Research Study 207). London: Home Office.

Winchester, S., & Jackson, H. (1982). *Residential burglary: The limits of prevention* (Home Office Research Study Number 74). London: Home Office.

Chapter 4
Which Security Devices Reduce Burglary?

Rebecca Thompson, Andromachi Tseloni, Nick Tilley, Graham Farrell, and Ken Pease

Abbreviations

CCTV	Closed-circuit television
CSEW	Crime Survey for England and Wales
CVS	Cadre de Vie et Sécurité
FAVOR	Familiarity, Accessibility, Visibility, Occupancy, Rewards
ONS	Office for National Statistics
SIAT	Security Impact Assessment Tool
SPF	Security Protection Factor
WD	Window and door locks
WDS	Window locks, door locks and security chains
WIDE	Window locks, internal lights on a timer, double door locks and external lights on a sensor

4.1 Introduction

Domestic burglary is costly in both human and financial terms (Dinisman and Moroz 2017). Individuals can attempt to protect their homes from burglary in a number of ways. Some may have informal agreements with neighbours to look after

R. Thompson (✉) · A. Tseloni
Quantitative and Spatial Criminology, School of Social Sciences, Nottingham Trent University, Nottingham, UK
e-mail: becky.thompson@ntu.ac.uk

N. Tilley
Jill Dando Institute, Department of Security and Crime Science, University College London, London, UK

G. Farrell
School of Law, University of Leeds, Leeds, UK

K. Pease
Criminology and Social Sciences, University of Derby, Derby, UK

© Springer Nature Switzerland AG 2018
A. Tseloni et al., *Reducing Burglary*,
https://doi.org/10.1007/978-3-319-99942-5_4

each other's homes or have house-sitting arrangements with friends and relatives. Some protective factors relate to features of the property and its surroundings, for example, culs-de-sac. Others relate to physical security measures installed at the home itself. The focus of this chapter is on such physical security devices.

In 2018, the market for home security devices in the UK was worth approximately £139.8 million (Mintel 2014). This may have been driven, in part, by discounted home insurance premiums conditional on installation of a burglar alarm. Insurance companies also routinely ask in proposal forms about the types of locks installed on external doors and windows. Burglar alarms and high-quality external locks are thus overtly preferred by insurers. However, it is unclear to what extent such preferences are evidence-based. Knowing if particular security devices are effective and, if so, when, where and how, is of great importance.

This chapter examines the protective effect of seven security devices (in every possible combination) to establish the most effective physical security configuration. This information is of direct benefit to householders, private landlords, housing associations, local authorities, government bodies, victims of crime, police forces, victim support organisations and insurance, security and building companies. The chapter extends the work of Tseloni et al. (2014) and Tseloni et al. (2017) in comparing security device effectiveness against domestic burglary with entry and attempted burglary separately. It offers a new perspective in relation to the distinctive mechanisms performed by different security devices. The findings seek to fill an important research gap with immediate policy applicability.

The structure of this chapter is as follows. Previous work on the *availability*[1] of security devices is discussed followed by an outline of existing research on the *effectiveness* of such devices. That different security devices have distinctive mechanisms is proposed followed by a discussion of findings which seek to test this proposition. The chapter ends with a discussion of the main findings and their potential policy implications.

4.2   Previous Work on Security Availability

It is reasonable to conjecture that, as offenders adapt, technology evolves and lifestyles change, the most commonly installed security devices also change as a consequence. Previous research on the availability of different security devices over time is relatively limited, with the exceptions of work by the Office for National Statistics (ONS) (2013, 2016) and Tseloni et al. (2017). The ONS uses data from the Crime Survey for England and Wales (CSEW) to estimate the proportion of

---

[1] Here, availability *does not* mean available to purchase from a shop. We use the term 'availability' throughout this and the following chapter to signify a security device was at least available *to use* by the householder if not actually used at the time of the incident. Over the time period studied, the data was not available to establish whether devices that require 'activation' of some kind (e.g. burglar alarms) were *in use* at the time of the incident.

households with particular security devices in their homes. Since 1995, there has been a statistically significant increase in the proportion of households with some form of security (ONS 2016). For the purposes of their reports, the ONS (2013) often groups household security into the following four categories:

- *No security* – the household has none of the security devices asked about in the CSEW.
- *Less than basic security* – the household is without both window *and* door locks but has some other devices.
- *Basic security* – the household has double locks/deadlocks on at least some external doors *and* locks on windows.
- *Enhanced security* – the household has window *and* door locks plus at least one additional security device.

According to the 2012/2013 CSEW, 75 percent of households had at least basic security (ONS 2013). Table 4.1 provides examples of the types of security combination (measured by the CSEW) that would fit under each of the above categories. ONS (2016) figures suggest there have been statistically significant increases in the uptake of individual devices since 1995, including window locks, security lighting, burglar alarms and double door locks/deadlocks.

This information is useful but focuses upon the availability of security devices (e.g. the proportion of households with a burglar alarm) irrespective of what other devices are also in place or upon relatively generic security categories (e.g. 'basic' and 'enhanced'). It is important to estimate the uptake of *specific* device *combinations* as failure to do so risks masking any underlying (and important) changes in security availability.

Table 4.1 Examples of CSEW security device combinations that fit the ONS (2013) 'basic/ enhanced' categories

Classification	Examples of device groupings that fit the relevant criteria		
Less than basic	W	DE	WIE
	D	WB	DES
	WS	WES	WI
	WE	E	DB
	DS	WEB	WIEB
	S	B	
Basic	WD		
Enhanced	WDE	WID	WIDSB
	WDS	WIDESB	WDC
	WDEB	WDESB	WIDECB
	WDES	WIDES	WDECB
	WIDEB	WDSB	WDEC
	WIDE	WIDB	WDSC
	WDB	WIDS	

Note: *B* burglar alarm, *C* CCTV, *D* double door locks or deadlocks on external doors, *E* external lights on a sensor, *I* internal lights on a timer, *S* security chains, *W* window locks. Only the most popular (>100 households) combinations are shown here

In an attempt to determine whether security played a part in the fall in burglary with entry, Tseloni et al. (2017) explored the availability and effectiveness of security devices over time (between 1992 and 2011/2012). They found the availability of most security devices increased just before and during the period of the drop in burglary.[2] There was a dramatic fall in the proportion of households with 'no security' around the same time. This figure has remained relatively consistent at around 5 percent since 1998 (see also Chap. 8). The proportion of households with more than one security device increased over the period of the crime drop (with the exception of combinations including security chains). The combination of window locks and double locks or deadlocks on external doors was the most widespread security configuration. Part of this may be due to the increasingly commonplace installation of double glazed windows and doors (Tilley et al. 2015a; Tseloni et al. 2017).

4.3   Previous Work on Security Device Effectiveness Against Burglary

The preceding section reported that security devices have become a more common feature in homes, but what do we know about their effectiveness? Householders, private landlords and housing associations can access advice from a range of organisations and individuals regarding how to protect their homes against burglary. Unfortunately, not all advice is based upon sound research. The accuracy of such advice, and the research upon which it may (or may not) be based, is rarely questioned. Recommendations are often made with little reference to the evidence upon which they are formed. Many examples of this can be found, for instance: 'alarms are undoubtedly the most effective deterrent against burglary' (Metropolitan Police 2017, p. 9) and 'visible burglar alarms will deter opportunist burglars and increase the security of your home' (Age UK 2017).

Previous research on security device effectiveness has tended to utilise data from one of three sources: (1) victimisation surveys, (2) accounts of imprisoned or active offenders or (3) evaluations of large-scale burglary reduction initiatives. The main findings from each of these sources will be discussed in turn.

4.3.1   Victimisation Survey Data

Speaking to victims of crime can provide unparalleled insight into their experiences and is a useful source to assess risk and determine patterns. The most detailed source of victimisation data in England and Wales is the CSEW. It collects information from a sample of the population on a wide range of crime- and justice-related topics (see ONS 2017). In relation to security, this survey asks respondents whether their

---

[2] According to the Crime Survey for England and Wales, domestic burglary levels peaked in 1993 and have since declined (Tseloni et al. 2017; ONS 2018a).

household has particular security devices, such as a burglar alarm, double locks or deadlocks on external doors, window locks and CCTV. The list of devices from which respondents could choose changed considerably between the 1996 and 1998 sweeps but has remained relatively consistent since then (see Appendix Table 4.7). Data from more recent sweeps include information regarding nine devices.[3] Research utilising this data generally focuses upon the protection conferred by categories such as 'basic security' (ONS 2013) or the presence or absence of specific devices *without consideration of other devices in place at the same time.*

Existing CSEW-based research finds that households with more security had lower burglary rates (Mayhew et al. 1993; Budd 1999; Flatley et al. 2010; ONS 2013; Tseloni et al. 2014, 2017). Conversely, households with none of the security devices listed in the CSEW were twice as likely to become victims of burglary compared to those with at least 'basic security' (see Table 4.1; ONS 2013). In two landmark burglary studies utilising victimisation survey data, Budd (1999, 2001) found security devices were strongly associated with a reduction in victimisation risk. In 1998, households without any of the security devices measured by the survey were at greatest risk (whilst controlling for a range of other factors). 'Basic' security measures (i.e. locks on external doors and windows) were particularly good at reducing risk, whilst adding burglar alarms, security lights and window grills offered additional protection (Budd 2001).

In a more recent study in pursuit of more detailed understanding of the relative efficacy of different security devices, Tseloni et al. (2014) analysed data from four sweeps of the CSEW (2008/2009–2011/2012). Put simply, the security devices installed in burgled homes were compared with non-burgled households. The most effective combination (that also provided the best value for money[4]) comprised window locks, internal lights on a timer, double locks/deadlocks on external doors and external lights on a sensor (inviting the mnemonic WIDE).

In an extension of the same team's previous work, Tseloni et al. (2017) explored the relationship between burglary with entry, security devices and burglars' modi operandi. This study found an increase in the availability of WIDE alongside a decline in the proportion of households without security over the period of the crime drop. The security increases were particularly prominent before, at the start and over the period of the steepest burglary fall in England and Wales (1992–2001/2002). Households with 'no security' were nearly eight times more likely (compared to the rest of the population) to be burgled in 2008/2009–2011/2012. It concludes there is strong evidence that security caused the decline in burglary in England and Wales in the 1990s. In sum, research based upon victimisation survey data suggests security is an important consideration in relation to burglary risk in that particular devices (in particular combinations) reduce the likelihood of becoming a victim (see also Chap. 8).

---

[3] These are burglar alarms, double locks or deadlocks on external doors, window locks, internal lights on a timer, external lights on a sensor, dummy alarms, security chains, window bars/grills and CCTV cameras. These nine devices are recorded in both the Victimisation and Crime Prevention modules.

[4] The highest total protection (by a small margin and ignoring outliers) was conferred by the following combination: window and door locks, security chains and CCTV (WDSC). In the authors' view, the second highest device configuration (WIDE) was both a cheaper and safer option given the high cost of CCTV and potential fire hazard posed by security chains (see Tseloni et al. 2014).

4.3.2 Offender Interviews

A growing body of research has explored security from the perspective of offenders (both imprisoned and active). This type of research generally uses semi-structured interviews, experimental scenarios or, more recently, simulation to determine what constitutes an attractive burglary target and to try better to understand the target selection process.[5] An example of research utilising offender interviews can be found in Chap. 3 of this book. Notwithstanding some limitations of obtaining information in this way (for example, offenders rationalising decision-making after the fact rather than accurately recalling the original event) (Wright and Decker 1994; Cromwell and Olson 2004; Roth and Roberts 2017) interesting findings have emerged from these studies.

There is consensus that there are rational elements to offender decision-making when selecting a property to burgle (Cornish and Clarke 1986; Hearnden and Magill 2004; Nee and Meenaghan 2006; Cromwell and Olson 2009). Offenders will mostly avoid particular targets if they determine the costs outweigh the expected rewards (Bennett and Wright 1984; Wright and Decker 1994; Cromwell et al. 1991). In relation to target selection, research of this kind highlights the importance of a number of specific environmental cues. The authors of the present chapter have devised the acronym FAVOR to summarise the environmental cues deemed most salient within previous research (see Table 4.2). FAVOR refers to the following five key offender considerations: Familiarity, Accessibility, Visibility, Occupancy and Rewards.

Decades of research with both incarcerated and active offenders have highlighted the relative importance of these 'FAVOR-able' environmental cues in selecting a target (Cromwell et al. 1991; Homel et al. 2014; Hearnden and Magill 2004; Nee and Meenaghan 2006; Clare 2011; Roth and Roberts 2017). Of particular relevance with regard to security are accessibility and occupancy; security devices can render a property more difficult to access (e.g. door and window locks) as well as make a property appear occupied (e.g. lights on a timer). We here review the literature with regard to offender perceptions of particular security devices.

4.3.2.1 Burglar Alarms

From an offender's perspective, burglar alarms may act as a visual deterrent by increasing the perceived risk of apprehension (Mayhew 1984). That said, there is mixed evidence regarding offender's attitudes towards alarms. The majority of studies have found alarms have deterred at least some offenders (Bennett and Wright 1984; Cromwell 1994; Maguire and Bennett 1982; Wright et al. 1995; Cromwell and Olson 2004). The presence of an alarm acted as a deterrent to 84 percent of

---

[5]Although not specifically utilising offender interviews, Langton and Steenbeck (2017) used Google Street View in an attempt to test the findings from this body of offender-based literature. They found ease of escape, accessibility and visibility were all positively related to burglary risk.

Table 4.2 The most common environmental cues considered by burglars in relation to burglary target selection as suggested by previous research (acronym FAVOR)

Familiarity	*The target household is close to the offender's home or in a familiar neighbourhood*
Accessibility	*The target household is easy to access from the street (and, conversely, has multiple escape routes), and it is easy to get inside*
Visibility	*The target household is not easily seen by passers-by and neighbours (sometimes referred to as surveillability)*
Occupancy	*The target household appears unoccupied (i.e. there is no car on the driveway or lights on in the house)*
Rewards	*The offender believes there are potentially high-value goods inside*

potential offenders in a study ($n = 82$) by Hearnden and Magill (2004). More recent research by Tilley et al. (2015b) suggests burglar alarms are associated with an *increased* risk of burglary victimisation (see also Chap. 8). There is evidence that an offender's level of experience may influence how alarms are viewed with more experienced offenders perceiving alarms, dogs and deadbolts as less of a deterrent than their less experienced counterparts (Clare 2011; Roth and Roberts 2017). Research by Armitage and Joyce (2015) suggests the quality of installation and type of alarm may influence offender perceptions (see also Chap. 3).

4.3.2.2   Locks

Window and door locks provide a physical barrier to entry (and, to some extent, can act as a visual deterrent). However, most previous research suggests that offenders are not particularly deterred by locks (Bennett and Wright 1984; Wright and Decker 1994; Wright et al. 1995). Hearnden and Magill (2004) found only 55 percent of offenders would be deterred by strong door and window locks (compared to 84 percent who were deterred by alarms). Some argue it might be the length of time taken to defeat a good lock that deters an offender, rather than the lock itself (Edgar and McInerney 1987).

4.3.2.3   Occupancy Cues/Proxies

Previous research suggests signs of occupancy (e.g. lights on inside the house, car on the driveway etc.) can be more important than physical security measures (Nee and Meenaghan 2006; Wright et al. 1995; Winchester and Jackson 1982; Roth and Roberts 2017; Snook et al. 2011). In Hearnden and Magill's (2004) study, 84 percent of offenders said they would be deterred if they believed someone was at home. Again, experience matters, with less experienced offenders more likely to be deterred by signs of occupancy (Clare 2011; Nee and Taylor 1988; Roth and Roberts 2017). As offenders gain experience, they may learn strategies to overcome such cues and/or assess the risk as low enough to discount.

Collectively, this body of research suggests that offenders draw upon their experience and expertise in the commission of offences (Nee and Meenaghan 2006; Bennett and Wright 1984; Nee and Taylor 1988; Wright et al. 1995; Roth and Roberts 2017). Nee and Meenaghan (2006, p. 935) view the burglar as a 'rational, "expert" agent' who relies upon a set of responses to past experiences. These learned responses will vary across individuals and may explain some of the mixed findings in relation to the importance of particular cues. 'Burglars select targets based on the unique combination of situational variables that each property/criminogenic environment presents' (Nee and Taylor 2000, p. 57). A range of environmental cues interact in any given situation. In the context of this chapter, it seems security is one of a number of potential considerations that (some) offenders take into account when deciding to select a target.

4.3.3 Large-Scale Initiatives

Over the past 30 years, a number of initiatives have attempted to reduce domestic burglary in the UK, for example the Kirkholt Burglary Prevention Project (Forrester et al. 1988), the Safer Cities Programme (Ekblom et al. 1996) and the Crime Reduction Programme (in particular the Reducing Burglary Initiative) (Tilley et al. 1999). Chap. 2 of this book provides an overview of national burglary reduction initiatives. In short, many of these initiatives drew on situational crime prevention principles, in particular, target hardening. Routinely employed as a burglary reduction strategy (Hamilton-Smith and Kent 2005), target hardening generally involves installing security devices to increase the effort an offender has to make to commit an offence or deter them altogether (Cornish and Clarke 2003).

Evaluations of these initiatives suggest that burglary victimisation risk can be reduced through improvements to household security (Forrester et al. 1990; Tilley and Webb 1994). The Kirkholt Burglary Prevention Project (Forrester et al. 1988; Pease 1991) offered security upgrades (with special attention to repeat victims), replaced coin meters, created Cocoon Neighbourhood Watch schemes and offered debt management advice to offenders. Burglary incidence in that area reduced by 75 percent over the three-year study period (Forrester et al. 1990). In particular, burglaries declined in households where security was installed but not in others.

In a similar vein, the Safer Cities Programme set up over 500 schemes to prevent burglary (Ekblom et al. 1996). Many of the projects involved target hardening/security upgrades (e.g. door and window improvements, alarms and security lighting). Target hardening reduced burglary under all conditions – although the best approach was a combination of target hardening and community-oriented action (see also Chap. 2).

In reviewing the large body of existing research, we conclude the evidence on the effectiveness of anti-burglary security devices is somewhat mixed. Data from the CSEW would suggest that window and door locks are a particularly effective burglary prevention measure whereas the message from offender interviews is less clear. It seems offender experience matters in determining which security devices deter. The evidence in relation to burglar alarms is also slightly conflicting with more recent studies suggesting they increase burglary risk. An explanation for these mixed results might be that grouping security devices together, or viewing devices in isolation, makes it difficult to identify the precise protective role of a range of different security combinations. Another explanation may relate to the wide range of devices available and their varying quality (Tilley et al. 2015a). This chapter presents findings from a project[6] which attempted to address some of these limitations.

4.4 Data and Methods

This analysis utilises every sweep of the CSEW between 2008/2009 and 2011/2012. Given the changing nature of crime and the evolving nature of our responses to it, the number and type of security devices examined in the CSEW has changed over time (albeit only minor changes since 1998) (see Appendix Table 4.7). The Crime Prevention and Victimisation modules for the latest survey analysed here, 2011/2012, included questions about the availability of nine household security devices.[7]

Information regarding the availability of particular security devices at the time of the incident was gathered (via the Victimisation module) from (the majority of) respondents whose home was burgled. For this analysis, if a victim reported more than one burglary within the survey reference period, their security at the time of the *first* burglary was retained for analysis. The security devices in non-burgled households were also extracted from the survey. This allows comparison of the security available in burgled households with non-burgled households. To increase the potential number of homes with any possible security combination from the list of devices, the various sweeps of the CSEW were merged to form one composite data set covering the period 2008/2009 through 2011/2012. More information regarding sample selection can be found in Appendix A.

This work employs the Security Impact Assessment Tool (SIAT) methodology developed and described by Farrell et al. (2011) in their study of vehicle security effectiveness. Put simply, the overall likelihood of becoming a victim of burglary is calculated for two distinct groups: (1) Households with no security[8]; and (2) Households with a particular security device (or combination of devices). The like-

---

[6]The project was funded by the Economic and Social Research Council (ESRC) Secondary Data Analysis Initiative and sought to answer: 'which burglary security devices work for whom and in what context'? (ES/K003771/1).

[7]Dummy alarms and window bars/grills are not included in the subsequent analysis due to their low (and diminishing) prevalence over time. Therefore, a total of seven devices are explored in this chapter.

[8]In this research, the term 'no security' should be taken to mean 'none of the CSEW listed devices' (see Appendix B in Chap. 5).

lihood that burglary will occur for each respective group is compared to the overall likelihood of burglary, which results in an odds ratio. The resulting metric is referred to as the Security Protection Factor (SPF). The SPF provides an indication as to the level of protection conferred by a security device (or combination of devices) compared to no security. If the security devices offer no protection relative to homes with no security devices, the proportions burgled would be the same (and the SPF value would be 1). The larger the SPF value from 1, the greater the level of protection provided by the device(s) under consideration.

One of the distinctive aspects of this research lay in the calculation of an SPF value for every possible combination of security devices listed in the CSEW. In the discussion of security device effectiveness, only devices (and combinations) available to a minimum of 50 households are reported. Another unique element of the research lies in the separate estimation of SPFs for burglary with entry and attempted burglary. The next sub-section will explain why we took this approach.

4.4.1 Why Examine Attempted Burglary and Burglary with Entry Separately?

There were a number of reasons behind our decision to analyse burglary with entry separately from attempted burglary. In his examination of attempted property crime, Farrell (2016) found the decline in attempted domestic burglary was delayed by a number of years (when compared with the fall in burglary with entry). This delay was seen as consistent with particular offenders trying, but failing, to commit burglary when faced with effective security (ibid). 'The time lag between the beginning of attempts and burglary with entry falls indicates that burglaries fell due to target characteristics encountered after the target had been selected (such as unanticipated guardianship in the form of security) rather than offenders' decisions not to target properties' (Tseloni et al. 2017, p. 3). Farrell (2016) proposes that the most easily deterred were younger, less experienced offenders. Added to this, Tseloni et al. (2017) report that the decline in burglary from the mid-1990s was mainly comprised of a dramatic drop in burglaries involving forced entry (e.g. overcoming security by forcing locks or breaking glass) as opposed to unforced entry (e.g. entering via an open window) which has remained relatively consistent (Tseloni et al. 2017). Collectively, the evidence suggests there are a number of 'distinct patterns of offending behavior and decision making' and that 'the delays are consistent with some offenders continuing to attempt crime but being thwarted by improved security' (Farrell 2016, p. 26). Here we examine this issue further.

To this end, we examine the role of security in relation to attempted burglary separately from burglary with entry. In Chap. 7 of this book, Sourd and Delbecque agree that '…it is essential to take into account attempts, not only as a failed burglary, but also as evidence of the targeting of the housing unit by a perpetrator' (pp. 195–220). They contend that burglary should be viewed as a three-step process (targeting, forced entry, theft) rather than as a single event. Using data from the French Victimisation Survey (the Cadre de Vie et Sécurité (CVS)), Sourd and Delbecque estimate the risk of becoming a victim at each of the three stages whilst controlling for a range of environmental and dwelling-specific factors (including security). The security devices examined include: alarms, security doors (a door which is reinforced with an internal steel plate/steel bars and can have multiple locks), cameras and digital locks. Their results reinforce their assertion that burglary should be seen as a three-step process as the environmental and housing unit-specific variables do not exert the same effect at each stage. They find the protective effect of security devices is generally greater during the forced entry stage.

Here, we build upon their work and that by Farrell (2016) and Tseloni et al. (2017) by developing and testing hypotheses about the role of security amongst the entire population of homes (i.e. victims of burglary, attempted burglary and non-victims) using a wider range of security data from the CSEW. In a similar vein to Sourd and Delbecque, we suggest that a home may fall into one of three categories – untargeted, targeted or penetrated. An *untargeted* home is one that is either never considered by a burglar or considered but rejected (i.e. the burglar makes no attempt to enter). A *targeted* home is considered a worthwhile target and entry is attempted. A property is *penetrated* should the burglar obtain entry (and in some cases, steal items). There are thus two transition points, which we term 'deter' and 'thwart'. In order for a home to move from untargeted to targeted, the burglar must not be *deterred* from selecting the property (in this example, by visible security devices). For a home to move from targeted to penetrated, entry must not be *thwarted* (e.g. by physical security devices or capable guardians) (see Diagram 4.1). An attempted burglary suggests the burglar was not deterred but was thwarted. By contrast, burglary with entry suggests the burglar was neither deterred nor thwarted. We suggest that some security devices are designed predominantly to deter and some to thwart. Here, we also draw upon Cornish and Clarke's (2003) 25 techniques of situational crime prevention, in particular increasing the effort (e.g. through target hardening) and increasing the risks (e.g. surveillance potential is increased through the use of security lighting). In other words, offenders may perceive the *risk* of continuing as too high and thus be deterred from targeting the property. If the offender concludes that the risk is not too high and the rewards are sufficient, an offender's entry may still be thwarted due to the *effort* required to enter the property (e.g. good-quality locks making entry more difficult).

Table 4.3 outlines the expected mechanism (deter or thwart) of the security devices examined here. For example, we hypothesise that burglar alarms are primarily designed to act as a visible deterrent to offenders as opposed to physically preventing them from gaining entry. By contrast, window and door locks are designed to prevent entry (or make entry more difficult). Previous research would lend support to this hypothesis in that a larger number of offenders claim to be deterred by burglar alarms than locks (see Sect. 4.3.2).

Different security devices may be more (or less) effective in providing protection at the two transition points (see Diagram 4.1). This chapter marks our first attempt to outline and tentatively test this proposition.

Table 4.3 Expected mechanism of different security devices

Security device	Dominant mechanism: deter or thwart?
Burglar alarm	*Deter*: a burglar alarm is designed to act as a visible deterrent. Its effectiveness relies upon the owner setting the alarm prior to leaving the house and a response if it sounds. It does not physically prevent someone from entering a property
CCTV	*Deter*: CCTV cameras are predominantly designed to act as a visible deterrent. They may increase the risk of being caught but do not physically prevent entry
Double locks or deadlocks	*Thwart*: double locks or deadlocks are designed to prevent entry or make entry more difficult
Dummy alarm	*Deter*: solely designed to act as a visible deterrent
External lights on a sensor	*Deter*: external lights may act as a visible deterrent. They are designed to increase surveillance by illuminating any passers-by
Internal lights on a timer	*Deter*: designed to act as a visible deterrent. They are used to make a house look occupied and therefore discourage burglars
Security chains	*Thwart*: chains cannot usually be seen from the outside and are designed to restrict access to a property. They rely on someone being in the property to be used. They are designed to prevent entry or make entry more difficult
Window bars/ grills	*Thwart*: their dominant function is to prevent access to a property through the window
Windows that require a key	*Thwart*: window locks are designed to prevent entry or make entry more difficult

4.5   Results

First, we present the results relating to the protection afforded by different combinations[9] of security devices against burglary with entry and attempted burglary in order to ascertain the importance of accessibility and occupancy (two of the FAVOR-able burglary cues outlined in Section 4.3.2). We then discuss our findings in relation to the specific deter/thwart mechanisms.

4.5.1   FAVOR-able Cues: Accessibility and Occupancy

As mentioned in Sect. 4.3.2, two FAVOR-able cues (accessibility and occupancy) are deemed to be of particular relevance with regard to security. In order to assess the importance of accessibility and occupancy, the ten most effective security device combinations are presented in Fig. 4.1 for burglary with entry and Fig. 4.2 for attempted burglary (only statistically significant results are presented within both figures). The security combinations have been ranked according to their

---

[9]The SPFs for individual devices (in isolation) are not presented here given that a small (and declining) proportion of the population only have single security devices. These results are available upon request.

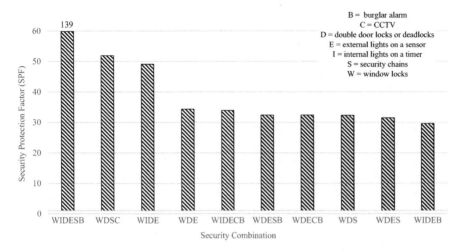

Fig. 4.1 Security Protection Factors against burglary with entry for the ten security device combinations with the highest SPFs (data taken from the 2008/2009–2011/2012 sweeps of the Crime Survey for England and Wales) (significant in burglary with entry, *p*-value <0.05) (capped at 60)

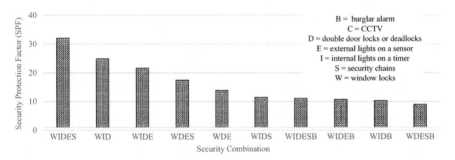

Fig. 4.2 Security Protection Factors against attempted burglary for the ten security device combinations with the highest SPFs (data taken from the 2008/2009–2011/2012 sweeps of the Crime Survey for England and Wales) (significant in attempted burglary, *p*-value <0.05)

effectiveness (SPF value) in descending order across the *x*-axis. In order to clarify how SPFs are interpreted, let us examine the second shaded bar in Fig. 4.1 (as the first bar is deemed to be an outlier). It denotes that, in 2008/2009–2011/2012, households with window locks, double door locks/deadlocks, security chains and CCTV cameras (WDSC) (and no other security devices) had just over fifty times the level of protection against burglary with entry compared to those with no security.

The findings in both Figs. 4.1 and 4.2 show that effectiveness does not increase linearly with the number of devices. Combinations of fewer devices often offer the same or, indeed, greater levels of protection than combinations of many devices. However, for both burglary with entry and attempted burglary, the 'top 10' device combinations all consist of at least three devices. Window and door locks form the basis of every combination for both burglary with entry and attempted burglary. External lights on sensors and internal lights on timers feature in the majority of combinations. Interestingly, the most effective devices are not exactly the same for burglary with entry as they are for attempted burglary.

Ignoring the SPF outlier (WIDESB), the most effective combination against burglary with entry was window locks, double door locks/deadlocks, security chains and CCTV cameras followed by 'WIDE' (window locks, internal lights on a timer, door double locks/deadlocks and external lights on a sensor) with an SPF of 49. For attempted burglary, the most effective combination was WIDE plus security chains (WIDES) with an SPF of 32. In the majority of cases, the addition of a burglar alarm generally leads to a reduction in protection against burglary with entry and attempted burglary (Tseloni et al. 2014; Tilley et al. 2015b).

4.5.2 Which Security Devices Deter and Which Thwart?

SPF values for attempted burglary can be difficult to interpret. One could argue that the security devices, to some extent, performed their function in preventing entry into the property (albeit they didn't stop the offender from *trying* to get in). In other words, if you are a victim of attempted burglary, the security devices may not have deterred the burglar, but they may have played a part in thwarting entry. As discussed previously, different security devices may be more (or less) effective in providing protection at the 'deter' or 'thwart' transition points (see Diagram 4.1). The way SPFs are calculated conflates deter and thwart thus making it difficult to draw any specific inferences in relation to these mechanisms.

This section reports the findings from our initial attempt[10] to address these limitations and more explicitly test the deter/thwart mechanisms. The population of interest here is all households (i.e. untargeted, targeted and penetrated).[11] The data are taken from the 2008/2009 to 2011/2012 CSEW sweeps. For this preliminary test, we calculate the odds of different events occurring, e.g. the number of households which experienced an event (a burglary with entry) divided by the number of those which did not experience the event (no burglary with entry). First, we calculate the odds of experiencing a burglary with entry for a range of different security combinations, followed by calculation of the odds of attempted burglary. We then use those two odds values to generate an odds ratio. This provides us with an initial (albeit somewhat crude) measure of association and an indication as to whether there are particular security device combinations which are more effective at thwarting entry. From a practical perspective, establishing if security devices have distinct mechanisms is useful as it is far better to deter offenders at the first hurdle (or 'tran-

---

[10] The findings in this section are preliminary. Work is ongoing to incorporate a larger number of data sets and carry out more advanced statistical analysis.

[11] Two alternative calculations were also tried (see Sect. A.5 in Appendix A).

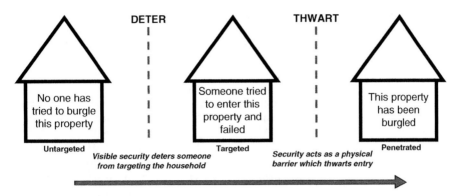

Diagram 4.1 Conceptual drawing of 'deter' and 'thwart' mechanisms

sition point' in relation to Diagram 4.1), given the damage often caused in attempts and the anxiety caused to victims who find out that someone has tried to break in.

Table 4.4 shows the odds of burglary with entry and the odds of an attempted burglary for different security combinations as well as the odds ratio comparing the two. Results are presented for device combinations where the sample of victims exceeds 50. For combinations where there were fewer than 50 victims (of burglary with entry or attempt), we can infer one of two things: (1) that the particular security device combination is not prevalent amongst the general population or (2) that the security devices had a strong deterrent effect (i.e. the house was not targeted, e.g. the WIDE combination).

In order to aid interpretation, we will provide a worked example. The first row of results shows the odds of a household with window and door locks plus security chains (WDS) experiencing a burglary with entry is 0.67 times that of the odds of experiencing an attempted burglary (in other words, the odds of experiencing a burglary with entry are lower than the odds of an attempt when the household has WDS). This was calculated as follows. First, the odds of experiencing a burglary with entry were calculated for households with WDS: the total number of burglary with entry victims with WDS ($n = 38$) (column 2) was divided by the total number of households with WDS who did not experience a burglary with entry ($n = 2705$) (column 3) giving 0.014 (column 4). The odds of experiencing an attempted burglary given WDS were then calculated: the total number of attempted burglary victims with WDS ($n = 56$) (column 5) was divided by the total number of households with WDS who did not experience an attempted burglary ($n = 2687$) (column 6) giving 0.021 (column 7). The odds of experiencing burglary with entry with WDS (0.014) were then divided by the odds of experiencing an attempted burglary with WDS (0.021) to give an odds ratio of 0.67 (column 8). If a security combination has a value of less than one in the eighth column, it suggests the odds of burglary with entry are lower than the odds of an attempt with that particular combination, i.e. the combination appears to be effective in thwarting burglary with entry. To take another

Table 4.4 What are the odds of burglary with entry and the odds of attempted burglary given a particular security device combination? With that security combination, are the odds of being a burglary with entry victim higher than attempted burglary? (CSEW, 2008/2009–2011/2012)

Security device(s)	Burglary with entry (BWE) victims (count)	Not a victim of BWE (count)[a]	Odds of experiencing BWE	Attempted burglary (ATT) victims (count)	Not a victim of ATT (count)[b]	Odds of experiencing ATT	Odds ratio (BWE to ATT)
WDS	38	2705	0.014	56	2687	0.021	0.67
WS	24	654	0.037	26	652	0.040	0.93
WDE	43	3264	0.013	37	3270	0.011	1.18
DS	27	436	0.062	23	440	0.052	1.19
WDEB	53	1900	0.028	40	1913	0.021	1.33
D	145	760	0.191	105	800	0.131	1.46
DB	31	105	0.295	21	115	0.183	1.61
WD	192	5189	0.037	111	5270	0.021	1.76
W	120	1645	0.073	69	1696	0.041	1.78
WDB	73	1382	0.053	39	1416	0.028	1.89
E	36	206	0.175	19	223	0.085	2.06
S	70	251	0.279	38	283	0.134	2.08
B	106	106	1.000	52	160	0.325	3.08
Total (across all combinations)	**2245**	**35,171**	**0.064**	**1356**	**36,060**	**0.038**	**1.68**

Note: Results shown for combinations with >50 victims of burglary with entry or attempted burglary

[a]Using WD as an example, this is calculated by subtracting the total number of burglary with entry victims with WD from the total population with WD. Therefore, this value includes non-victims *and* victims of attempted burglary with WD. This is justified because here we are interested in the odds of a *specific* event (burglary with entry) occurring

[b]Using WD as an example, this is calculated by subtracting the total number of attempted burglary victims with WD from the total population with WD. Therefore, this value includes non-victims *and* victims of burglary with entry with WD. This is justified because here we are interested in the odds of a *specific* event (attempted burglary) occurring

example, in the case of WD, an odds ratio of above 1 (in this case, 1.76) means the odds of burglary with entry are higher than the odds of an attempt. The SPF value for WD would suggest this combination is, generally, effective. This result suggests WD may stop the majority of properties being targeted (evidenced by the effective SPF), but when they are, having only WD does not consistently thwart entry. This may be due to the widespread prevalence of WD and the greater potential number of lower-quality locks in the population.

The only combinations which have lower odds of burglary with entry when compared with attempt are WDS and WS. The largest 'thwart' effect was found for window and door locks plus security chains. This is entirely consistent with our predictions outlined in Table 4.3 – for all other devices the dominant mechanism was predicted to be to deter (except for window bars/grills which were not included in any of the analysis due to their low (and diminishing) prevalence over time). The findings regarding security chains are also somewhat unsurprising given the emphasis upon occupancy within previous literature. Security chains are generally not visible from the outside of a property. Added to this, a person usually has to be inside the house for this device to be used. An offender may target the property but be thwarted upon discovery of a door secured by a security chain signalling that someone is at home. However, as discussed in previous papers, there are issues regarding security chains posing a potential fire safety hazard. The odds of entry (when compared to attempt) were highest for single devices in isolation, namely, burglar alarms, security chains and external lights on a sensor. This initial analysis supports our assertions regarding the deter/thwart mechanisms, and thus more advanced statistical analysis is warranted in order to draw firmer conclusions.

4.6 Discussion and Conclusion

A number of findings reported here warrant particular attention. Window and door locks consistently form the foundation of the most effective security combinations (see Figs. 4.1 and 4.2). In addition, all combinations in the 'top 10' contain at least three devices (i.e. window and door locks plus at least one other device). This supports the 'basic/enhanced' ONS (2013) security classification, whereby basic security comprises double locks/deadlocks on at least some external doors and locks on windows. Enhanced security comprises window and door locks plus at least one other security device. The findings from this research show the exact devices to be added to the 'basic' specification of window and door locks. Adding to the combination of window and door locks, security chains and CCTV cameras (WDSC) or internal lights on a timer and external lights on a sensor (WIDE) can increase protection against burglary with entry by up to 51 and 49 times (respectively) compared to households with no security.

In relation to 'FAVOR-able' cues, the findings outlined in this chapter highlight the importance of restricting access (through the use of good-quality window and door locks and, notwithstanding, fire safety considerations, security chains), simulating occupancy and increasing visibility (through the use of lights on timers/sensors). The protective effect of simulating occupancy is in agreement with much previous research (see Section 4.3.2.3) and serves to reinforce the importance of capable guardianship (Cohen and Felson 1979). The wider project involved advanced statistical modelling of a number of potential proxy indicators of visibility (area type and type of accommodation) and potential reward (household income, number of cars, etc.) (see Chap. 5).

Some security devices perform automatically (requiring no formal activation/always 'ready to function', e.g. an external light on a sensor), whereas others rely upon human activation (e.g. using the security chain on an external door). In relation to WIDE, both forms of lights (once set up) should function automatically. Double door locks/deadlocks require a householder to remember to lock the door when the house is unoccupied. Similarly, window locks will only perform their security function if the window is shut and locked (which requires action upon the part of the householder). Police forces in England and Wales are routinely required to respond to burglaries of 'insecure' properties, particularly during warmer weather/summer months when windows are more likely to be left open. Suffolk Constabulary (2017) reported that 'more than a third of reported residential burglaries in April and May this year were as a result of an insecure door or window'. Research by Tseloni et al. (2017) also shows burglaries involving unforced entry have remained relatively stable over the period of the crime drop in contrast to the dramatic drop in burglaries involving forced entry. Householder vigilance is thus an increasingly important additional consideration in relation to security (Winchester and Jackson 1982; Hearnden and Magill 2004; Nee and Meenaghan 2006).

Perhaps the biggest contribution of this chapter relates to our two proposed security mechanisms – deter and thwart. We propose different security devices are designed to activate distinctive primary mechanisms. Some are designed to deter (i.e. for the offender to assess the risk of targeting the property as too high (Cornish and Clarke 2003), e.g. CCTV cameras) and some to thwart (i.e. to physically prevent entry/increase the effort (ibid), e.g. security chains). As anticipated, we find window locks, door locks and security chains exert the biggest 'thwart' effect of all device combinations. We conclude that the security combinations including some devices which deter and some which thwart appear to be most effective. For example, the combination of WIDE provides both a deterrent (internal and external lights which simulate occupancy and increase visibility) and a thwarting function (window and door locks in certain combinations physically prevent entry). Burglary is now generally committed by more experienced offenders who may be less deterred by particular security devices (see Section 4.3.2) (Farrell 2016; Farrell et al. 2015) highlighting the need for security combinations which both deter *and* thwart (to deter those less experienced and thwart the more experienced).

As mentioned, the analysis reported in relation to deter/thwart marks a tentative first step in the exploration of the effectiveness of security in relation to attempted burglary. Further, more sophisticated, analysis is required which could examine the household (including security), incident and area characteristics that make an attempted burglary more likely. Chapter 7 provides an example of this type of analysis using French data. Another example of this type of research can be found in Thompson's (2014) analysis of theft from the person and robbery. Here, logistic regression models were run whereby the likelihood that an incident would be 'attempted' as opposed to 'completed' was estimated taking into account a range of individual and incident factors. In other words, Thompson (2014) estimated whether particular characteristics of the individual or the incident increased the likelihood of items *not* being stolen from the person (as opposed to an incident where something was stolen). She found the incident was less likely to be an attempt if force, violence or a weapon was used, if the victim was aware of the incident and if they had contact with the offender. Incident characteristics held the greatest explanatory power compared to victim characteristics when modelling the likelihood of attempted victimisation against victimisation. This serves to highlight the importance of incident characteristics, the 'near causes' of crime (Tilley 2009) and situational crime prevention as an approach to reducing crime (Farrell 2016). This modelling approach could be extended to burglary with entry and attempted burglary in England and Wales to determine if there are particular security device combinations, incident or household characteristics that increase the likelihood of the crime to be thwarted.

To conclude, a number of policy implications arise from this research. In terms of sustaining the fall in volume crimes such as burglary, those most vulnerable to burglary with entry or attempted burglary should be given access to the most effective security. We know that particular households and areas are both more vulnerable to burglary and less likely to be protected by the most effective security (see Chap. 5). It is fairly well established that the crime falls have been unevenly distributed amongst different population groups and areas – property crime has become more concentrated amongst particular groups (e.g. social renters and repeat victims of crime) (Ignatans and Pease 2015, 2016; Hunter and Tseloni 2016; Tilley et al. 2011). Research conducted in the USA suggests that the financial cost is one of the primary reasons for not installing additional security (Roth 2017). Table 4.5 presents the demographic characteristics of the households with the most prevalent device combination, window and door locks, as well as the characteristics of households with WIDE and no security. It shows that households with the most effective security are fundamentally different to those with none. Compared to households with WIDE and WD, those with no security are more likely to be comprised of lone parents on lower incomes in rented properties. On the other hand, residents in households with WD and WIDE are both less likely to have children under 16 and more likely to earn a higher household income.

Overall, this research indicates that a more detailed and useful picture is painted when looking at specific device ownership rather than security availability more generally. It is therefore important to consider different 'security packages' and their relative effectiveness in order to provide more accurate crime prevention advice. As is highlighted in the Home Office's (2016) Modern Crime Prevention Strategy, opportunities to commit crime should be removed or designed out. It is clear from the review of the literature and the findings of this project that particular security combinations offer an effective way to do this.

Table 4.5 Descriptive statistics (2008/09–2011/12) for no security, WD and WIDE populations

Characteristics	No security (%)	WD (%)	WIDE (%)
Number of adults (two adults)			
One adult	51.3	33.8	22.5
Three or more adults	13.5	16.9	13.3
Children under 16 in household (no children)			
Children	35.3	28.5	19.3
Lone parent (not a lone parent)			
Lone parent	14.9	6.0	1.5
Ethnicity of head of household (White)			
Black	3.6	2.3	1.3
Asian	5.3	3.3	1.1
Mixed, Chinese or other	2.8	1.8	0.8
Household income (£20,000–£29,999)			
£4999 and under	10.1	4.2	1.7
£5000–£9999	17.8	12.3	5.3
£10,000–£19,999	24.3	22.7	19.0
£30,000–£49,999	11.1	17.5	20.2
£50,000 or more	7.0	10.1	16.0
No income/no info	18.8	17.0	19.3
Tenure (owner)			
Social rented sector	30.9	16.4	3.8
Private rented sector	25.9	12.4	4.8
Area type (rural)			
Inner city	13.2	7.8	3.1
Urban	64.6	68.0	66.4
Region (South East)			
North East	6.8	6.2	6.8
Yorkshire and Humberside	7.7	6.8	7.5
North West	10.9	11.1	9.7
East Midlands	7.9	8.2	12.6
West Midlands	8.3	9.8	9.4
East	11.6	13.2	16.1
London	13.2	8.5	5.4
South West	10.8	13.0	11.4
Wales	12.1	10.9	8.0
Victim of burglary or attempted burglary	59.4	5.5	1.6
N	1348	4625	1418

Acknowledgement The authors would like to thank Dr. Puneet Tiwari for his useful comments during the writing of this chapter.

Appendix A

A.1   Introduction

The Crime Survey for England and Wales (CSEW) is a large-scale, face-to-face victimisation survey widely considered to be the most comprehensive long-term measure of crime trends available in England and Wales (Tilley and Tseloni 2016). The survey was first conducted in 1982 and from then was run approximately every 2 years until 2001, at which point it became a continuous survey.

The CSEW was, prior to April 2012, known as the British Crime Survey (BCS). From April 2012, responsibility for the survey moved from the Home Office to the Office for National Statistics (ONS). For consistency throughout the book, we refer to the survey as the CSEW.

A.2   Crime Survey for England and Wales Sample Selection

Since 2001/2002 the CSEW has used an annual rotating sample of between (approximately) 32,000 and 48,000 adults resident in England and Wales. The aim is to conduct at least 1000 core interviews in each police force area. The CSEW has sampled adults over the age of 16 since 1982. From January 2009, the survey was extended to include 10–15 year olds, although this data is not analysed within this book. Specific details regarding the sampling design can be found within TNS BMRB (2012).

A.3   Crime Survey for England and Wales Questionnaire Structure

The structure of the CSEW is relatively complex. In general, it consists of a number of core modules asked of the whole sample (e.g. socio-demographic details, experiences of the criminal justice system, etc.), a set of modules asked of different sub-samples (e.g. crime prevention and security (the exact topics vary each year)), self-completion modules (e.g. offending behaviour and drug use) and, where relevant, modules concerning crime victimisation (see Appendix Table 4.6).

As shown in Appendix Table 4.6, the CSEW is a rich data source, collecting data regarding the characteristics (both demographic and attitudinal) of individual respondents, their household and the area in which they live as well as whether they have been a victim of crime or anti-social behaviour. In relation to household security measures, detailed information is collected from the majority of burglary victims and a randomly selected subset of the total sample (this subset comprises approximately 10,000 respondents in each sweep). Information regarding the availability of

Appendix Table 4.6 Modules of the 2011–2012 CSEW questionnaire and subset of respondents who were asked each module

Questionnaire module	Core sample
Household grid	All
Perceptions of crime	All
Screener questionnaire	All
Victim modules	All victims
Performance of the criminal justice system	All
Experiences of the criminal justice system	All
Mobile phone and bicycle crime	All
Module A: Experiences of the police	Random 25% – group A
Module B: Attitudes to the criminal justice system	Random 25% – group B
Module C: Crime prevention and security	Random 25% – group C
Module D: Ad hoc crime topics	Random 25% – group D
Plastic card fraud	Random 75% (groups B, C, D)
Mass marketing fraud	All
Anti-social behaviour	Random 25% – group A
Demographics and media consumption	All
Self-completion module: Drugs and drinking	All aged 16–59
Self-report offending behaviour	Random 25% – group B
Self-completion module: Domestic violence, sexual victimisation and stalking	All aged 16–59[a]

Table taken from TNS BMRB (2012, p. 15)
[a]Questions on stalking were put to a random 50% (groups C and D); questions on attitudes to domestic violence were put to a random 25% (group D)

security devices at the time of the incident is gathered (via the Victimisation module) from the majority of respondents whose household was burgled. The list of devices from which respondents could choose changed considerably between the 1996 and 1998 sweeps but has remained relatively consistent since then (see Appendix Table 4.7).

A.4   Limitations

A.4.1   Security Information Is Not Available for All Burglary Victims

Information about the home security devices in place at the time of the burglary is not available for a small minority of victims who reported at least three unconnected crime incidents of higher seriousness than burglary (according to standard CSEW offence classification - see Hales et al. 2000) during the reference period. By way of

Appendix Table 4.7 Household security measures over time, as measured by the CSEW (in both the Victim and Non-Victim Forms) 1992–2011/2012. Note: Adapted from Tseloni et al. (2017)

	1992	1994	1996	1998	2000	2001	01/02	02/03	03/04	04/05	05/06	06/07	07/08	08/09	09/10	10/11	11/12
Burglar alarm	✓	✓	✓	✓	✓	✓	✓	✓	✓	✓	✓	✓	✓	✓	✓	✓	✓
Dummy box	✓	✓	✓	✓	✓	✓	✓	✓	✓	✓	✓	✓	✓	✓	✓	✓	✓
Double locks/deadlocks	✓	✓	✓	✓	✓	✓	✓	✓	✓	✓	✓	✓	✓	✓	✓	✓	✓
Security chains/bolts				✓	✓												
Security chains						✓	✓	✓	✓								
Security chains/doorbars										✓	✓	✓	✓	✓	✓	✓	✓
Bars/metal grill/bar door																	
Window locks	✓	✓	✓	✓	✓	✓	✓	✓	✓	✓	✓	✓	✓	✓	✓	✓	✓
Indoor lights				✓	✓	✓	✓	✓	✓	✓	✓	✓	✓	✓	✓	✓	✓
Outdoor lights				✓	✓	✓	✓	✓	✓	✓	✓	✓	✓	✓	✓	✓	✓
Lights	✓	✓	✓														
Window bars/grilles		✓	✓	✓	✓	✓	✓	✓	✓	✓	✓	✓	✓	✓	✓	✓	✓
CCTV camera														✓	✓	✓	✓
Dog (provides security)	✓																✓

illustration, details about home security are not solicited from a multiple victim of three unrelated incidents of assault or robbery whose home has also been burgled. The security of the most vulnerable population is therefore unknown due to survey limitations.

A.4.2   Victims of Both Attempted Burglary and Burglary with Entry Are Excluded

The unit of analysis here is the household. Therefore, when a victim reported more than one burglary incident, their home security availability at the time of the *first* burglary (during the survey's reference period) was retained for analysis. A minority of cases where a respondent experienced both an attempted burglary and a burglary were, however, excluded. Using the 2008/2009–2011/2012 data as an example, in a small number of cases (which make up 0.17% of the total sample, 1.6% of all burglary victims, 2.6% of victims of burglary with entry or 4.4% of victims of attempts), a respondent experienced both an attempted burglary and a burglary (separate incidents not considered to be part of a series). For the purposes of this analysis, security device availability was measured at the time of interview for non-victims and at the time of the *first* incident for victims. It was therefore necessary to establish when each incident happened in order to ascertain which victimisation happened first – the burglary or the attempt. Data regarding the month in which each incident happened was originally established for nine cases from the 2011/2012 sweep. Of the nine, four respondents first experienced an attempted victimisation and two burglaries with entry. With regard to the remaining three cases, both incidents happened in the same month. Therefore, we were unable to ascertain which incident happened first. As a result, because they constitute a small proportion of the total sample, cases where a respondent experienced both an attempt and a burglary with entry were excluded from this analysis.

A.5   Alternative Deter/Thwart Calculations

Using WD as an example, the population of interest in Table 4.4 was calculated by subtracting the total number of burglary with entry victims with WD from the total population with WD. Therefore, this value includes non-victims *and* victims of attempted burglary with WD. This is justified because here we are interested in the odds of a *specific* event (burglary with entry) occurring. However, one could argue the events (burglary with entry and attempted burglary) are not independent – both burglary with entry and attempted burglary households have been targeted, and therefore their security did not deter in the first place. The calculations in Appendix Tables 4.8 and 4.9 offer an alternative means of calculation (where the population includes only victims) which incorporates the events' potential non-independence. In Appendix Table 4.8, our population is victims with particular security

Appendix Table 4.8 Alternative deter/thwart risk calculations (where the population selected comprises only *victims* with particular security combinations) (CSEW, 2008/2009–2011/2012)

Security device(s)	Burglary with entry (BWE) victims (count)	Attempted burglary (ATT) victims (count)	Total victims (BWE + ATT) (*total*) (count)	BWE/*total*	ATT/*total*	Relative risk (BWE relative to ATT)
WDS	38	56	94	0.404	0.596	0.68
WS	24	26	50	0.480	0.520	0.92
WDE	43	37	80	0.538	0.463	1.16
DS	27	23	50	0.540	0.460	1.17
WDEB	53	40	93	0.570	0.430	1.33
D	145	105	250	0.580	0.420	1.38
DB	31	21	52	0.596	0.404	1.48
WD	192	111	303	0.634	0.366	1.73
W	120	69	189	0.635	0.365	1.74
S	70	38	108	0.648	0.352	1.84
WDB	73	39	112	0.652	0.348	1.87
E	36	19	55	0.655	0.345	1.89
B	106	52	158	0.671	0.329	2.04
Total victims (across all combinations)	**2245**	**1356**	**3601**	**0.623**	**0.377**	**1.65**

Note: Results shown for combinations with >50 victims of burglary with entry or attempted burglary

Appendix Table 4.9 Alternative deter/thwart odds calculations (where the population selected comprises all victims) (CSEW, 2008/2009–2011/2012)

Security device(s)	Burglary with entry (BWE) victims (count)	Attempted burglary (ATT) victims (count)	BWE/*total* BWE *victims* (BWE2)	ATT/*total* ATT *victims* (ATT2)	Odds ratio (BWE2/ ATT2)
WDS	38	56	0.017	0.041	0.41
WS	24	26	0.011	0.019	0.56
WDE	43	37	0.019	0.027	0.70
DS	27	23	0.012	0.017	0.71
WDEB	53	40	0.024	0.029	0.80
D	145	105	0.065	0.077	0.83
DB	31	21	0.014	0.015	0.89
WD	192	111	0.086	0.082	1.04
W	120	69	0.053	0.051	1.05
S	70	38	0.031	0.028	1.11
WDB	73	39	0.033	0.029	1.13
E	36	19	0.016	0.014	1.14
B	106	52	0.047	0.038	1.23
Total victims (across all combinations)	**2245**	**1356**	–	–	–

Note: Results shown for combinations with >50 victims of burglary with entry or attempted burglary combinations installed in their homes. Here, we are comparing the proportion of burglary with entry victims with (for example) WD, with the proportion of attempted burglary victims with WD to calculate their relative risk. In other words, are victim households with WD at greater risk of becoming a victim of burglary with entry or attempted burglary? Note, the results are almost identical to Table 4.4.

In Appendix Table 4.9, our population includes all victims. Here, we are comparing the odds of having each security combination for burglary with entry victims with the odds of attempted burglary victims having the same combination. In other words, are burglary with entry victims more likely (than attempted burglary victims) to have (by way of example) WD? Note that here the results are slightly different. The top two 'thwarting' device combinations are still window locks, door locks and security chains (WDS) followed by window locks and security chains (WS). However, the results in Appendix Table 4.9 suggest there are a larger number of 'thwarting' device combinations. All 'thwarting' combinations (bar one, WS) include door locks which is consistent with our expectations outlined in Table 4.3. Although Appendix Table 4.9 presents a larger number of thwarting combinations, the results are consistent (across Table 4.4 and Appendix Tables 4.8 and 4.9) in that window locks, door locks and security chains (in various combinations) are deemed to thwart.

A.6   More Information

For more details regarding CSEW methodology, questionnaires and topics covered, see Hough and Maxfield (2007), Flatley (2014) and the various CSEW Technical Reports (TNS BMRB 2012). For more information on crime statistics more generally, please see the ONS (2018b) user guide.

References

Age UK. (2017). Crime prevention. http://www.ageuk.org.uk/home-and-care/home-safety-and-security/crime-prevention-/security-in-your-home/. Accessed 16 June 2017.

Armitage, R., & Joyce, C. (2015). Why my house? Exploring offender perspectives on risk and protective factors in residential housing design. *National Architectural Liaison Officer's Conference*. February 2015, Stratford Upon Avon (Unpublished).

Bennett, T., & Wright, R. (1984). *Burglars on burglary: Prevention and the offender*. Aldershot: Gower.

Budd, T. (1999). *Burglary of domestic dwellings: Findings from the British Crime Survey* (Home Office Statistical Bulletin Issue 4/99). London: Home Office.

Budd, T. (2001). *Burglary: Practice messages from the British Crime Survey* (Briefing note 5/01). London: Home Office.

Clare, J. (2011). Examination of systematic variations in burglars' domain-specific perceptual and procedural skills. *Psychology, Crime & Law, 17*(3), 199–214.

Cohen, L. E., & Felson, M. (1979). Social change and crime rate trends: A routine activity approach. *American Sociological Review, 44*(4), 588–608.

Cornish, D. B., & Clarke, R. V. G. (1986). *The reasoning criminal*. New York: Springer.

Cornish, D. B., & Clarke, R. V. G. (2003). Opportunities, precipitators and criminal decisions: A reply to Wortley's critique of situational crime prevention. In M. Smith & D. B. Cornish (Eds.), *Theory for situational crime prevention, Crime prevention studies* (Vol. 16, pp. 41–96). New York: Criminal Justice Press.

Cromwell, P. (1994). Juvenile burglars. *Juvenile and Family Court Journal, 45*, 85–92.

Cromwell, P., & Olson, J. N. (2004). *Breaking and entering: Burglars on burglary*. Belmont: Wadsworth.

Cromwell, P., & Olson, J. N. (2009). The reasoning burglar: Motives and decision-making strategies. In P. Cromwell (Ed.), *In their own words: Criminals on crime* (pp. 42–56). Oxford: Oxford University Press.

Cromwell, P., Olson, J. N., & Avary, D. (1991). *Breaking and entering: An ethnographic analysis of burglary*. London: Sage.

Dinisman, T., & Moroz, A. (2017). *Understanding victims of crime*. London: Victim Support.

Edgar, J., & McInerney, W. (1987). Locks: Attacks and countermeasures. Technical notebook: How to defeat a lock and how to keep a lock from being defeated. *Security Management, 37*, 61–63.

Ekblom, P., Law, H., & Sutton, M. (1996). *Safer cities and domestic burglary* (Home Office Research Study 164). London: Home Office.

Farrell, G. (2016). Attempted crime and the crime drop. *International Criminal Justice Review, 26*(1), 21–30.

Farrell, G., Tseloni, A., & Tilley, N. (2011). The effectiveness of vehicle security devices and their role in the crime drop. *Criminology and Criminal Justice, 11*(1), 21–35.

Farrell, G., Laycock, G., & Tilley, N. (2015). Debuts and legacies: The crime drop and the role of adolescence-limited and persistent offending. *Crime Science, 4*, 1–10.

Flatley, J. (2014). British crime survey. In G. Bruinsma & D. Weisburd (Editors in Chief). *Encyclopedia of criminology and criminal justice (ECCJ)* (pp. 194–203). New York: Springer.

Flatley, J., Kershaw, C., Smith, K., Chaplin, R., & Moon, D. (2010). *Crime in England and Wales 2009/10: Findings from the British Crime Survey and Police Recorded Crime*. London: Home Office.

Forrester, D., Chatterton, M., & Pease, K. (1988). *The Kirkholt burglary prevention project, Rochdale* (Crime Prevention Unit Paper 13). London: Home Office.

Forrester, D., Frenz, S., O'Connell, M., & Pease, K. (1990). *The Kirkholt burglary prevention project: phase II* (Crime Prevention Unit Paper 23). London: Home Office.

Hales, J., Henderson, L., Collins, D., & Becher, H. (2000). *2000 British Crime Survey (England and Wales): Technical report*. London: National Centre for Social Research.

Hamilton-Smith, N., & Kent, A. (2005). The prevention of domestic burglary. In N. Tilley (Ed.), *Handbook of crime prevention and community safety* (pp. 417–457). Devon: Willan Publishing.

Hearnden, I., & Magill, C. (2004). *Decision-making by house burglars: Offenders' perspectives*. London: Home Office.

Home Office. (2016). *Modern crime prevention strategy*. London: Home Office.

Homel, R., Macintyre, S., & Wortley, R. (2014). How house burglars decide on targets: A computer-based scenario approach. In B. Leclerc & R. Wortley (Eds.), *Cognition and crime: Offender decision-making and script analyses (crime science series)* (pp. 26–47). London: Routledge.

Hough, M., & Maxfield, M. (2007). *Surveying crime in the 21st century, Crime prevention studies* (Vol. 22). New York: Criminal Justice Press.

Hunter, J., & Tseloni, A. (2016). Equity, justice and the crime drop. The case of burglary in England and Wales. *Crime Science, 5*(3). https://doi.org/10.1186/s40163-016-0051-z.

Ignatans, D., & Pease, K. (2015). Distributive justice and the crime drop. In M. Andresen and G. Farrell (Eds.), *The criminal act: Festschrift for Marcus Felson* (pp77–87). London: Palgrave Macmillan.

Ignatans, D., & Pease, K. (2016). On whom does the burden of crime fall now? Changes over time in counts and concentration. *International Review of Victimology, 22*(1), 55–63.

Langton, S. H., & Steenbeck, W. (2017). Residential burglary target selection: An analysis at the property-level using Google Street View. *Applied Geography, 86*, 292–299.

Maguire, E. M. W., & Bennett, T. (1982). *Burglary in a dwelling: The offence, the offender and the victim*. London: Heinneman Educational Books.

Mayhew, P. (1984). Target hardening: How much of an answer? In R. V. G. Clarke & T. Hope (Eds.), *Coping with burglary: Research perspectives on policy* (pp. 29–44). Boston: Kleuer Nijhoff.

Mayhew, P., Aye Maung, N., & Mirrlees-Black, C. (1993). *The 1992 British Crime Survey* (Home Office Research Study 132). London: HMSO.

Metropolitan Police. (2017). Advice on burglary prevention. http://212.62.21.14/Site/crimepreventionbumblebee. Accessed 16 June 2017.

Mintel. (2014). Home Security – UK – May 2014. http://academic.mintel.com/display/679647/. Accessed 12 Mar 2018.

Nee, C., & Meenaghan, A. (2006). Expert decision making in burglars. *British Journal of Criminology, 46*, 935–949.

Nee, C., & Taylor, M. (1988). Residential burglary in the Republic of Ireland. In M. Tomlinson, T. Varley, & C. McCullagh (Eds.), *Whose law and order* (pp. 82–103). Galway: The Sociological Association of Ireland.

Nee, C., & Taylor, M. (2000). Examining burglars' target selection: interview, experiment or ethnomethodology? *Psychology, Crime & Law, 6*(1), 45–59.

Office for National Statistics. (2013). *Chapter 3 – Burglary and home security*. Office for National Statistics. http://webarchive.nationalarchives.gov.uk/20160105160709/http://www.ons.gov.uk/ons/dcp171776_340685.pdf. Accessed 16 June 2017.

Office for National Statistics. (2016). *Focus on property crime: Year ending March 2016*. Office for National Statistics. https://www.ons.gov.uk/peoplepopulationandcommunity/crimeandjustice/bulletins/focusonpropertycrime/yearendingmarch2016. Accessed 16 June 2017.

Office for National Statistics. (2017). *Crime and justice methodology*. London: Office for National Statistics.

Office for National Statistics. (2018a). *Crime in England and Wales: Year ending December 2017*. https://www.ons.gov.uk/peoplepopulationandcommunity/crimeandjustice/bulletins/crimeinenglandandwales/yearendingdecember2017. Accessed 7 June 2018.

Office for National Statistics. (2018b). *User guide to crime statistics for England and Wales*. https://www.ons.gov.uk/peoplepopulationandcommunity/crimeandjustice/methodologies/userguidetocrimestatisticsforenglandandwales#crime-survey-for-england-and-wales-csew. Accessed 14 June 2018.

Pease, K. (1991). The Kirkholt project: preventing burglary on a British public housing estate. *Security Journal, 2*(2), 73–77.

Roth, J. J. (2017). The role of perceived effectiveness in home security choices. *Security Journal, 31*(3), 708–725.

Roth, J. J., & Roberts, J. J. (2017). Now, later, or not at all: Personal and situational factors impacting burglars' target choices. *Journal of Crime and Justice, 40*(2), 119–137.

Snook, B., Dhami, M. K., & Kavanagh, J. (2011). Simply criminal: Predicting burglars' occupancy decisions with a simple heuristic. *Law and Human Behavior, 35*, 316–326.

Suffolk Constabulary. (2017). Over 1/3 burglaries are because homes are left insecure. https://www.suffolk.police.uk/news/latest-news/over-13-burglaries-are-because-homes-are-left-insecure. Accessed 16 June 2017.

Thompson, R. (2014). *Understanding theft from the person and robbery of personal property victimisation trends in England and Wales, 1994–2010/11*. PhD: Nottingham Trent University.

Tilley, N. (2009). *Crime prevention*. Devon: Willan Publishing.

Tilley, N., & Tseloni, A. (2016). Choosing and using statistical sources in criminology – What can the Crime Survey for England and Wales tell us? *Legal Information Management, 16*(2), 78–90.

Tilley, N., & Webb, J. (1994). *Burglary reduction: Findings from Safer Cities schemes*. London: Home Office.

Tilley, N., Pease, K., Hough, M., & Brown, R. (1999). *Burglary prevention: Early lessons from the Crime Reduction Programme.* (Crime Reduction Research Series Paper 1). London: Home Office.

Tilley, N., Tseloni, A., & Farrell, G. (2011). Income – Disparities of burglary risk and security availability over time. *British Journal of Criminology, 51*(2), 296–313.

Tilley, N., Farrell, G., & Clarke, R. V. (2015a). Target suitability and the crime drop. In M. Andresen & G. Farrell (Eds.), *The criminal act: The role and influence of routine activities theory* (pp. 59–76). London: Palgrave Macmillan.

Tilley, N., Thompson, R., Farrell, G., Grove, L., & Tseloni, A. (2015b). Do burglar alarms increase burglary risk? A counter-intuitive finding and possible explanations. *Crime Prevention and Community Safety, 17,* 1–19.

TNS BMRB. (2012). *The 2011/12 Crime Survey for England and Wales technical report volume one.* http://doc.ukdataservice.ac.uk/doc/7252/mrdoc/pdf/7252_csew_2011-2012_technicalreport.pdf Accessed 14 June 2018.

Tseloni, A., Thompson, R., Grove, L., Tilley, N., & Farrell, G. (2014). The effectiveness of burglary security devices. *Security Journal,* 1–19.

Tseloni, A., Farrell, G., Thompson, R., Evans, E., & Tilley, N. (2017). Domestic burglary drop and the security hypothesis. *Crime Science, 6*(3), 1–16.

Winchester, S., & Jackson, H. (1982). *Residential burglary: The limits of prevention.* London: Home Office.

Wright, R., & Decker, S. (1994). *Burglars on the job: Street life and residential break-ins.* Boston: Northeastern University Press.

Wright, R., Logie, R., & Decker, S. (1995). Criminal expertise and offender decision making: An experimental study of the target selection process in residential burglary. *Journal of Research in Crime and Delinquency, 32,* 39–53.

Chapter 5
Household- and Area-Level Differences in Burglary Risk and Security Availability over Time

Andromachi Tseloni and Rebecca Thompson

Abbreviations

CSEW	Crime Survey for England and Wales
DWP	Department for Work and Pensions
FAVOR	Familiarity, Accessibility, Visibility, Occupancy, Rewards
HMOs	Houses in multiple occupation
HRP	Household Reference Person
ONS	Office for National Statistics
RH	Reference household
WD	Window and door locks
WDS	Window locks, door locks and security chains
WDSC	Window locks, door locks, security chains and CCTV cameras
WDE	External lights on a sensor, window and door locks
WIDE	Window locks, internal lights on a timer, double door locks and external lights on a sensor

5.1 Introduction

It is well-established that crime concentrates amongst people (victims and offenders) and places (O et al. 2017; Lee et al. 2017; Farrell 2015; Sherman et al. 1989; Wolfgang et al. 1972; Weisburd 2015). In their seminal work, Trickett et al. (1992) found households in the highest crime areas of England and Wales experienced ten times more property crimes (burglaries, thefts from dwelling, criminal damage and vehicle crimes) than those in the lowest crime areas. In addition, those victimised experienced crime more frequently (ibid). Thus, in high-crime areas, there are fewer victims but more crimes per victim than would be predicted if crime victimisation was random (Trickett et al. 1992;

A. Tseloni (✉) · R. Thompson
Quantitative and Spatial Criminology, School of Social Sciences, Nottingham Trent University, Nottingham, UK
e-mail: andromachi.tseloni@ntu.ac.uk

© Springer Nature Switzerland AG 2018
A. Tseloni et al., *Reducing Burglary*,
https://doi.org/10.1007/978-3-319-99942-5_5

Osborn and Tseloni 1998). Household burglary is no exception. Few households experience burglary in any given year,[1] but previous research found 38 percent of all burglaries in 1 year were suffered by the 18 percent most burgled households (Tseloni and Pease 2005). This suggests there may be households and areas which would particularly benefit from preventive measures. Chapters 4 and 8 explore the average (across all households and areas) effectiveness of security devices. Window and door locks (WD) formed the basis of all the most effective security device combinations. The most efficacious configurations were found to be window and door locks, security chains and CCTV cameras (WDSC) followed by window and door locks plus internal and external security lighting (WIDE),[2] then external lights on a sensor, window and door locks (WDE) and window and door locks (WD) (see also Chap. 8; Tseloni et al. 2014, 2017). In the interests of brevity, only results relating to WIDE will be discussed here. Other results are available from the authors upon request.

The availability of effective security may differ across (or be conditioned by) population sub-group and area type. Different household types (e.g. owner-occupiers, renters), area types (e.g. inner city, urban or rural) and regions provide the *context* within which security works (or doesn't work) against burglary. To put it more simply, households may suffer high burglary risk either because they do not have effective security or, although they have it, it does not protect them enough. In the latter case, the presence of (generally) effective security may not protect certain households because they have disproportionately high burglary risks due to, for example, proximity to potential offenders or housing design and estate layout. Therefore, knowing who has effective security and how this is linked to their specific burglary risk is important information for both theory and policy. Domestic burglary has fallen considerably in England and Wales (by 64 percent from 1993 to 2011/2012) and elsewhere (by nearly 30 percent internationally from 1995 to 2004) (Tseloni et al. 2010; 2017; Chaps. 1 and 2 of this book) arguably as a result of widespread adoption of effortless and nonintrusive security (Farrell et al. 2014; Tilley et al. 2015; Chap. 8 of this book). The question is whether the burglary falls and security upgrades have been similar across different contexts.

The current chapter aims to examine:

(a) The relationship between the risk of becoming a victim of burglary and the odds of owning the most effective security (both nationally and in context)
(b) How, if at all, this relationship has changed (both nationally and in context) over the period of the crime drop

---

[1] The study (which relied on data from 1999) found 3.5 percent of households in England and Wales were victims of burglary.

[2] The highest total protection (by a small margin and ignoring outliers) was conferred by WDSC. This combination was not analysed here due to the high cost of CCTV and the potential fire hazard posed by security chains. In addition, unlike the other security measures, security chains can only be used if the occupants are present in their dwelling.

Context in this study refers to population groups of different demographic and socio-economic characteristics, defined, for example, by ethnicity, household composition, tenure, income, number of cars and type of area of residence. As will be seen in the ensuing discussion, this is the first study that addresses the above research aims. It thus fills an important gap in the literature on the crime drop that informs theory and offers policy lessons for sustaining and expanding the burglary falls to groups that have not experienced a fall in burglary and/or producing further falls.

The outline of this chapter is as follows. After presenting the relevant theory (Sect. 5.2), it will first review what is already known about (the relationship between) burglary risk and security availability in different contexts as well as which population groups experienced the largest burglary falls (Sect. 5.3). In particular, Sect. 5.3 is split into five parts. It will review previous research findings on which types of households experience the highest burglary rates (Sect. 5.3.1) and which have adequate security (Sect. 5.3.2), the limited evidence on both burglary risk *and* security availability in context (Sect. 5.3.3), any changes over the last 20 years and whether the same or different households have been at greater risk than others over time (Sect. 5.3.4). Section 5.3 will conclude by outlining how the present work expands current knowledge on whether burglary concentration has changed over the period of the sharp crime drop, 1993–2004/2005 (Sect. 5.3.5). The research aims with respect to the national picture of the over time relationship between security and burglary are addressed in Sect. 5.4.2 after a brief introduction of the data and methodology (Sect. 5.4.1). The fifth part of the chapter (Sect. 5.5) investigates the same relationship in context – with respect to selected household (Sects. 5.5.2, 5.5.3, 5.5.4, 5.5.5 and 5.5.6) and area (Sect. 5.5.7) types – after an introduction to this selection (Sect. 5.5.1). The penultimate Sect. 5.6 summarises the findings considering what information they hold about the distributive justice of the crime drop. The chapter ends with a discussion of the new evidence for crime prevention theory and policy (Sect. 5.7). Specific details of the data, methodology and statistical results are given in Appendix B.

5.2 Theoretical Framework

Testing to what extent (and how) burglary risks can be predicted has been founded on criminological theories of lifestyle, routine activities (for a combination of both theoretical propositions, see Gottfredson 1981) and social disorganisation (Shaw and McKay 1942). A household may be an attractive burglary target for a number of reasons:

(a) The physical features of a property (including physical security) and its immediate surroundings
(b) The household's socio-economic characteristics, such as household composition and income
(c) The household's routine activities, such as whether they are away from home a lot

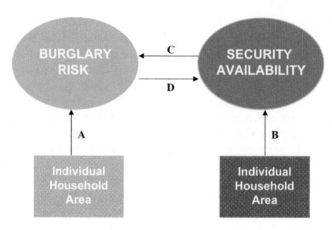

Diagram 5.1 The association between burglary risk and security availability considering contextual influences

(d) The population profile of the neighbourhood

(e) The interplay of all the above

These features collectively determine a household's exposure to potential burglars in the absence of capable guardians and can be used to predict their risk of victimisation and expected number of burglaries within a particular time period. The physical features of properties and their surroundings, households' composition, their routine activities and area profile collectively relate to a number of the FAVOR-able cues often considered by burglars in the commission of an offence (see Chap. 4): Familiarity with the area, Accessibility and Visibility of the household and whether the property appears Occupied. In relation to the final cue, burglars' perceived Rewards may be gauged by household and area socio-economic profile (Chap. 3). Jointly they shape burglars' perceptions about effort and detection risk. Therefore they offer one explanation of victimisation based on population heterogeneity: the set of stable characteristics that makes a household an attractive burglary target (Tseloni 1995).[3]

To a large extent, the same household and area attributes – factors (b) to (e) – may also determine the quantity and quality of security present in a home. In effect, security availability is shaped by the same *causes* as burglary risk whilst, at the same time, is also one of the factors which *influences* burglary risk. Diagram 5.1 clarifies the role of security. As Diagram 5.1 shows, individual, household (including routine activities) and area characteristics and their interactions influence both

---

[3] Event dependence and spells are additional explanations of crime (re-)victimisation. The former refers to how victimisation history alters future victimisation risk and may proxy unobserved population heterogeneity, in other words stable differences that may become known to offenders after their first encounter (Tseloni and Pease 2004; Tseloni 2014). The latter arguably indicates offenders' movements which are analogous to animal foraging that make entire areas riskier for a short period (Johnson 2014).

burglary risk and the availability of physical security in the property. In addition, the latter, security, exerts a direct effect on the chances of becoming a burglary victim.[4] Therefore, household and area characteristics influence burglary risk both directly and indirectly via the mediating factor of security.

5.3   Previous Research Evidence

5.3.1   Burglary Risks in Context

This section will provide a brief overview of previous research which has explored the risk factors of burglary victimisation, addressing the following question:

Who is at risk of being burgled and where do they live?

There is a fairly well-established body of research evidence regarding what makes a household an attractive burglary target (Kennedy and Forde 1990; Rountree and Land 1996; Osborn and Tseloni 1998; Tseloni 2006). Chapters 3 and 4 discussed how the physical features of a domestic property and its surroundings can deter potential burglars (or thwart them) from breaking in. Security devices are often an integral part of a property with the aim of preventing burglary. As outlined in Chap. 4, however, there are a number of additional cues which can play an important role in target selection: socio-economic characteristics and routine activities of the household, the population profile of the area of residence and their interplay. The fact that different households have a different risk of victimisation and that there is variation in the number of expected incidents over a particular time period is termed population heterogeneity (Tseloni 1995).

Research carried out using Crime Survey for England and Wales (CSEW, formerly known as the British Crime Survey, BCS) data between 1992 and 2000 (Ellingworth et al. 1997; Osborn and Tseloni 1998; Tseloni 2006, 2014) shows that household types with high burglary risk include:

- Those who have previously experienced burglary or car theft and/or a household member who has been a victim of assault in the previous 4 years
- Those with two or more cars
- Lone parents
- Social renters (analogous to public housing tenants in the USA)
- Those living in inner-city areas
- Householders who have recently (within 1–2 years) moved into the area

---

[4]The opposite direction of effect, whereby an initial burglary prompts security uptake is equally plausible but is not examined here. Here, we analyse data relating to the first burglary reported by victims (within the recall period) and any security in place *prior* to this incident. The data hold no specific information about incidents that occurred earlier than the recall period (see Chap. 4, Sect. 5.4.1 and Appendix A).

In addition, households experiencing high burglary risk are often located in areas with a certain demographic and socio-economic profile, delineated by a low number of cars per household, a high percentage of youths (5–24 years old), high population density and high poverty (which is a composite factor indicating areas characterised simultaneously by high percentages of lone parents, households without a car, those renting from the local authority and a high mean number of people per room but low percentages of non-manual workers and households owning their homes) (Ellingworth et al. 1997; Kershaw and Tseloni 2005; Osborn and Tseloni 1998; Tseloni 2006, 2014). The same picture of burglary vulnerability exists cross-nationally, allowing for country-specific spatial distribution of the population (Rountree and Land 1996).[5] It should be stressed that all the above risk factors exhibit independent (evidenced via statistical modelling that accounts for group composition) effects on burglary victimisation.

5.3.2   Security Availability in Context

Security devices in residential properties are often installed in response to a real or perceived risk of burglary (Van Dijk and Vollaard 2012). This section explores previous empirical findings with regard to:

Who has anti-burglary security devices and where do they live?

To a large extent the same household and area attributes relating to burglary risks can also determine the quantity and quality of security available in the home (Tilley 2012). In addition, burglary experiences (or indeed those of neighbours and friends) may influence households' decisions to increase the number and/or quality of anti-burglary devices in their homes (Lewakowski 2012). Budd (1999) examined household security measured at different points for burglary victims and non-victims employing the 1998 CSEW; for the former, what counted was any security in place at the time of the first burglary reported to the survey whereas for non-victims the data referred to security at the time of interview. Distinguishing whether security was in place prior to a burglary taking place is essential for avoiding reverse causality problems when examining security and burglary (see next two subsections). To our knowledge, with the exception of pioneering work by Budd (1999), on which more will follow below, and our own collaborative research (Hunter and Tseloni 2016; Tilley et al. 2011; Tseloni and Thompson 2015), the relationship between household attributes and physical security has been unexplored.

---

[5] For example, the vast spaces in the USA allow affluent households to live in gated neighbourhoods and/or very far away from places accessible via public transport to potential burglars. Therefore, household affluence is equivalent to area affluence and a protective factor against burglary in this country. This is very much in contrast with the well-established finding that household affluence is a risk factor whereas area affluence protects against burglary in England and Wales (Tseloni et al. 2002, 2004; Tseloni 2006).

A relatively limited body of previous research has shown that security is unevenly distributed across the population (Budd 1999; Tilley et al. 2011; Tseloni and Thompson 2015). 'Ownership of security devices varies greatly among different types of household…Young households, households with one adult and children, and economically disadvantaged households are particularly likely to have low levels of security' (Budd 1999, p. 38). Quite reasonably, income (or lack thereof) is likely to be a contributing factor to having good quality (or, indeed, any) security devices. For example, in 1998, households earning more than £50,000 per annum were 80 percent more likely to have, what was termed by Budd (1999), a *high level of security* or, by Tilley et al. (2011), *enhanced security* (at least three security devices including door and window locks) than households in the lowest income bracket (less than £5000 per annum). This differential in enhanced security between the highest and lowest income households has remained after the crime drop: households earning more than £50,000 per annum were roughly 60 percent more likely to have enhanced security and 50 percent less likely to have no or inadequate security compared to the less affluent in 2005/2006 (Tilley et al. 2011).

Particular areas have also been found to have an unequal uptake of household security. In 1998, areas of social housing (council estates), those with a high level of disorder and inner cities had higher than the national average proportion of households with less than two security devices. The same was true for two regions: Wales and the North East (Budd 1999, p. 39). However, the findings from these two UK studies (Budd 1999; Tilley et al. 2011) rely on bivariate associations (contingency table analyses) and therefore do not account for group composition which means these apparent income and area differences in levels of security may mask other unobserved causal factors.[6]

5.3.3 *Security Availability* and *Burglary Risk in Context*

Evidence from two countries to date confirms that security has been the main driver behind the burglary drop over the last two and a half decades. The widespread availability of security measures of increasing effectiveness adopted by private households in England and Wales and the mandatory burglary proofing of new housing in the Netherlands have been shown to relate to burglary falls in the respective countries (Tseloni et al. 2017; Vollaard and Van Ours 2011). However, the question of whether these security induced burglary reductions are similar across different contexts (i.e. particular population groups and areas) remains. This question is theoretically important given the uneven distribution of both burglary risk and security availability in the population outlined in the two previous sections. It is also of urgent practical importance: it may explain the almost flat trajectory of burglary in

---

[6] To our knowledge, apart from the studies in this book (Chaps. 5 and 7), the only exceptions are Tseloni (2011) and Lewakowski (2012). The former is a conference presentation. The full text of Lewakowski's (2012) excellent Master's thesis does not seem to be widely available. Therefore, both will not be further referred to in this book.

England and Wales over the last 10 years and therefore inform how to affect further burglary declines and avoid future increases.

Budd's (1999) pioneering work estimated the burglary risk differentials across security levels (as defined below) taking into account the socio-economic characteristics of respondents and their areas of residence, what is termed *group composition*, using data from the 1998 CSEW. She differentiated across security levels as follows (see also Chap. 4, Table 4.1): 'Households with low level security are those with only window locks or deadlocks. Those with high security have [additionally] a burglar alarm, security lights or window grilles. [Most important, s]ecurity was [recorded] at time of interview for non-victims and at time of incident for victims' (Budd 1999, p. 82). She found that those with no or low security had, respectively, 628 percent and 77 percent higher risk of experiencing a burglary compared to households with high levels of security (but otherwise identical socio-economic and area profile). To our knowledge all other research that has examined security as a predictor of burglary (in addition to socio-economic factors) provided inconclusive evidence due to potential reverse causality stemming from measuring security irrespective of the timing of the burglary.[7]

5.3.4 Who Has Benefited the Most (or, Conversely, Drew Negligible Benefits) from the Reduction in Burglary Risk and the Increase in Security Availability?

Burglary rates increased considerably from 1981 (when the first crime survey data became available in England and Wales) to 1993 (see Chaps. 1 and 2, Figs. 1.8 and 2.1). From 1993, burglary rates fell to the lowest levels ever recorded. In England and Wales, the number of recorded incidents '…dropped by 64 percent and the percentage of burgled households fell from 7 to 2.1 per 100 households between 1993 and 2011/12' (Tseloni et al. 2017, p. 3). From 2004/2005 the dramatic fall stalled at about 700,000 incidents per year or 1.9 per 100 households with no statistically significant year-on-year variation until the time of writing (Tseloni et al. 2017; ONS 2017a). Sect. 5.3.1 reviewed previous evidence pertaining to the question of who is affected by burglary. Given the fall in burglary since 1993, however, it is arguably more pertinent to find out who benefited the most (or, conversely, drew negligible benefits) from the steep decline.

There is no evidence to date to dispute that all population groups suffer fewer burglaries (and crime in general) now than before the crime drop (Hunter and Tseloni 2016; Ignatans and Pease 2016). However, the fall in crime was *uneven*

---

[7] Similarly to Budd (1999), a number of studies have examined the effect of burglary prevention measures as an additional predictor of burglary along with household and area factors (Tseloni 2006; Wilcox et al. 2007; for a more comprehensive list, please see Vollaard and Van Ours 2011), but unlike Budd (1999) they overlooked whether security was installed before or after the burglary. This omission confounds the direction of causality since households may have adopted security as a result of a previous burglary (Vollaard and Van Ours 2011).

across different population groups and areas. Crime concentration, in particular in relation to property crime, has increased during the crime drop (Ignatans and Pease 2015, 2016). The same household types that were at greater risk of burglary when crime rates were rising in England and Wales continue to suffer the bulk of burglaries. This is now at an even higher rate (relative to others) than before the crime drop.

The relative incidence of burglaries significantly increased between 1993 and 2008/2009 for households with any of the following characteristics: single adult, lone parent, non-White, social renters and inner-city residents (Hunter and Tseloni 2016). Households with a combination of the previous attributes have experienced compounded increases in their relative burglary risk and incidence over the period of the crime drop. Another study comparing 2005/2006 burglary risks to 1995 found that the burglary victimisation gap between the most affluent (earning at least £30,000 per annum) and the poorest households (earning less than £5000 per annum) in England and Wales widened, whilst the gap in enhanced security availability narrowed moderately (Tilley et al. 2011).

Households on an average income (£20,000–£29,999) have generally always been found to have the lowest burglary risk. Looking at a range of other household characteristics, owner-occupiers, two-adult households and those with one car benefited the most from the crime drop experiencing higher than average burglary risk falls (Hunter and Tseloni 2016). Considering the question of equity in relation to the crime falls (Rawls 1999), with the exception of households with no car, 'all socio-economic groups heavily burdened by burglaries in 1993 (lone parents, those in social housing, households earning at least £50,000[8] and inner-city residents) experienced inequitable burglary falls' (Hunter and Tseloni 2016, p. 11). 'Unlike the general picture of widening burglary divides with respect to population socio-economic classifications, burglary incidence rates became more comparable across regions in England and Wales during the crime drop, resulting in a more equal regional distribution' (Hunter and Tseloni 2016, p. 10).[9]

Security devices in residential properties are often installed in response to a real or perceived risk of burglary (Van Dijk and Vollaard 2012). From 1992, when the first crime survey data on security appear, to 1998, the availability of security lights more than doubled, window locks increased by almost 50 percent, whereas significant increases also occurred in double door locks and burglar alarms (Budd 1999). The proportion of residential properties with more than one security device, and especially effective ones, such as lights and locks, roughly doubled from 1992 to 2011/2012 (see also Chap. 8). Conversely those with no security (or just a single device) decreased by roughly two thirds over the same period. Almost the entire population of England and Wales (95 percent) now has some form of security in their homes (Tseloni et al. 2017, see also Chap. 8). As seen in Chap. 3, Secured by Design

---

[8] The relative increase in burglary risk of households earning £50,000 or more per year is minimal. A caveat here is that this income group is compared to those earning at least £30,000 in the pre-burglary fall comparative year (1993) since it did not exist as a separate income category.

[9] Examining the distributive aspect of the crime drop in Sweden in relation to offenders' characteristics, Nilsson et al. (2017) reported increased crime concentration amongst the less affluent population groups as a result of inequitable falls in acquisitive crime.

requirements which were introduced in 1989 (and greatly expanded since 1998) in the UK may have contributed to the widespread adoption of security since the early 1990s (see also Armitage and Monchuk 2011). Burglary security has become more effective in deterring or thwarting burglaries (see Chaps. 4 and 8 of this book, and Tseloni et al. 2017). Given the rises in combinations of security devices installed in households since the early 1990s (Tseloni et al. 2017), it would be useful to ascertain who adopted this more effective security and, conversely, who has not.

To our knowledge, previous work on changes in the distribution of security availability over the period of the crime drop focused only on income population groups (Tilley et al. 2011). Rather counter-intuitively however the lowest income group (of less than £5000 per annum) had the highest (a) basic or enhanced security uptake and (b) enhanced security effectiveness (measured via Security Protection Factors; see Chap. 4) in preventing burglary, whereas the security improvements of the most affluent were the second highest (Figures 4 and 11 in Tilley et al. 2011). One caveat needs to be entered. The study's start year of 1997 may impact upon the findings. This is some 4 years after the burglary peak. We know that major security prevalence increases had occurred prior to the burglary peak (Tseloni et al. 2017), and, as already mentioned, the gap in burglary risk between the most affluent (defined in that year as those on at least £30,000 per annum) and the poorest households in England and Wales increased between 1995 and 2005/2006 (Figure 6 in Tilley et al. 2011).

5.3.5 Limitations of Previous Research

Budd's (1999) pioneering analysis successfully overcame the potential reverse causality pitfalls of cross-section data (which refer to a single point in time without any follow-up interviews of participants) relating to the time sequence between security installation and burglary. The current work therefore follows in her steps by measuring security at the point of the (first) burglary incident for burglary victims (within the survey reference period) and at the time of interview for non-burglary victims (Tseloni et al. 2014; Chap. 4 of this book). Budd's (1999) research however has the following six limitations:

(i) It defines security levels in broad terms whereby different combinations are measured within the same category, thus impeding specifying the effectiveness of specific devices or combinations.

(ii) It does not consider the fact that security installation is in itself dependent upon household and area demographic and socio-economic characteristics.

(iii) It refers to one point in time, 1997, and, naturally at the time it was undertaken, it could not have employed the rich pre- and post-crime drop time series data now available.

(iv) Relatedly, it lacks information concerning whether and, if so how, the relationship between security and burglary has changed over time (particularly over the period of the crime drop purely on account of the study's timing).

(v) It does not include information regarding how the relationship between security and burglary risk is mediated by household and area factors.

(vi) It cannot draw conclusions regarding which security devices are most effective in preventing burglaries and which population groups deserve to be the focus of target hardening initiatives.

To our knowledge, the only published research that has examined both burglary risk and security availability trends over the period of the crime drop is Tilley et al. (2011) which shares some of the same limitations outlined above. That study relied on the same broad security categories as Budd (1999) – for a comparison of security groupings of this body of research and the current study see Chap. 4, Table 4.1. It investigated bivariate associations across income groups, which did not measure the possible latent effects on either burglary risk or security availability due to group composition (i.e. an omitted variables problem, Green 1997).[10] Further, although their data was spread over a period longer than a year, the study examined a narrow timeframe.

The work reported in Chaps. 4, 5 and 6 of this book has sought to address these important research gaps. Chapter 4 provided a detailed assessment of the effectiveness of a range of security devices and their combinations, largely addressing limitation (i) above. This showed that the most efficacious security combinations all include window and door locks; adding security chains and CCTV cameras (WDSC) or internal and/or external security lighting (WIDE/WDE/WID) offers the most promising results. The current chapter addresses limitations (ii), (iii), (iv) and (v). Finally, Chap. 6 and the conclusion focus on overcoming limitation (vi).

5.4 Effective Security Availability and Burglary Risks During the Crime Drop

5.4.1 Data and Methodology

The previous sections of this chapter highlighted that both burglary and security are unevenly distributed across population groups and areas. The evidence is ample and consistent with regard to household types with elevated burglary risks. However, previous research on who owns burglary security is limited and methodologically crude. Although there is strong evidence that significant increases in the uptake of household security devices are responsible for the burglary drop, the relationship between security and burglary across population socio-economic groups and areas, i.e. in context, has mostly been unexplored to date. In addition, previous evidence on how security is related to burglary within specific contexts is limited to income groups and, largely due to research design limitations, inconclusive (arrows C and D from Diagram 5.1). Therefore whether (and how) burglary risk and security availability are related within specific contexts and their respective trajectories over the period of the crime drop presents a knowledge gap which the following discussion addresses. Before answering the above questions, a short note on data and methodology is given below.

---

[10] Bivariate analysis does not consider group composition and entails omitted variables problems (Green 1997). This is a potentially serious limitation since the socio-economic profiles of low- or high-income groups may differ widely between 1997 and 2005/2006 (see also Sect. 5.5.5).

Burglary risk is defined here as the likelihood of experiencing at least one attempted burglary or burglary with entry. So, unlike Chap. 4, the two burglary types are examined together. This is mainly to ensure there is an adequate number of respondents for analysis from each population group but it also has a theoretical basis. Together attempted burglary and burglary with entry victimisation give households where physical security neither deterred nor thwarted an offender. The current study employed data from all CSEW sweeps from 1994 to 2011/2012 aggregated in five sets as follows: 1994 and 1996, 1998 and 2000, 2001/2002 to 2004/2005, 2005/2006 to 2007/2008 and 2008/2009 to 2011/2012.[11] The data come from CSEW respondents who completed the Crime Prevention module (or the relevant Follow-Up Questionnaire in the pre-2001/2002 sweeps) which includes information on security devices in the household. Details of CSEW survey design, sample selection and questionnaire structure is discussed in the Appendix of the previous chapter (Appendix A in Chap 4). Three separate data sets, one for each effective security combination (WD, WDE and WIDE[12]), were used for each period. They consisted of all burglary victims and non-victim respondents who reported no security and those who had WD, WDE and WIDE security availability, respectively. The analysis discussed in this Sect. 5.4 refers to all three data sets of respective effective security combination and no security. The respective sample sizes for all combinations and time periods (15 in total) are given in Appendix Table 5.4.

In the interest of brevity, population group-specific findings for burglary risk and effective security availability will only be discussed for the most effective of the combinations: WIDE. Other results are available upon request. WIDE security availability compares the number of households with window locks, internal lights on a timer, double locks/deadlocks and external lights on a sensor with those that have no security. Therefore, findings on population group-specific burglary risk and security availability discussed in Sect. 5.5 are based on CSEW data from all CSEW burglary victims and non-victim respondents who reported having no security or having WIDE. How this sample compares to the entire aggregated CSEW data sets and the population in England and Wales over time is discussed in Sect. 5.5.1. Detailed discussion of the data, methodology and modelling strategy is provided in Appendix B.1.

In order to estimate jointly the probabilities of burglary victimisation and effective security availability for specific contexts, bivariate (with two dependent variables) logit regression models were estimated via the computer software MLwiN (Snijders and Bosker 1999; Rasbash et al. 2009). The associated burglary risk and security availability regression equations include a large number of independent variables informed by the routine activities theory of crime (Cohen and Felson 1979; Felson 2002) and, to a lesser extent, by social disorganisation theory (Shaw and McKay 1942). Together they denote measurable household and area of residence characteristics that delineate different contexts and are outlined in Sect. 5.5.

---

[11] As seen in Chap. 1, Fig. 1.1, the first two sets give estimates of 1993/1995 and 1997/1999 crime rates, respectively. The remaining data sets gauge crime rates of the corresponding financial years, for example, the 2001/2002 to 2004/2005 CSEW measures crimes from April 2001 to March 2005.

[12] The highest total protection (by a small margin and ignoring outliers) was conferred by WDSC. This combination was not analysed in relation to different population groups for the reasons explained at the beginning of this chapter.

Details of the respective sample sizes (Appendix Table 5.3) and estimated models (Appendix Table 5.5) are given in Appendix B.

Regression analysis is a statistical technique that allows group composition to be considered. In regression models based on data with more than one qualitative variable, such as the rich set of socio-economic and routine activity information from the CSEW, each such variable must be denoted with respect to a reference category (Johnston 1984). The set of all reference characteristics gives a fictitious household: the reference household (RH). The RH in this study refers to two adults without children, of professional social class, earning £20,000–£29,999 per annum, owning their home, having two cars and living for 10 or more years in the same detached house in a rural area of the South East. In addition, the respondent/Household Reference Person (HRP)[13] is of White ethnicity, and (for all models except those from the 1994 to 1996 CSEW data) their house is left unoccupied for 7 or more hours on a typical weekday[14] (see Appendix A.1.1). In this study, the RH is (by construct) expected to have the lowest burglary risk. The next subsection investigates the association between burglary risk and security availability during the crime drop in England and Wales nationally and for the RH.

5.4.2   Burglary Risk and Effective Security Correlation During the Crime Drop

This section presents findings relating to the following three questions:

(RQ_1) Is there a negative relationship between burglary risk and security availability, i.e. are burglary risks lower when there is more security available?

(RQ_2) Has the relationship between burglary risk and effective security ownership changed over the period of the crime drop and, if so, how?

(RQ_3) How is the relationship between burglary risk and effective security affected by context and has it changed over the period of the crime drop?

Both RQ_1 and RQ_2 examine the national average (also called unconditional, i.e. without consideration to context) correlation between burglary risk and effective security (RQ_1) over time (RQ_2) from the research aims of Sect. 5.1. This, in principle, should be negative since more and better security would be expected to reduce burglary by increasing the (perceived or real) time and effort to break in and increasing detection risk. Figure 5.1 shows the trajectory of the (estimated national average) correlation between burglary risk and the three security combinations, WD, WDE and WIDE, from 1993 to 2012 (illustrated via arrow C from Diagram 5.1[15]).

---

[13] Ethnicity refers to the respondent except for the 2008/2009–2011/2012 period when the CSEW started measuring ethnicity of the household reference person (HRP) – previously termed 'Head of Household' (see Appendix A.1).

[14] House occupancy was not measured in the early sweeps.

[15] Figure 5.1 gives the relationship depicted via arrow C of Diagram 5.1 for multiple time periods based on findings in Appendix Table 5.4.

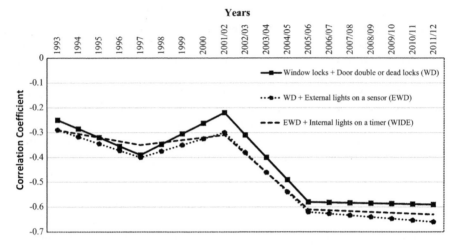

Fig. 5.1 National average correlation between burglary risk and the availability of effective security combinations, WD, EWD and WIDE, over the period of the crime drop (1996–2011/2012 CSEW data). Note: The y-axis values refer to the national average (unconditional) correlation estimated from joint logit empty models of burglary risk and availability of respective WD, EWD and WIDE effective security combinations from five aggregated CSEW data sets, 1994–1996, 1998–2000, 2001/2002–2004/2005, 2005/2006–2007/2008 and 2008/2009–2011/2012. The first two models for each security combination refer to years 1993–1996 and 1997–2000, respectively. The in-between years' correlation estimates have been interpolated from the values given by the models of adjacent periods

Burglary risk and effective security availability are indeed negatively correlated, meaning that greater presence of effective security (particularly the WDE and WIDE combinations) corresponds to lower burglary risks. It is striking that their negative relationship became stronger at two points: the start of the burglary drop, from 1993 to 1997, and subsequently in 2005/2006 when the burglary drop stalled (Tseloni et al. 2017; see also Chap. 2 of this book). In 2005/2006 the negative correlation between burglary risk and effective security availability became (at least) twice as strong compared to that of 2001/2002 and has remained stable (at roughly −0.6) since then. This is so across all three security combinations. The trajectories in Fig. 5.1 imply an increase in security effectiveness or a stronger reliance on security to prevent burglary at a period where overall the average burglary trends have been stationary (neither falling nor increasing) which arguably masks divergent burglary trends across different contexts. This suggests that the main reason behind the burglary falls is better physical security and provides further evidence supporting the security hypothesis for the crime drop (Chap. 8; Farrell et al. 2014; Tseloni et al. 2017).

The third question (RQ₃) asks whether incorporating context (in the form of household and area characteristics that affect each, i.e. arrows A and B in Diagram 5.1) influences this negative correlation between effective security availability and burglary risk. At one extreme, once who owns effective security is determined, there should be no 'left over' (residual or unexplained) correlation between security and burglary. In this case their relationship (arrow C in Diagram 5.1) is entirely a function of mediating household and area characteristics (arrows A and B in Diagram 5.1). At the other extreme their correlation may be unaffected by context, either because it is uniform across contexts or because context is inadequately defined. In the former instance, security would reduce burglary risk to the same extent regardless of where it operates, for example, be it a rural or urban household, in a detached house with children or a non-pensioner, single adult flat. Alternatively, if context is inadequately defined, the characteristics under arrows A and B in Diagram 5.1 would be irrelevant. In reality, the truth lies somewhere in between.

The negative correlation between security availability and burglary risk becomes weaker (particularly for combinations of more devices and in the periods 1997–2000 and after 2005/2006) when the analysis incorporates context. However, it does not disappear – far from it. Indeed context-specific analysis does not eliminate the negative correlation between burglary and security. This implies that their relationship persists beyond the (CSEW measured) mediating household and area characteristics and arguably depend upon unmeasured aspects. These may include more detailed area characteristics, for example, neighbourhood social networks, physical layout and design features of the house/flat and area of residence, as well as connectivity to other places (see also Chap. 3). For economy of space, the context-specific correlations between burglary and each of the three effective security combinations (WD, WDE and WIDE) are shown in Appendix Table 5.4. Appendix B.2 discusses these in more detail.

5.5   Effective Security Availability and Burglary Risks in Context over the Period of the Crime Drop

5.5.1   General Remarks, Population Groups and Their (National Average) Burglary Risks

Having painted the national picture with regards to the key research aims (a) and (b) from the preamble to this chapter, the remainder of this chapter focuses on the relationship between burglary risk and availability of WIDE security for key household and area types, i.e. in context, during the crime drop. To reiterate, this section discusses:

(a) The relationship between the risk of becoming a victim of burglary and the odds of owning the most effective security in different contexts
(b) How, if at all, this relationship has changed for each population group over the period of the crime drop?

The only previous study to examine population-specific over-time burglary victimisation found the relative burglary incidence rates experienced by single adults, lone parents, households from ethnic minorities, social renters and inner-city residents increased over the crime drop, but any differences across regions narrowed over the same period (Hunter and Tseloni 2016). That study also found that households own-ing one car benefited from the burglary drop more than those with none or two or more cars. There is no previous work on population-specific security changes dur-ing the burglary drop, except for examining income groups (Tilley et al. 2011). This select body of previous work furnishes a list of six socio-economic classifications to examine here: ethnicity, household composition, tenure, household income, number of cars owned or regularly used by the household and area type. These are just a few of a large number of household characteristics that were included in the wider study this chapter draws from. Focusing on a select list of characteristics is however necessitated for economy and justified by previous research evidence.

The entire list of household types examined and how they fared with regard to burglary risk and availability of the WIDE security combination can be found in Appendix A.1 and A.3, respectively. Specifically, the following discussion does not include population groups with respect to social class, accommodation type, length of residence at current address, hours home left unoccupied and region. Age of the HRP is included via quadratic function (Tseloni 2006, see Appendix Table 5.5) and shows that burglary risk decreases and the availability of WIDE security increases with age. This is a rather consistent finding throughout the period of the crime drop.

Table 5.1 shows the percentage of the population and households in England and Wales that belonged in each group of interest here across the three Censuses from 1991 to 2011. With the exception of percentage of households with children under 16 years old, the socio-economic landscape of the country changed over the 20-year period. The two starkest differences between 1991 and 2011 are evident in relation to households living in private rented accommodation and lone parents, both of which almost doubled within this period. In 2011 the population was also more ethnically diverse, more households were using two or more cars, whilst fewer households had no car compared to 1991. These changes provide the backdrop against which each population group's effective security and burglary risk trajecto-ries will be discussed in the following sections.

Table 5.1 also includes sample percentages of each population group of interest across the five CSEW aggregate data sets employed here. Appendix Table 5.3 gives the same information for all the characteristics included in the complete models. With the exception of number of adults in the household and again children, all other population groups' share of the employed CSEW samples varies over time. Population groups with increasing presence in more recent CSEW samples include lone parents, private renters, households with an annual income of at least £30,000, those who do not disclose or have any information about income and households with at least three cars. Conversely there were fewer social renters, households with an annual income less than £20,000, households without a car and inner-city resi-dents in more recent CSEW sweeps.

Overall, the presence of different population groups in the CSEW samples over time agree with the unambiguous trends of the general population recorded by the 1991, 2001 and 2011 Census (see Table 5.1). In two instances however it may be an artefact of changes in CSEW sampling methodology. The diminishing share of

Table 5.1 Census and employed CSEW sample percentages of selected population subgroups

Household characteristics (reference category)	Census 1991[h]	Employed CSEW sample 1994–1996	1998–2000	Census 2001[h]	Employed CSEW sample 2001/2002–2004/2005	2005/2006–2007/2008	2008/2009–2011/2012	Census 2011[h]
Ethnicity of respondent[a] (White)								
Black	1.8	2.7	1.4	2.2	1.4	1.8	2.4	3.1
Asian[b]	2.9	2.3	2.5	4.4	1.5	1.8	3.2	3.6
Mixed, Chinese or Other[c]	1.2	1.0	2.1	2.2	1.3	2.4	1.8	3.9
Number of adults (two adults)								
One adult	30.6	32.4	38.3	36.5	32.5	37.1	36.6	37.4
Three or more adults	17.7	16.6	12.1	–	14.9	15.4	13.4	–
Children under 16 in household (no children)								
Children	28.0	28.3	27.5	29.5	25.8	29.6	27.1	29.1
Lone parent (not a lone parent)								
Lone parent	3.7	5.0	6.5	6.5	5.6	8.3	8.0	7.2
Tenure (owner)								
Social rented sector[d]	23.0	26.2	28.6	19.1	17.9	21.8	17.0	17.6
Private rented sector[e]	9.3	4.7	11.0	11.9	10.0	12.0	15.1	18.1
Household income (£20,000–£29,999)[f]								
£4999 and under	–	21.0	15.8	–	6.9	7.5	5.8	–
£5000–£9999	–	20.5	18.8	–	13.5	12.2	11.4	–
£10,000–£19,999	–	26.9	28.1	–	19.8	19.6	21.6	–
£30,000–£49,999	–	11.3	10.2	–	15.6	17.3	15.7	–
£50,000 or more	–	–	3.4	–	6.3	8.7	11.6	–
No income information	–	6.2	8.5	–	23.3	22.0	19.1	–
Number of cars owned/used in last year (two cars)								
No car	32.4	30.6	28.4	26.8	22.0	23.3	22.3	25.6
One car	43.7	43.6	47.7	43.6	44.6	40.9	44.1	42.2
Three or more cars	4.2	5.0	4.4	5.9	7.3	8.6	8.4	7.4

(continued)

Table 5.1 (continued)

Area of residence	Census 1991	Employed CSEW sample		Census 2001	Employed CSEW sample			Census 2011
		1994–1996	1998–2000		2001/2002–2004/2005	2005/2006–2007/2008	2008/2009–2011/2012	
Area type (rural)[g]								
Inner city		21.6	16.3		7.2	8.2	8.0	81.5
Urban		54.9	60.5		61.7	64.1	65.5	

[a]Ethnicity of Household Reference Person (HRP) in 2008/2009-2011/2012 and for consistency ethnicity of Head of Household in the 2011 Census
[b]Indian/Pakistani/Bangladeshi
[c]Includes the Census ethnicity categories: All Mixed/Chinese/Other Asian/All Other
[d]Rented from a housing association/council (local authority)
[e]Private landlord or letting agency/rented with a job or business/living rent free/other
[f]No Census data on household income. For UK – not just for England and Wales – median disposable income trajectory in equivalised prices indexed to 1970 see ONS (2018) and to 1977 see ONS (2017b). Based on the latter report, the UK disposable income was estimated as follows: 1991 £18,763, 1992 £18,387.50, 1993 £17,937.50, 1994/1995 £18,137.50, 1995/1996 £18,412.50, 1996/1997 £19,337.50, 1997/1998 £19,850.00, 1998/1999 £20,550.00, 1999/2000 £21,437.50, 2000/2001 £22,162.50, 2001/2002 £22,987.50, 2002/2003 £23,787.50, 2003/2004 £24,287.50, 2004/2005 £24,987.50, 2005/2006 £24,900.00, 2006/2007 £25,375.00, 2007/2008 £25,687.50, 2008/2009 £25,300.00, 2009/2010 £25,600.00, 2010/2011 £25,337.50 and 2011/2012 £24,975.00
[g]Census data distinguish between number of households in rural and urban areas
[h]ONS Crown Copyright Reserved – data based on tables downloaded from Nomis on 18 January and 6 February 2018
– Not possible to identify from available 2001 and 2011 Census tables

inner-city households in the post 2001/2002 samples may arguably reflect the change in CSEW sampling methodology introduced in that year (Tilley and Tseloni 2016) rather than a general population trend, but, as the Census does not differentiate between households in inner-city and urban areas (see Table 5.1), this remains a hypothesis. It is also hard to justify the U-shaped changes in the percentages of ethnic minorities across samples after 1996 (the last year that the survey included an ethnic minority booster sample).[16]

Prior to discussing burglary risks and effective security availability in context (as derived from the statistical models), it is worth looking at the national average burglary risk estimates obtained from bivariate cross-tabulations with each household attribute of interest and area type for two reasons. First, the ONS and previously the Home Office have published these in their Crime Statistics reports and currently in the ONS/BBC online crime risk calculator (Tseloni and Pease 2017). Secondly, the available data and therefore our analysis are based on a subsample (albeit randomly

---

[16] 'The CSEW uses a stratified multistage cross-section sample design, which (a) had over-representation of inner city [defined inconsistently over time] constituencies until 1998 and has had over-represented low-density areas since 2001/2002, as well as (b) included ethnic minority booster samples until 1996' (Tilley and Tseloni 2016, p. 83).

selected) of burglary non-victims[17] which, although over-represents burglary victims, does not compromise this analysis because the focus is on *relative* population groups' burglary risk. For comparison, Table 5.2 shows the national average burglary risk estimates (second column) obtained from bivariate cross-tabulations with each household attribute of interest and area type, respective odds ratios in comparison to the base category of each attribute (third column) and their respective sample sizes (percentages) in the entire 2008/2009–2011/2012 CSEW aggregate data (first column). With few exceptions, the employed CSEW samples in this study (Table 5.1) are effectively identical to the ones from the entire CSEW data in Table 5.2.[18]

Ignoring compositional effects, the highest burglary risk in 2008/2009–2011/2012 was faced by the following household types (with respect to the characteristics selected for discussion in this work): HRP of Mixed, Chinese or Other ethnicity; single adult households; those with children; lone parents; social renters; households earning under £5000 per year; households without a car; and those living in inner cities of England and Wales.

The following Figs. 5.2, 5.3, 5.4, 5.5, 5.6, 5.7, 5.8 and 5.9 illustrate the relative burglary risk (via dotted bars) and WIDE security combination availability (via lined bars) across different population groups with respect to ethnicity, household composition, tenure, household income, number of cars owned or regularly used by the household and area of residence type over time. *Dotted* bars depict relative burglary risks,[19] whereas *lined* bars represent relative WIDE security availability.[20] Any missing bars for one or more time periods within each population group indicate that the respective characteristic had essentially identical burglary risk and/or security availability to the RH (see Appendix B.1.3).

---

[17] As seen in Appendix A in Chap. 4, to avoid respondent fatigue and minimise survey costs, not all CSEW respondents complete the Crime Prevention module which includes essential home security information (Tseloni et al. 2014). In addition, the Crime Prevention module each year was administered to a proportion of the total CSEW sample which varied (half the sample in the early 1990s and a quarter in recent years, see Appendix Table 4.6 in Chap. 4 for details). As a result, the sample employed here over-represents burglary victims to a different extent over the period of the crime drop (see also Appendix Sect. B.3 and Appendix Table 5.6). This is necessitated by the fact that security availability for the entire CSEW sample does not exist. Therefore the employed CSEW sample in this study is the only available data on anti-burglary security devices in dwellings in England and Wales.

[18] This confirms the randomness of the selection of respondents to the Crime Prevention module. The two samples may differ with respect to the percentage of lone parents, three or more adult households, private renting households and those earning under £5000 or at least £50,000 per year. These differences may, however, not be statistically significant. In addition, the percentage of lone parents and private renting households in the employed 2008/2009–2011/2012 CSEW sample are closer to the 2011 Census than the ones from the entire sample.

[19] Strictly speaking Figs. 5.2, 5.3, 5.4, 5.5, 5.6, 5.7, 5.8 and 5.9 and Appendix Table 5.5 show the odds of burglary – the ratio of the likelihood of being burgled over the complement probability of not being burgled (Long 1997, p. 51). Since burglary victimisation is a rare event, the odds approximate the risk. Therefore, for convenience they are referred to as burglary risk in the ensuing discussion.

[20] WIDE security availability is contrasted to no security in Figs. 5.2, 5.3, 5.4, 5.5, 5.6, 5.7, 5.8 and 5.9 and Appendix Table 5.5. For this reason, it is denoted here as a binary outcome ignoring all other possible security combinations. The sample selection implications are touched upon in Table 5.2, Appendix Sect. B.3 and Appendix Table 5.6.

Table 5.2 Sample percentages and burglary risk of selected population groups via bivariate (contingency tables) analysis of the entire CSEW data 2008/2009–2011/2012

Household characteristics and area of residence type	2008/2009–2011/2012		
	Entire CSEW Sample %	Burglary risk	Odds ratio to the base
Ethnicity of respondent[a]			
White (base)	93.2	1.9	1.0
Black	2.0	3.1	1.6
Asian	3.3	3.5	1.8
Mixed, Chinese or Other	1.6	4.0	2.1
Number of adults			
One adult	31.9	2.6	1.6
Two adults (base)	52.4	1.6	1.0
Three or more adults	15.7	1.9	1.2
Children under 16 in household			
No children (base)	73.7	1.6	1.0
Children	26.3	2.9	1.8
Lone parent			
Not a lone parent (base)	95.3	1.8	1.0
Lone parent	4.7	5.7	3.2
Tenure			
Social rented sector	16.0	3.3	2.2
Private rented sector	11.6	2.8	1.9
Owner (base)	72.5	1.5	1.0
Household income			
£4999 and under	3.8	3.8	2.1
£5000–£9999	11.4	2.5	1.4
£10,000–£19,999	20.4	1.9	1.1
£20,000–£29,999 (base)	14.2	1.8	1.0
£30,000–£49,999	17.7	1.7	0.9
£50,000 or more	14.0	2.0	1.1
No income information	18.5	1.7	0.9
Number of cars owned/used in last year			
No car	19.3	3.0	1.9
One car	42.7	1.8	1.1
Two cars (base)	29.2	1.6	1.0
Three or more cars	8.9	1.7	1.1
Area type			
Inner city	7.8	3.3	3.7
Urban	66.3	2.2	2.4
Rural (base)	25.9	0.9	1.0

[a]Ethnicity of Household Reference Person (HRP) in 2008/2009–2011/2012

Population groups are given in the horizontal *x*-axis which also denotes the CSEW years examined for each group. One population group from each classification however is the reference characteristic. As mentioned, the RH delineated previously (Sect. 5.4.1) comprises the entire set of these reference characteristics. The burglary risk and security presence of the RH serves as a baseline; as already outlined it cannot be individually estimated (see Sect. 5.4.1 and Appendix B.3) and thus coincides with the *x*-axis at value 1 of the *y*-axis of the following set of graphs.

The vertical *y*-axis shows the estimated burglary risk and security availability relative to the reference category of each population classification. A horizontal line at value 1 denotes the baseline burglary risk and WIDE security availability compared to none, i.e. those for the RH. Values above 1 show higher burglary risk or security availability of the respective population group compared to the RH. Values below 1 give a lower risk or availability equal to the difference between the estimated value (bar height) and 1. For example, if households in group A have a burglary risk of 0.5 and effective security availability of 2 compared to reference group B, then households in group A have half the risk and double the protection of group B.

Last but by far not least, readers should note that the bar heights in Figs. 5.2, 5.3, 5.4, 5.5, 5.6, 5.7, 5.8 and 5.9 represent the magnitude of the estimated odds ratios. Bars without shading indicate that the respective odds ratio is not statistically significant at the conventional p-value of 0.05 or lower. To distinguish between burglary and security availability non-statistically significant odds ratios (white bars), the outline of bars for burglary are thicker than for security availability. The following discussion focuses on statistically significant relationships of population subgroups with respect to burglary risk and WIDE security availability. Readers interested in more exact statistical significance of the estimated effects may further consult the models' estimated coefficients in Appendix Table 5.5.

5.5.2   Effective Security and Burglary Risk Across Ethnic Groups

According to the latest (at the time of writing) Census in 2011, the majority (86.0 percent) of the usual resident population in England and Wales was White (Office for National Statistics 2012). The second largest ethnic group was Indian (2.5 percent) followed by Pakistani (2.0 percent). According to the same source, England and Wales has become more ethnically diverse since the 1991 Census (ibid). This is obvious in the figures of Table 5.1.[21] The group with increasingly higher presence is Mixed,

---

[21] Unlike the earlier years, Table 5.1 shows the percentage of 'head of households' – using the Census terminology – rather than population from each ethnic group in the 2011 Census for consistency with the 2008/2009–2011/2012 CSEW models that depict HRP's ethnicity. For this reason, the percentage presented here of White 'heads of household' (89.5 percent) is higher than that of the White population (86.0 percent) in England and Wales. Interestingly the percentage of Mixed, Chinese or Other (3.9 percent) and Asian (Indian, Pakistani or Bangladeshi, at 3.6 percent) 'heads of household' is lower than that of the respective population groups (5.4 and 5.3, respectively).

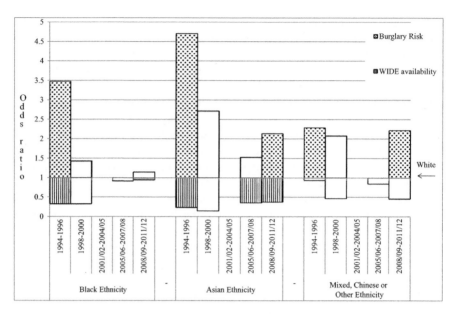

Fig. 5.2 Burglary risk and WIDE security availability (odds ratios) of ethnic minority groups in comparison to White ethnicity, 1993–2011/2012

Chinese or Other which between 1991 and 2011 rose from 1.2 to 5.4 percent of the population. In the latest Census, 3.9 percent of 'heads of household' were Mixed, Chinese or Other. Ethnicity is highly relevant to crime victimisation, in particular burglary. In recent years burglars have been found to target Asian households which traditionally keep gold jewellery in their homes (The Guardian 2012; Metro 2017).

In order to examine relative burglary risk and effective security availability by ethnic groups, data from the CSEW regarding the ethnicity of the respondent (or HRP where available[22]) were used. Respondents were categorised into one of four groups: Black (referring to individuals of Black/African/Caribbean ethnicity); Asian (defined as individuals with an Indian, Pakistani or Bangladeshi background); Mixed, Chinese or Other; or White.[23] Figure 5.2 shows the relative burglary risk and the presence of WIDE across the first three ethnic groups in comparison to the reference category (White) over time.

At the beginning of the burglary drop, 1993–1996,[24] all ethnic minority groups were facing a significantly higher likelihood of burglary victimisation (see the first

---

[22] The estimates refer to respondent's ethnicity pre-2008/2009 and the Household Reference Person's ethnicity in the 2008/2009–2011/2012 models when this information became available in the CSEW. This inconsistency does not, however, compromise the findings since respondent's and Household Reference Person's ethnicity is highly associated in the CSEW (Appendix Table 5.7).

[23] Ethnicity categories have been selected on the basis of consistency with earlier CSEW sweeps and adequate sample sizes.

[24] Although the data come from the 1994 and 1996 CSEW, as mentioned, they refer to victims' security availability and burglaries that occurred in the respective previous calendar years, and therefore the first bars refer to the period 1993–1996. Similarly, the 1998–2000 data cover the period 1997–2000.

dotted bar of each bar-stack in Fig. 5.2) relative to White respondents. With the exception of Mixed, Chinese or Other, ethnic minority households also had significantly lower availability of effective security than White households (see the first lined bar of each bar-stack in Fig. 5.2). The relationship between effective security and burglary risk across ethnic groups was as expected, insofar that those with the lowest chances of having WIDE faced the highest burglary risks. Asian respondents faced nearly four times higher burglary risks alongside being 76 percent less likely to have the WIDE security combination than White respondents. Black respondents came second, facing a two and half times higher burglary risk and being 67 percent less likely to have WIDE security than White respondents. Finally, Mixed, Chinese or Other respondents had just over double the burglary risk of White respondents but were no less likely to have WIDE security.

The burglary risk and effective security profile across ethnic groups has changed since the mid-1990s. Households of Black origin have faced similar burglary risks and had similar levels of effective security as White households since 1997 (see previous footnote). By contrast, the heightened burglary risk and lower availability of effective security in relation to White respondents has been sustained for the other two ethnic minority groups, especially in the latest years examined here, 2008/2009–2011/2012. Households with an Asian or Mixed, Chinese or Other HRP had over double the burglary risk of households with a White HRP and, respectively, 62 and 54 percent lower availability of effective security.[25]

Therefore, with respect to ethnic differences in both burglary risks and effective security availability over the course of significant national burglary falls, Black and White households have completely converged, Asian households have remained worse off than White households, but to a lesser degree than before, whereas there has been no improvement for Mixed, Chinese or Other households. It might also be worth stressing that Asian households faced increasingly higher burglary risks than Black households during the same period.

5.5.3 Effective Security and Burglary Risk with Respect to Household Composition

Over a quarter (29.1 percent) of households in England and Wales have children under 16 years old according to the 2011 Census – a relatively stable figure since the 1991 Census (28.0 percent; see Table 5.1). With the exception of lone parent households which have increased – from 3.7 percent in the 1991 Census to 6.5 and 7.2 in

---

[25] Asian households had nearly three times the burglary risk of White households and 85 percent less effective security in the period 1997–2000, as well as about 65 percent lower odds of WIDE security in the years from 2005/2006 to 2011/2012.

the 2001 and 2011 Censuses, respectively – all other household types in terms of size and composition have remained stable over time (see Table 5.1).[26]

Household composition – specified here by the number (one, two or three or more) of *adults* and whether there are *children* in the household, as well as whether it is a *lone parent* household – is an important factor in determining the risk of burglary. Large family units and extended families living in the same household would naturally be presumed to have increased presence at home. However, households of three or more adults may also be comprised of students or single young adults sharing a house, also referred to as houses in multiple occupation (HMOs). HMOs are attractive burglary targets for a number of reasons: they often contain multiple gadgets (as each person living there has her or his own laptop, mobile phone, etc.). HMOs often also have low home occupancy during the day as well as during the evenings and weekends since young adults may leave the house more frequently and leave the house empty for longer periods (Bernasco 2009). They may also be more likely to unintentionally offer easy access to intruders by leaving windows open and doors unlocked assuming another household member would be in the house and/or take responsibility to lock them. For these reasons, local councils and the police have particularly focused on HMOs for target hardening via a mixture of increased security requirements in terms of security devices available in the property and public awareness initiatives (Nottinghamshire Police 2013).

Other household types potentially vulnerable to burglary would be single, non-pensioner adult households due to potentially low levels of presence at home and households with children which may offer access to potential burglars by leaving doors and/or French windows leading to the garden open for the children to come in and out of the house. The combination of the two above elements is found in lone parent households (single adults with children under 16 years old). Many such households have priority in local council housing (for more on this type of housing, please see next subsection) policies and thereby become social renters. Indeed, 23 percent of council housing tenants are lone parent households (Wilson and Barton 2017). Furthermore, lone parents suffer exponentially more property crimes than others if they live in deprived areas, whereas in average or affluent areas of residence, they experience similar crime rates to others (Tseloni 2006). Therefore, it is crucial to separate any lone parent household composition effect from any social housing effect (see next Sect. 5.5.4) on burglary risk and security availability.

Figure 5.3 shows the relative burglary risk and availability of WIDE for single and three or more adult households relative to two-adult households over time. Figures 5.4 and 5.5 present the same information for households with children compared to those without and for lone parents compared to non-lone parents, respectively.

---

[26] It was not possible to ascertain the number (percentages) of two adult and three or more adult households in the publicly available 2001 and 2011 Census data; therefore the two categories have been collapsed into two or more adult households. The increase in single adult households presented in Table 5.1 across the three Censuses reflects the rise in the percentage of lone parent households.

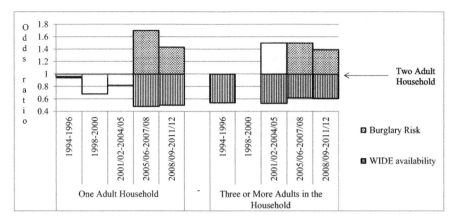

Fig. 5.3 Burglary risk and WIDE security availability (odds ratios) of single and more than two-adult households in comparison to two-adult households

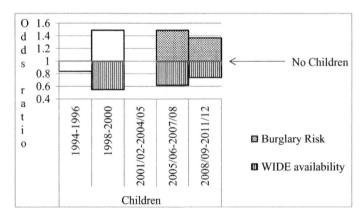

Fig. 5.4 Burglary risk and WIDE security availability (odds ratios) of households with children under 16 years old in comparison to households without children

The number of adults in the household was not a risk factor for burglary for the earlier time periods, 1993–2004/2005, but it has become so since mid-2000. In the respective periods of 2005/2006–2007/2008 and 2008/2009–2011/2012, single adult households were 70 and 43 percent more at risk of burglary with 50 percent less effective security than two-adult households. During these two periods three or more adult households were 50 and 39 percent at higher burglary risk with almost 40 percent lower availability of effective security than the RH. It is interesting that until 2004/2005, despite having a 45 percent lower presence of effective security, three or more adult households were not at statistically significant higher burglary risk than two-adult households. By contrast the relationship between burglary risk and security was as expected in relation to single adult households; they had both similar effective security and burglary risk as two-adult households up until 2005/2006 (see Appendix Table 5.5).

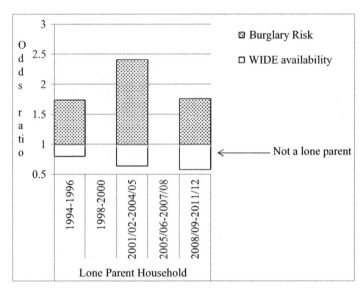

Fig. 5.5 Burglary risk and WIDE security availability (odds ratios) of single adult with children under 16 years old households in comparison to other households

The presence of children in the household was not a risk factor for burglary victimisation or lower levels of effective security at the start of the crime drop (1993–1996). This however changed after this period. Households with children faced almost 50 percent higher burglary risk and had 45 and 38 percent lower effective security presence than others in the years 1997–2000 and 2005/2006–2007/2008, respectively. Their elevated risk fell slightly to 39 percent as did their lack of effective security (to 24 percent) in comparison to households without children in the most recent period examined, 2008/2009–2011/2012.

Lone parents faced about 75 percent higher likelihood of burglary than others both at the start of the crime drop (1993–1996) and the final period (2008/2009–2011/2012) examined here and twice as much in 2001/2002–2004/2005 but did not exhibit any (statistically significant) lower availability of effective security than others. This elevated risk is over and above that of single adult households and those with children, especially in the most recent period (2008/2009–2011/2012). Incorporating these interactions, the overall burglary risk of lone parent households is almost two and a half times greater than that of a two-adult household without children.[27] By contrast, lone parents have 79 percent lower WIDE security availability than a two-adult household with no children.

---

[27] This is calculated from Appendix Table 5.5 as the product of the respective burglary odds for one adult, children and lone parent from the penultimate column, $1.43 \times 1.37 \times 1.76$. The result is 3.45 which is 245 percent higher than the RH [$100 \times (3.45–1)$]. In a similar manner, but using the respective figures in the last column, the odds of WIDE availability compared to no security is 0.21 ($=0.50 \times 0.74 \times 0.58$) which is 79 percent lower than the RH [$100 \times (0.21–1)$].

Therefore, in recent years with largely constant burglary rates, two-adult house-holds have achieved significantly lower burglary risks and higher presence of effective security than others. Households with children are at higher burglary risk and have less effective security than others since 1998, but this gap seems to have narrowed in recent years. Finally, lone parents have relatively high risks of burglary that have remained fairly constant and (non-statistically significant) lower levels of effective security availability than others over time.

5.5.4 *Effective Security and Burglary Risk with Respect to Household Tenure*

Householders may own their home (either outright or with the help of a loan or mortgage), rent from private landlords or rent from local authorities (including social landlords managing public housing, as defined in the CSEW). Local authorities provide housing to priority need population categories, effectively low-income households with additional needs, such as whether there are children or household members with a disability (Wilson and Barton 2017). This type of housing in the UK is currently termed 'social rented' accommodation, formerly known as 'council housing', and it is similar to 'public housing' in the USA, or 'Habitations à loyer modéré' in France. Social rented housing in the UK has undergone unprecedented changes in the last three decades on a number of fronts, including housing design, layout, availability, spatial concentration and location, as well as how the general public views it (Ginsburg 2005; The Guardian 2017; Tunstall 2011).

The percentage of households in social rented accommodation has decreased steadily from 23.0 percent in the 1991 Census to 19.1 and 17.6 percent in the 2001 and 2011 Census, respectively (see Table 5.1). Conversely private renting has become more common in England and Wales; the percentage of households in privately rented housing almost doubled from 9.3 percent in 1991 to 18.1 percent in 2011, mirroring the reduction not just in social rented but also owner-occupied (outright or with a mortgage or loan) accommodation, especially from 2001 to 2011 (see Table 5.1). As discussed in Chap. 2, many of the anti-burglary initiatives of the Home Office in England and Wales sought to prevent burglaries in social housing. Social renters indeed have experienced and, as will be seen shortly, continue to experience a high volume of burglaries (Osborn and Tseloni 1998; Tseloni 2006). In recent years UK policy attention has turned towards crime problems and the increased risk faced by tenants in privately rented accommodation (Higgins and Jarman 2015). In the late 2000s private renters were the most likely tenure group to be in fuel poverty, whilst social renters were not worse off than owner-occupiers according to this indicator of relative poverty (Tunstall 2011).

Figure 5.6 shows burglary risk and availability of the WIDE security combination for social and private renting households in relation to owner-occupiers over time. Social renters had 33 percent higher burglary risk and 61 percent lower avail-

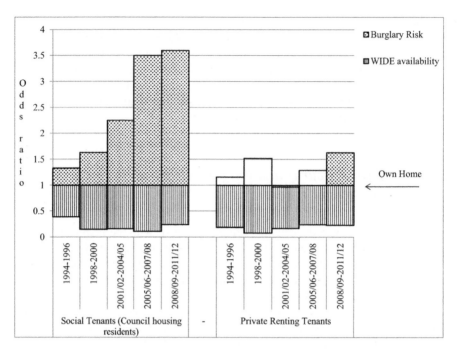

Fig. 5.6 Burglary risk and WIDE security availability (odds ratios) of households in rented from local authorities or private landlords accommodation in comparison to households living in their own (outright or mortgaged) home

ability of effective security than households who owned their homes at the start of the burglary drop. Since 1996, the gap between social renters and owner-occupiers on both fronts has widened. Indeed by 2005/2006–2007/2008, by which time almost all of the national burglary fall had been achieved, social renters were at 250 percent higher risk of burglary and had 89 percent lower availability of the WIDE security combination than owner-occupiers. Effective security availability amongst social renters has slightly improved in the most recent years examined here, 2008/2009–2011/2012; they are 76 percent less likely than owner-occupiers to have the WIDE security combination. However, the burglary risk gap between the two household types has not narrowed.

In comparison to owner-occupiers, private renters have also lacked effective security over the entire period examined here (at about 80 percent lower odds to be WIDE secured). This gap did not affect their burglary risk which was comparable to that of the RH until the most recent period studied, 2008/2009–2011/2012, when private renters faced 63 percent higher burglary risk.

To summarise, the gap in burglary risk between social renters and owner-occupiers became almost 8 times wider during the burglary drop, especially after 2004/2005, when national burglary rates stabilised. During the same period and despite the considerable investment in target hardening (for a detailed overview, see Chap. 2), social renters' shortfall in effective security compared to owner-occupiers widened by one and a half times (to 2005/2006–2007/2008). The improvement in the security gap of social renters from 89 percent in 2005/2006–2007/2008 to 76 percent in 2008/2009–2011/2012 was a welcome one but appeared ineffective in lowering risk. By contrast, the private renters' security gap did not coincide with (statistically significant) increased burglary risks over time, except in the last period examined.

5.5.5 Effective Security and Burglary Risk with Respect to Annual Household Income

Home ownership and income are linked but their relationship is not linear. For example, elderly households may live on low incomes in their own house, whilst by contrast highly paid young professionals may rent. Physical security can be costly and therefore it cannot be afforded equally by all. For this reason, previous research has focused upon security availability and burglary risks across income groups (Tilley et al. 2011). Other characteristics of households, for example, tenure or household composition, may however be more critical than affordability in the decision regarding whether to obtain anti-burglary devices. They would remain confounded without examining them in tandem with income. To give another example, social renters, as discussed earlier, are also low-income households, but not all low-income households are social renters. We therefore need to be able to separate how much burglary risk and effective security absence are directly linked to low income or social housing which refers to a different, albeit linked, attribute: tenure. At the other end, it is reasonable to assume that high-income households would own their homes; however, not all owner-occupiers are on a high income (e.g. pensioners), whilst high-income households may be renting through a private landlord. Household income, in theory, is also related to the number of (working) adults in the household since more household members at work would theoretically bring in more money. Again, the relationship is not straightforward as some high-income households are single high earners, whilst 10 percent of working-age adults in working families live in relative low income, i.e. 60 percent below the UK median income (DWP 2017). Therefore, we need to be able to distinguish between income effects on burglary and security from those of other factors, such as tenure and household composition.

The CSEW asks respondents to place their annual household income within income bracket bands of £5000 which we merged into the following categories: 'under £5000', '£5000–£9999', '£10,000–£19999', '£20,000–£29999', '£30,000–£49999' and '£50,000 or more'. People are commonly reluctant to disclose this information, even in non-precise terms, or if the survey respondent is not the HRP, she or he may genuinely not know how much the entire household income is, especially in HMO. For these reasons, the income variable has an unusually high proportion of missing information in the CSEW data, which has been retained here and presented as an additional income category, labelled 'No income/No information'.

The fact that the CSEW data include income in monetary rather than relative values presents two problems. First the category '£30,000–£49999' used to be, in earlier sweeps, '£30,000 or more', whilst the last ('£50,000 or more') did not exist until the 1998 CSEW. More importantly, the same monetary values have quite different purchasing power and imply diverse affluence levels over time, especially over the 20-year period examined here. Indeed, as seen in the next paragraphs, the ONS publishes UK income trend estimates indexed at a specific starting year of the ONS series and in equivalised prices (via the modified OECD scale) to aid over time comparisons. Thus, the ONS figures overcome the problem of the variable purchasing power of nominal income, whilst they highlight the need to be cautious in interpreting over time comparisons of CSEW income groups.

It is no surprise that data consistent with the CSEW income groups do not exist in the Census (see Table 5.1). The ONS provides information on mean and median (earned by the poorest 50 percent of households) disposable (after taxes and benefits) income in the UK (that includes Northern Ireland and Scotland in addition to England and Wales) over time standardised at 1970 equivalised prices (ONS 2018).[28] According to this source the median disposable income fell about 5 percent between 1991 and 1993, the peak year of the burglary rates, but it had recovered to previous levels by 1996 and rose almost consistently until 2007 by 37 percent overall. From 2007 to 2012 however it dropped again by 5 percent. The actual value in British pounds of the median income is not provided in this publication (ONS 2018). According to an earlier ONS (2017b) report, the median household disposable income grew from £12,500 in the financial year ending 1977 to £25,700, £26,700 and £27,200 in the financial years ending 2008, 2016 and 2017, respectively.[29] Using the ONS figures as a guide, we estimated (via multiplying the original ONS given value for 1977 with the annual indexed value and dividing by 100) that during the period of this study, the median UK disposable income ranged from £18,763 in 1991 to £24,975 in 2011/2012 (see relevant note of Table 5.1). Considering that middle and upper income groups in principle pay more taxes than subsidies received, their disposable income would be lower than their nominal one. Therefore, these estimates arguably suggest that the median UK income is included within the range of the base income bracket of '£20,000–£29999' per annum in the CSEW analyses offering therefore an intuitive interpretation of the following findings.

Figure 5.7 shows burglary risk and the availability of WIDE security for households on 'under £5000', '£5000–£9999', '£10,000–£19999', '£30,000–£49999' and '£50,000 or more' in comparison with households earning '£20,000–£29999' per annum.

Unsurprisingly households on the lowest annual income, 'under £5000', face higher burglary risks (albeit not always) and have lower effective security availability than middle-income households '£20,000–£29999'. Their effective security

---

[28] Disposable income is expected to be lower than the annual household income for households at a minimum of low middle income which do not receive benefits and pay taxes.

[29] These figures have been adjusted for inflation, deflated to 2016/2017 prices using the Consumer Prices Index, which includes owner-occupiers' housing costs and changes in household composition over time (ONS 2017b).

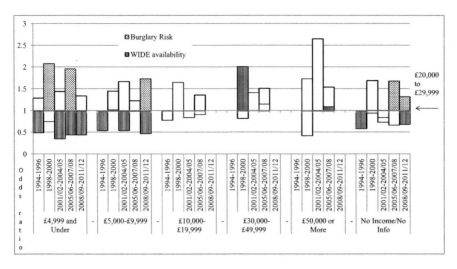

Fig. 5.7 Burglary risk and WIDE security availability (odds ratios) of non-affluent and affluent households in comparison to average income households

deficit has remained relatively constant over time. In the most recent period examined (2008/2009–2011/2012), households in the second lowest income group, '£5000–£9999', were 73 percent more at risk of burglary and had 53 percent lower availability of effective security than those on '£20,000–£29999'. Similarly, households with 'No income/No information' were 32 percent more likely to be burgled (roughly half of their 2005/2006–2007/2008 even more elevated risk) and less likely to have WIDE security. Households in the low middle-income group, '£10,000–£19999', have effectively similar burglary risks and effective security availability as the RH. Those in the top income groups, '£30,000–£49999' and '£50,000 or more', had higher WIDE security than those earning '£20,000–£29999', up until 2001/2002–2004/2005 – the year that burglary stopped falling significantly – but since then had (statistically) similar levels of effective security. They also faced the same (without significant differences) burglary risk as households with roughly median income throughout the period examined.

There is therefore an inverse relationship between availability of the WIDE security combination and burglary risk both for the low-income households and for those which did not provide income information to the survey. Considering these three population groups as one, their burglary risk gap, compared to those with upper middle income, '£20,000–£29999', has narrowed over time, whilst their security deficit has remained relatively unchanged. Apart from up to a 165 percent higher availability of effective security during the years of sharp burglary drop (1997–2004/2005), the two upper income groups ('£30,000 or more' and '£50,000 or more') did not differ from middle-income households. This implies that over time, as security became more affordable (Van Dijk and Vollaard 2012), their advantage in effective security disappeared, whereas they have not been any more or less burgled than middle-income groups in the time period from 1993 to 2011/2012.

5.5.6 Effective Security and Burglary Risk with Respect to Household Car Ownership

The final household characteristic to be examined is the number of cars the household owned or had use of. Cars parked outside a home are both visible signs of occupancy and affluence shaping burglars' perceptions of interruption/detection risk and target attractiveness, respectively. The 2011 and 1991 Censuses show that over the period of the crime drop, car availability increased – more households have at their disposal at least two cars and fewer are carless.[30] Interestingly, the percentage of households with one car has remained about the same (at roughly 42 to 44 percent) over the 20 years examined in this study. This Census information, although referring to car ownership rather than ownership or usage, confirms car availability trends observed across suites of CSEW data (see Table 5.1).

Car availability may signal affluence: homes with many, especially expensive, cars parked in their driveway may be targeted by burglars. The security hypothesis for the crime drop suggests that manufacturer inbuilt car security initiated the crime drop which was then generalised into other crime types (Farrell et al. 2011, 2015; Tseloni et al. 2010). Since new cars had become harder to steal by hotwiring (car thieves' modus operandi in the 1980s), breaking into the owners' home to steal the car keys may be one way of carrying out car theft. In theory therefore cars, as both indicators of affluence and targets, are expected to be positively related to burglary risk and security availability. By contrast, a car on the driveway may indicate the home is occupied and therefore actually serve as a deterrent.

In this study households were initially classified into four types: those with no, one, two or three or more cars (where two cars formed the reference category). In preliminary statistical modelling however, households with three or more cars had no statistically different burglary risk or WIDE security availability than two-car households. For this reason, the last two categories are merged here, and Fig. 5.8 shows the relative burglary risk and WIDE availability of households with no or one car in comparison to the base of at least two cars over time (see Appendix B.1 for the modelling strategy and Appendix Table 5.3 for more detailed results).

Confirming the earlier theoretical expectation above, households with no car have between 50 (in 1998–2000) and 74 percent (in 2001/2002–2004/2005) lower odds of WIDE security than households with two or more cars, and this shortfall has remained significant throughout the crime drop (see also Appendix Table 5.5). They did not however have any significantly higher burglary risk except for the most recent period at 52 percent compared to two-car households (2008/2009–2011/2012). One-car households were less secured in two early periods (by 38 and 41 percent in 1996–1998 and 2001/2002–2004/2005, respectively) of the crime drop but did not

---

[30] Specifically, the percentage of households with no car fell from 32.4 percent in 1991 to 26.8 and 25.6 percent in 2001 and 2011, respectively (see Table 5.1). Households with at least three cars almost doubled from 4.2 percent in 1991 to 7.4 percent in 2011, whilst two-car households increased by roughly a quarter.

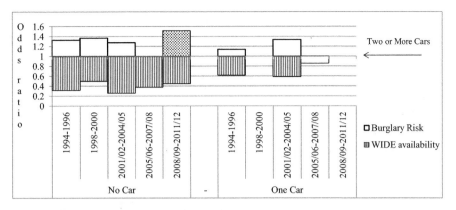

Fig. 5.8 Burglary risk and WIDE security availability (odds ratios) of no or one-car households in comparison to households with two or more cars

have any significantly higher burglary risk than those with more cars. Therefore, in relative terms, lack of WIDE security is associated with higher burglary risks only for carless households and in the most recent period. The next discussion moves away from household classifications to examine burglary and security across different area types.

5.5.7 Effective Security and Burglary Risk by Area of Residence

This section focuses on burglary risks and the availability of WIDE security for households residing within inner cities or urban areas in comparison to those in rural areas. In the 1994–1998 CSEW samples, inner cities[31] were over-represented (Tilley and Tseloni 2016). As seen in Chap. 2, most burglary target hardening has previously been directed towards non-rural areas. In terms of population statistics, the 1991 and 2001 Census did not provide area type information on households – the only information comes from the 2011 Census which registered 81.5 percent of households in England and Wales living in urban areas with the remaining 28.5 percent in rural areas (see Table 5.1). A possible explanation for this lack of earlier data is that area classification definitions have undergone a number of changes over time (Pateman 2011).

Figure 5.9 shows the relative burglary risk and availability of WIDE security for inner-city or urban households in comparison to those in rural areas over time.

---

[31] Defined as constituencies where at least one of the following applies: their population exceeds 50 persons per hectare, fewer than 54 percent of households are owner-occupiers or fewer than 1 percent of household 'heads' are classified as professional or managerial (Hales and Stratford 1997, 1999).

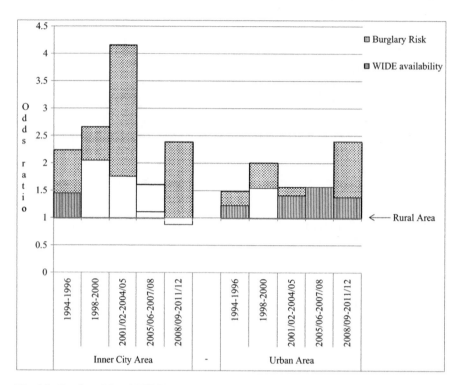

Fig. 5.9 Burglary risk and WIDE security availability (odds ratios) of households in inner-city or urban areas in comparison to rural areas

The picture here is partly different from what has been reported so far in this chapter: households living in urban areas had between 23 and 57 percent higher availability of effective security than those in rural areas throughout the examined period. However, they were also at higher risk of burglary by between 49 and 140 percent with this risk increasing in the latest period (2008/2009–2011/2012) examined.[32] Inner-city residents had significantly higher availability of effective security only in the early years, 1993–1996, whilst they have had sustained higher burglary risks by between 124 percent, in 1993–1996, and 316 percent, in 2001/2002–2004/2005, compared to rural households.[33]

Therefore, non-rural households have faced increasing relative burglary risks over the period of the crime drop and stable availability of effective security compared to residents of rural areas. Specifically, after the burglary drop (2008/2009–

---

[32] The respective estimates of the effects of urban residence were not statistically significant at the conventional 0.05 p-value in two instances: (a) for WIDE security presence in the 1997–2000 and (b) for burglary risk in the 2005/2006–2007/2008 models.

[33] The inner city parameters for effective security availability in the models after 1996 and for 2005/2006–2007/2008 burglary risk are not statistically significant at the conventional 0.05 p-value.

2011/2012), urban households were worse off than before (1993–1996) despite a small improvement in their availability of WIDE security. Inner-city households had about the same relative burglary risks before and after the crime drop. They experienced however substantial risk increases between 1997 and 2004/2005 (the period of national burglary falls). Inner-city households' advantage in WIDE security availability (compared to rural households) gradually disappeared. What is unique in the above results (compared to those reported in the previous subsections) is that elevated relative burglary risk coincides with moderately elevated (or at least not reduced) relative availability of effective security. This counter-intuitive finding will be discussed in the following sections.

Regional differentials in burglary risk and the availability of effective security which are not presented here (interested readers can find them in Appendix Table 5.5) confirm the above observation. For example, in 1993–1996 London had both higher burglary risk and greater presence of effective security than the rest of the South East, which is predominantly rural or urban with significant rural parts (Official Statistics 2013). Meanwhile Wales, which is also predominately rural, has had lower effective security throughout the examined period and no greater (or even lower in 1997–2000) burglary risk than the South East. The concluding section discusses further the burglary prevention implications of these findings.

5.6   Security-Driven Burglary Drop and Distributive Justice

The current chapter examined the overall national correlation between burglary risk and effective security availability and how it informs contextual falls. More availability of effective security (particularly WDE and WIDE combinations) decreases household burglary risk but not to the same extent over time – since 2004/2005 the (negative) connection of burglary risk with effective security has become stronger than in earlier years. This coincides with the period that the national burglary rate has plateaued after a period of significant decline (see Chaps. 1, 2 and 8). Stagnant trends at a national level are the combination of uneven trends, increasing for some groups and falling for others which overall cancel each other out. The flat national burglary rate and a higher correlation with effective security (observed since 2004/2005) suggest divergent burglary trends across contexts with varying levels of security. Indeed, 'households with "no security" have experienced a fourfold increase in their relative burglary risk during the crime drop' (Tseloni et al. 2017, p. 7). After a period of considerable (between 56 and 72 percent) contraction from 1992–1996 to 2008/2009–2001/2002, the proportion of 'no security' households in England and Wales has remained at around 5 percent (Tseloni et al. 2017; Chap. 8 of this book). Therefore, since 2004/2005 the gains in burglary falls for population groups with effective security have been negated by rises affecting those without it.[34]

---

[34] Conversely the overall burglary drop between 2001/2002 and 2004/2005, a period of effectively stable 'no security' levels, suggests that the benefits of burglary falls experienced by effectively secured households exceeded the burglary risks of 'no security' households.

With the confidence afforded by the current and previous research that security uptake and upgrades have reduced burglary over time (at least) in England and Wales (Tseloni et al. 2017; Chaps. 5 and 8 of this book), the next important question is whether these changes were similar across all segments of the population. As already pointed out, the burglary fall has been reversed for some population groups since 2004/2005 in the context of a stable burglary rate overall. It is of no surprise therefore that the overall burglary fall was uneven across population groups from the mid-1990s (Hunter and Tseloni 2016; Ignatans and Pease 2016; Tilley et al. 2011).

From a distributive justice perspective, the burglary drop and effective security uptake should be *equitable*. Most empirical studies are concerned with *horizontal equity*, i.e. the proportional distribution of resources or policy outputs across population groups or neighbourhoods on the basis of an agreed criterion (Lucy et al. 1977). Horizontal equity however is concerned with outputs not *outcomes*. In the context of ensuring that (relative to others) those at high risk and low availability of effective security at the outset experienced the sharpest burglary reduction (and steepest security rise), we are concerned with *vertical equity*, in other words 'the unequal, but equitable, treatment of unequals' (Mooney and Jan 1997, p. 80).[35] The next paragraphs (a) summarise the findings on the burglary drop and effective security uptake across different population groups (b) to identify which households experienced (in)equitable trends and, therefore, where prevention needs to be directed.

Hunter and Tseloni (2016) looked at whether the burglary fall had been equitable by comparing two points in time: the distribution of crime rates at a point before the crime drop (1993) and the distribution at a point after the crime drop (2008/2009). They found that:

(a) Burglary victimisation rates divides widened, and burglary is relatively more concentrated on households of non-White ethnicity HRP, single adults, lone parents, renters, without a car and inner-city residents.
(b) The burglary drop was inequitable for lone parents, renters and inner-city residents; by contrast it was to some extent (compared to two-car households but not in relation to those with one car) equitable for households without a car.

The analysis reported in this chapter extended the above research by (i) examining all years during the entire crime drop period at the time of analysis, 1993–2011/2012, (ii) across comparable population groups over time and (iii) putting security availability trends in the picture.[36] The results here largely confirmed and further refined the findings from previous research. To summarise, the burglary drop was uneven across different contexts:

(a) The burglary victimisation divide *widened* to disadvantage single or three or more adult households, with children and/or lone parents, private and social

---

[35] Throughout the discussion that follows, we will refer to vertical (in)equity as defined here.

[36] The current study is however inferior to the previous one in that it looks at burglary risk (rather than number of burglaries experienced) and employs CSEW subsamples, both limitations necessitated from the need to investigate security availability.

renting households, households earning an annual income of £5000–£9999 or without income information, those without a car and residents of urban areas. It remained *stable* for households with Mixed, Chinese or Other ethnicity HRP and inner-city residents and *narrowed* for households of Asian HRP (although widened in comparison with Black HRP), whilst it disappeared for households of Black HRP.

Since data from all years were analysed in some cases, the changes between the end years are not the cumulation of incremental changes towards the overall outcome but follow a non-linear trajectory. For example, the victimisation divide trend has a U-shape in the case of households without a car, an inverse U-shape for lone parents and short ups and downs for inner-city and urban households (for details see Sect. 5.5).

(b) Focusing on those with significantly higher burglary risks in the mid-1990s, the burglary drop was inequitable for (in order of magnitude) social renters (who experienced continuous incremental deterioration over subsequent time periods), urban residents (who however experienced some equitable drop in the periods from 1997 to 2007/2008), inner-city residents, lone parents[37] and households of Asian HRP in relation to Black HRPs.

This study examined the same issues with regards to trajectories in the availability of effective (WIDE) security across population groups during the crime drop. As with burglary falls, security uptake was not uniform, and, as seen, income was not the only barrier to obtaining effective protection.

(a) The gap in WIDE security availability *widened* to the detriment of households with a HRP of Mixed, Chinese or Other ethnic origin, single adult households, with children and/or lone parents, social renters, households with an annual income of less than £10,000, inner-city residents and, compared to urban, rural residents; remained essentially *unchanged* maintaining the gap disadvantaging three or more adult households and private renters; and *narrowed* for households of Asian HRP (but widened with regard to households with a Black HRP), without a car and no income information, whilst it was *eliminated* for those with a Black HRP and, from 2005/2006, one-car households.
(b) The WIDE security combination uptake during the crime drop was inequitable for households of Asian HRP with regard to Black HRP, social renters (considerably so until 2007/2008), households earning an annual income of less than £10,000 and rural households in comparison to urban.

As Ignatans and Pease (2016) have pointed out, crime concentration increased after the crime drop. So, what do the above findings tell us about the relationship between burglary risk and effective security availability in context? Our findings suggest

---

[37] Inner city residents and lone parents experienced more unjust burglary drops up until 2004/2005 than in comparison to the last period (2008/2009–2011/2012). Counting all the effects that make up the overall lone parent effect (single adult + children + lone parent) rather than just the interaction shows that lone parents have experienced the second most unjust (least) burglary drops after social renting households.

burglary risks follow the same pattern, having become more concentrated on those who increasingly or constantly lack (relative to others) effective security, namely:

- Households with a HRP of Asian or Mixed, Chinese or Other ethnicity
- Single or three plus adult households, with children, and/or lone parents
- Social and private renters
- Low-income (under £10,000 per year) households
- Inner-city residents

During the crime drop, these households experienced the least reductions and since 2004/2005 possible increases in burglary victimisation. Household types with a combination of two or more of the above characteristics have exponentially higher burglary risk and lower WIDE security availability than the individual percentage differences reported in this chapter.

5.7   How Can Crime Prevention Redress the Uneven Burglary Drop and Reignite Overall Falls?

Widespread target hardening in recent years (see also Chaps. 2 and 3) has strengthened the critical role of effective anti-burglary devices (see also Sect. 5.4). This is in line with the security hypothesis for the crime drop (Farrell et al. 2011, 2014; Chap. 8 of this book). The security hypothesis which explains both crime rises and falls (and derives from opportunity theories of crime (Clarke 2012)) to date has been confirmed cross-nationally for car crime falls (Farrell and Brown 2016) and domestic burglary (Tseloni et al. 2017; Vollaard and Van Ours 2011). Can the security hypothesis explain the uneven crime drop across different contexts?

The current chapter examined the relationship between the risk of becoming a victim of burglary and the odds of owning the most effective security across different socio-economic groups and how, if at all, this relationship has changed over the period of the crime drop. It thus provided the first piece of evidence that the security hypothesis explains the uneven burglary falls that disadvantaged households with Asian or Mixed, Chinese or Other ethnicity HRP; single or three or more adult households, with children and/or lone parents; social and private renters; low-income (under £10,000 per year) households; and inner-city residents. Over the crime drop periods, burglars' activity has targeted the above household types, which lack adequate physical security and thus may be perceived as easily accessible targets. Therefore, the security hypothesis for the crime drop also explains burglary falls across different contexts.

Few exceptions however exist, one of which might be seen to negate the above general conclusion. Households without income information, those without a car and residents of urban areas have experienced increasing burglary risks (relative to others) without any reduction in their relative availability of effective security. Despite the narrowing gap the first two groups have still *considerably* lower security than the RH. Therefore, the findings in relation to households without car or income information do not contradict the security hypothesis.

The findings for urban households are distinct. Both risk and availability of effective security have increased in urban areas: households are more burgled and WIDE secured than rural households over the entire crime drop period. This finding contradicts the security hypothesis. One perhaps worthy observation is that urban households' security advancement was below their risk increase. Specifically, their security uptake (65 percent) was roughly one third of their burglary risk rise (186 percent) over the 20 years. In addition, the latter occurred predominately in the last period examined here which coincides with the latest economic downturn (2008/2009–2011/2012).

Both observations indicate that the key factor for this finding is urban households' close proximity and exposure to potential burglars (Felson 2002; Wiles and Costello 2000; Winchester and Jackson 1982; Chaps. 2, 3, and 4 of this book). Previous research has shown that a specific population group's risk can differ across areas over and above the exposure afforded by household and personal characteristics (Tseloni 2006; Tseloni and Pease 2015).[38] In this vein, the effectiveness of household security can be moderated or intensified by the aggregate area exposure to crime (Wilcox et al. 2003). Therefore, in this case the success of otherwise effective security is arguably weakened in urban areas which offer a particularly familiar and accessible environment to potential burglars (Chap. 4).

The above beg the question of what should be done to re-enact the burglary fall which stalled after over 10 years of significant reductions. The obvious policy recommendation to tackle burglary concentration this study offers is that crime prevention agencies engage in public awareness campaigns (see Chaps. 6 and 9), paying particular attention to the set of vulnerable household types listed earlier and their landlords. Further, the central government (Ministry of Housing, Communities and Local Government currently in the UK) and/or local authorities should offer incentives, including grants, for upgrading poor home physical security (see Chaps. 2, 3 and 6). In particular, owner-occupied dwellings by households of the ethnicity, household composition, income (including no car households) and area type vulnerable categories, listed earlier, should take priority. Similar incentives for home upgrades have, to date, been successfully offered for improving home energy efficiency but have yet to be implemented on a large scale for crime-proofing (Skudder et al. 2017). In addition, local authorities and housing associations should look to upgrade the physical security of their own stock of housing and incorporate effective device combinations in their planning and building of new homes.

Finally, for households which, although have higher levels of effective security, are still at greater burglary risk than others, such as urban area residents, policies should look to supplement physical security with both formal (e.g. police patrols and preventive checks) and informal (such as developing a stronger sense of community and involving neighbours) surveillance as well as adaptations to the built environment (Chap. 3). Households living in areas accessible to burglars would require multiple target hardening techniques which combine technological, policing and community-based preventive aspects to counteract their already elevated risk.

---

[38] For example, Tseloni (2006) found that, although lone parents living in the highest crime areas experience over a quarter more property crimes than others, in average or low crime areas, they are no more at risk than others. Similarly, widowed people suffer significantly more personal crimes in high population density areas but less than other household types according to marital status in other areas (Tseloni and Pease 2015).

Acknowledgement The authors are grateful to Dr. James Hunter and Professor Nick Tilley for insightful comments. Any errors are the authors' responsibility.

Appendix B

B.1   Data and Methodology

B.1.1   Variables

Population group-specific findings of burglary risk and effective security availability will only be discussed for the most effective of the combinations: WIDE. WIDE security availability compares the number of households with window locks, internal lights on a timer, double locks/deadlocks and external lights on sensor with those that have no security. Other results are available upon request.

Burglary risk is defined here as the likelihood of experiencing at least one attempted burglary or burglary with entry. So, unlike Chap. 4, the two burglary types are examined together to ensure there is an adequate number of respondents for analysis from each population group. This has theoretical disadvantages as attempts are (often) a product of the presence of security devices that were effective in thwarting a burglar. However, they did not deter burglars in the first place, and therefore the two burglary types together relate to household characteristics and security effective in deterring or thwarting.

Fourteen sets of explanatory variables that may affect both burglary risk and security availability entered the models, which were (reference category in italics):

- Number of adults (16 years old or older) in the household (one adult, *two adults*, three or more adults)
- Presence of children (under 16 years old) in the household (children, *no children*)
- Lone parent (lone parent, *not a lone parent*)
- Ethnicity of the respondent and in 2008/2009–2011/2012 ethnicity of the HRP (Black, Asian, Mixed/Chinese/Other, *White*)
- Annual household income before tax (£4999 and under, £5000–£9999, £10,000–£19,999, *£20,000–£29,999*, £30,000–£49,999 or £30,000 or more in 1994–1996, £50,000 or more which exists since 1998, no income information)
- Tenure (social rented sector, private rented sector, *owner*)
- Number of cars owned/used by the household in the last year (no car, 1 car, *2 cars*, 3 + cars)
- Area type (*rural*, inner city, urban). The following are not discussed in the main text of Chap. 5:
- Social class, classifying households in three groups according to whether the Household Reference Person (HRP) is in routine (formerly manual or blue collar) or intermediate occupations, or never worked/not classified social class in comparison to *professionals*

- Accommodation type, classifying households in three groups according to whether they live in a semi-detached or terraced (row in the USA) house or flat/maisonette/other as opposed to a *detached house*
- Length of residence at current address, where the base of '*10 years or more*' is compared to residence for '1 to 2 years', '2 to 5 years' or '5 to 10 years'
- Hours home left unoccupied, a variable attempting to gauge household's guardianship if the home is left unoccupied 'less than 3 hours' or '3 to 7 hours' compared to '*7 or more hours*' on a typical weekday
- Region, contrasting North East (North in 1994–1996), Yorkshire and Humberside, North West, East Midlands, West Midlands, East (East Anglia in 1994–1996), London, South West and Wales to the *South East*
- Age of the Household Reference Person formerly termed 'Head of Household' (HRP) who is the individual in the household who owns or rents the accommodation

As seen in Chap. 4, the study employed all CSEW sweeps since 1992. A large number of household characteristics however are not available in the 1992 CSEW data set that currently exists in the UK Data Service – social class, number of hours home left unoccupied, area type and region. In addition, the routine activity variable – number of hours home left unoccupied – is not available in the 1994 and 1996 CSEW. To overcome the large number of missing household characteristics, two sets of models were estimated for this early period:

1. Based on the 1992–1996 CSEW data sets whilst including only those limited number of variables which were available
2. Based on the 1994–1996 CSEW data sets and including all variables consistently available over time (as shown in the above bullet points with the exception of hours home left unoccupied)

In this work the estimated model from the aggregate 1994–1996 (excluding 1992) CSEW data with the wider set of household characteristics and area type is reported in order to ensure that the results are consistent over time. The 1992–1996 model results with fewer explanatory variables are available upon request.

B.1.2   Data and Sample Sizes

The data are taken from the 1994 to 2011/2012 CSEW merged into five aggregate data sets and descriptive statistics of all variables used – household characteristics, area type, region, burglary risk and WIDE security availability – including sample sizes across the five aggregate CSEW data sets (1994–1996, 1998–2000, 2001/2002–2004/2005, 2005/2006–2007/2008 and 2008/2009–2011/2012) are given in Table 5.3. Prior to 2001, the full recall period was from 1 January of the year preceding interview until the date of interview – a period of about 14 months. For example, interviews for the 1996 BCS were conducted from January 1996 to June

Appendix Table 5.3 Descriptive statistics of household and area characteristics, burglary risk and WIDE security availability from the WIDE sample of the CSEW aggregate data sets (1994–2011/2012)

Characteristics (reference category)	1994–1996 (LWD)	1998–2000	2001/2002–2004/2005	2005/2006–2007/2008	2008/2009–2011/2012
Number of adults (two adults)					
One adult	32.4	38.3	32.5	37.1	36.6
Three or more adults	16.6	12.1	14.9	15.4	13.4
Children under 16 in household (no children)					
Children	28.3	27.5	25.8	29.6	27.1
Lone parent (not a lone parent)					
Lone parent	5.0	6.5	5.6	8.3	8.0
Social class of HOH (professional)					
Routine occupation	23.0	26.1	41.1	40.8	38.1
Intermediate occupation	54.8	39.8	18.0	19.9	19.1
Never worked/not classified	2.9	5.8	6.3	6.2	6.3
Ethnicity of respondent[a] (White)					
Black	2.7	1.4	1.4	1.8	2.4
Asian	2.3	2.5	1.5	1.8	3.2
Mixed, Chinese or Other	1.0	2.1	1.3	2.4	1.8
Household income (£20,000–£29,999)					
£4999 and under	21.0	15.8	6.9	7.5	5.8
£5000–£9999	20.5	18.8	13.5	12.2	11.4
£10,000–£19,999	26.9	28.1	19.8	19.6	21.6
£30,000–£49,999	11.3	10.2	15.6	17.3	15.7
£50,000 or more	–	3.4	6.3	8.7	11.6
No income information	6.2	8.5	23.3	22.0	19.1
Tenure (owner)					
Social rented sector	26.2	28.6	17.9	21.8	17.0
Private rented sector	4.7	11.0	10.0	12.0	15.1
Type of accommodation (detached house)					
Semi-detached	32.2	33.1	34.7	31.0	33.4
Terraced	31.0	31.2	27.0	28.7	27.1
Flat/maisonette/other	15.6	18.5	12.2	15.2	14.1
Hours home left unoccupied (7+ hours)					
Less than 3 hours	–	42.1	46.9	43.0	40.2
3–7 hours	–	31.7	27.8	31.6	32.0
Length of residence[b] (10+ years)					
12 months to 2 years	7.4	9.2	9.3	9.7	9.5
2 to 5 years	18.0	22.5	21.2	21.3	21.1
5 to 10 years	21.9	17.6	16.2	17.6	17.9
Number of cars owned/used in last year (two cars)					
No car	30.6	28.4	22.0	23.3	22.3
One car	43.6	47.7	44.6	40.9	44.1
Three or more cars	5.0	4.4	7.3	8.6	8.4

(continued)

Appendix Table 5.3 (continued)

Characteristics	1994–1996 (LWD)	1998–2000	2001/2002–2004/2005	2005/2006–2007/2008	2008/2009–2011/2012
Area type (rural)					
Inner city[c]	21.6	16.3	7.2	8.2	8.0
Urban	54.9	60.5	61.7	64.1	65.5
Region (South East)					
North East (North in 1994–96)	6.2	6.3	4.9	5.2	6.8
Yorkshire and Humberside	11.0	10.2	9.4	9.7	7.6
North West	9.8	10.7	9.0	10.4	10.3
East Midlands	9.0	6.8	9.1	10.4	10.3
West Midlands	10.4	10.5	10.0	9.2	8.9
East (East Anglia in 1994–96)	5.0	10.2	14.6	16.0	13.9
London	12.7	8.9	6.6	6.6	9.2
South West	8.7	11.5	11.9	10.4	11.1
Wales	5.9	10.3	11.2	9.1	10.0
Victim of burglary or attempted burglary	*17.5*	*30.4*	*15.3*	*32.7*	*29.8*
Security availability (no security)					
WIDE[d]	47.8	28.3	49.7	47.5	51.3
Age of HOH					
Mean	52.82	51.66	52.69	51.55	52.17
Standard deviation	17.07	16.49	16.56	16.32	16.40
Minimum	18	18	17	16	17
Maximum	99	91	93	95	94
N	*3998*	*812*	*1434*	*1570*	*2766*

Notes: [a]Ethnicity of Household Reference Person (HRP) in 2008/2009–2011/2012
[b]1994–1996 refers to length of residence in area rather than at the same address
[c]Inner-city areas were oversampled before the 2001/2002 CSEW (see Tilley and Tseloni 2016)
[d]This is not a reflection of higher levels of ownership in 1994–1996 but a data limitation. The list of security devices from which respondents could choose in 1992–1996 was much smaller (see Sect. 4.3.1 and Appendix Table 4.7 in Chap. 4); therefore although the data may suggest respondents had 'LWD', (a) they may also have had a number of other (unmeasured) devices in addition to LWD and (b) LWD could mean WIDE, WDE or WID. In other words, the smaller number of security device choices in 1992–1996 means the sample proportion for LWD is likely to be higher and incorporate additional combinations which were measured and not subsumed in WIDE from 1998 onwards

1996, with incidents therefore reported from January 1995 to June 1996. After 2001 and a move to continuous interviewing, the 'moving reference period' includes the current month plus the 12 months prior to the date of the interview.

The analysis reported in Chap. 5 requires information about security devices in both the 'general population' *and* burgled households. Some security devices are not strictly comparable over time due to changes in question wording. In addition, on a small number of occasions, device information was only collected for burglary victims and not the general population. For example, in the 1992–1996 sweeps, burglary victims were asked whether they had security chains, window bars/grills or dogs at the time of the event, but this information was not collected consistently from the general population sample. As a result, in the 1992–1996 sweeps individual devices for some victims may in fact represent these in combination with security chains, window bars or grills and/or dogs. The same devices may also exist unacknowledged within security combinations in the 1992–1996 data.

It should also be noted that the term 'no security' should be taken to mean 'none of the CSEW listed devices'. Therefore, strictly speaking 'no security' is not comparable over time except between the 1998 and the 2007/2008 CSEW sweeps. 'No security' in the 1992–1996 sweeps means no burglar alarm, no double locks, no window locks and no lights. From the 1998 sweep onwards, more categories were included so in addition to the previous list, 'no security' means no security chains, no indoor lights on a timer and no external lights on a sensor. Therefore 'no security' in 1992–1996 means something different to 'no security' in the following sweeps. This may, to some extent, explain the higher frequency of 'no security' in the earlier 1992–1996 CSEW sweeps in Appendix Table 5.3. The same argument in theory applies for the pre- and post-2008/2009 sweeps of the CSEW due to the introduction of questions about the availability of CCTV in the respondent's home, but the very small proportion of households with this device makes this issue negligible.

The samples for the statistical analyses reported in the current Chap. 5 are subsets of the data used in Chap. 4. In particular the samples of the analyses on the effectiveness of security against burglary nationally across England and Wales in Chap. 4 consist of all burglary victims and non-victims who (after being randomly selected) completed the Crime Prevention module (see Appendix A – Appendix Table 4.6, Chap. 4). To reiterate, information about household security availability of Crime Prevention module respondents who were also victims of burglary have been taken from the Victim Form. It therefore refers to the time of the first burglary rather than the time of the interview. For details please see Tseloni et al. (2014) and Appendix A in Chap. 4.

The data for the statistical modelling reported in the current Chap. 5 consist of all households (burglary victims and Crime Prevention module non-victim respondents) with effective (WD, WDE and WIDE) security availability and households with no security. For this reason, the sample sizes depend on the specific combination investigated. Indeed, the sample sizes examined in this work are dependent on the availability of effective security. The likelihood of having the respective security reduces from the most common effective combination of WD, which was present at roughly 14 percent of households, to the least common one, WIDE, which was found in 4 percent of the households in England and Wales between 2001/2002 and 2011/2012 (Tseloni et al. 2017). The later Appendix Table 5.4 gives the sample sizes of all data sets across the three (WD, WDE and WIDE) effective security combinations over the five sets of CSEW aggregate data. Comparing with the previous chapter's Table 4.5 – the last row of which gives the sample sizes of households with no security, WD and WIDE – it can be seen that the sample sizes for the 2008/2009–2011/2012 statistical models of WD (5973) and WIDE (2766), shown in the third and last rows, respectively, of the last column of Appendix Table 5.4, have been obtained by adding the first figure in the last row of Table 4.5 to the second and the third, respectively.

B.1.3   Statistical Model and Modelling Strategy

Models were run for all the effective security combinations: WD, WDE and WIDE from 1998 to 2000 onwards. The 1992–1996 group of sweeps used the following combinations, WD and LWD, due to the fact that the distinction between internal and external lights was not made during this period. As mentioned in the main text of Chap. 5, the results of the estimated models of burglary risk and WIDE security availability are discussed here, whereas results for WD and WDE are available upon request.

The statistical model used is the bivariate logit regression model (Snijders and Bosker 1999) of two associated outcomes – the likelihood of burglary and WIDE security availability – over a set of explanatory variables, here the household characteristics, area type and region outlined earlier. The choice of the statistical model is justified by the fact that both outcomes are in theory affected by the same household and area characteristics, whilst they are also interrelated. The model allows estimating the correlation of the likelihood of burglary and WIDE security availability in addition to the effects of the explanatory variables on each outcome. Further explanation of this statistical model within criminology (with respect to multiple fear of crime measurements) and a discussion on how to interpret its fixed and random parameters, including the correlation between outcomes, can be found in Tseloni and Zarafonitou (2008). The model across the five sets of data was estimated via the computer software MLwiN version 2.10 (Rasbash et al. 2009). For a complete guide go to http://www.bristol.ac.uk/cmm/software/mlwin/).

The modelling strategy included four stages as follows: for each of the five sets of aggregate data from 1994–1996 to 2008/2009–2011/2012, first a baseline model, whereby respective pairs of random constant terms are only estimated, to establish that burglary risk and WIDE security availability are overall correlated. The 'national average' figures in Appendix Table 5.4 give the estimated baseline correlations. The last row of figures in Appendix Table 5.5 show the baseline constant estimates. Secondly all variables in the next Appendix Table 5.3 were added to the baseline models to give over time estimates of all theoretically relevant effects on burglary risk and security availability. These models also provide a consistent profile of the RH over time. Not all estimated parameters however were statistically significant. In the third stage of the model fitting, characteristics with no statistically significant (at p-value greater than 0.10) parameter in both burglary

Appendix Table 5.4 Correlation (standard error of covariance) between effective security availability and burglary risk during the crime drop

	Aggregated Crime Survey for England and Wales Sweeps (reference periods)				
	1994–1996[a] (1993–1996)	1998–2000 (1997–2000)	2001/2002– 2004/2005	2005/2006– 2007/2008	2008/2009– 2011/2012
	Estimated correlation coefficient (standard error of covariance)				
	Presence of window locks and door deadlocks or double locks (WD) and burglary risk				
National average[b]	−0.25 (0.01)	−0.39 (0.02)	−0.22 (0.02)	−0.58 (0.01)	−0.59 (0.01)
Reference household[c]	−0.26 (0.01)	−0.36 (0.02)	−0.21 (0.02)	−0.55 (0.01)	−0.54 (0.01)
Sample size	5354	1555	3111	3104	5973
	Presence of window locks, door deadlocks or double locks and external lights on a sensor (WDE) and burglary risk				
National average[b]	−0.29 (0.01)	−0.40 (0.03)	−0.30 (0.02)	−0.62 (0.01)	−0.66 (0.01)
Reference household[c]	−0.26 (0.01)	−0.34 (0.03)	−0.29 (0.02)	−0.58 (0.01)	−0.59 (0.01)
Sample size	3998	1018	2096	2256	4312
	Presence of window locks, door deadlocks or double locks, internal lights on a timer and external lights on a sensor (WIDE) and burglary risk				
National average[b]	−0.29 (0.01)	−0.35 (0.03)	−0.31 (0.02)	−0.61 (0.01)	−0.63 (0.01)
Reference household[c]	−0.26 (0.01)	−0.24 (0.03)	−0.30 (0.02)	−0.54 (0.02)	−0.50 (0.01)
Sample size	3998	812	1434	1570	2766

[a]The 1994–1996 CSEW data include information on whether the household uses lights as a burglary prevention method, but there was no distinction between internal lights on a timer and external lights on a sensor. Therefore the combinations of WDE and WIDE are for this period LWD and strict comparisons between the period 1994 and 1996 and the following years impossible

[b]The national average correlation is the unconditional one which has been estimated via the baseline or empty model whereby only the random intercepts of the joint models of burglary risk and presence of effective security have been estimated

[c]The reference household refers to a two-adult household without children, professional social class, household income of £20,000–£29,999 per annum, owning their home and two cars and living in a detached house in a rural area of the South East. In addition, the respondent (and in 2008/2009–2011/2012 the Household Reference Person) is of White ethnicity, and the home is left unoccupied for 7 or more hours on a typical weekday. The correlations of burglary risk and effective security availability for the reference household across the three effective security combinations and the five time periods examined here have been estimated via a (two-equation) bivariate logit model which includes all statistically significant (at *p*-value less than 0.10) covariates and is shown in Appendix Table 5.5. They are identical to the ones from respective models including all covariates as delineated in Appendix Table 5.3

and security regressions were omitted from the model. To put it differently, household characteristics were retained on the basis of having at least one parameter at p-value lower than 0.10 in at least one outcome, burglary or security. Further in the fourth and final stage, individual categories with p-value greater than 0.10 were removed from the models only if they could be merged with the base category of their encompassing characteristic. The following example may clarify the third

Appendix Table 5.5 Estimated odds ratios, $Exp(b)$, of fixed effects of household characteristics, area type and region of England and Wales on joint burglary risk and WIDE security availability over time, 1993–2011/2012 – based on bivariate logit regression models of CSEW aggregate data sets (1994–2011/2012)

Household characteristics (base category)	1994–1996 (LWD)		1998–2000		2001/2002–2004/2005		2005/2006–2007/2008		2008/2009–2011/2012	
	Burglary	Security	Burglary	Security	Burglary	Security	Burglary	Security	Burglary	Security
	Exponentials of estimated fixed parameters, $Exp(b)$									
Number of adults (two adults)										
One adult	0.94	0.96	0.98	0.68	0.97	0.82	1.70^	0.48^	1.43^	0.50^
Three or more adults	0.96	0.54^	–	–	1.50*	0.53^	1.50**	0.62^	1.39**	0.61^
Children under 16 in household (no children)										
Children	0.85	0.84^	1.49*	0.55**	–	–	1.49**	0.62^	1.37**	0.74**
Lone parent (not a lone parent)										
Lone parent	1.74**	0.80	–	–	2.41^	0.64	–	–	1.76**	0.58*
Social class of HOH (professional)										
Routine occupation	0.94	0.61^	0.86	0.38^	0.73	1.37*	1.09	1.02	0.91	0.93
Intermediate occupation	0.84	0.76^	0.83	0.95	1.08	0.77	1.00	0.70**	1.08	0.86
Never worked/not classified	0.76	0.41^	0.37**	2.65**	1.12	0.90	0.75	1.08	1.20	0.51^
Ethnicity of respondent[a] (White)										
Black	3.48^	0.33^	1.43	0.33	–	–	0.97	0.92	0.95	1.15
Asian	4.71^	0.24^	2.72*	0.15*	–	–	1.53	0.36**	2.14^	0.38^
Mixed, Chinese or Other	2.29**	0.94	2.08	0.47	–	–	0.99	0.85	2.22**	0.46*
Household income (£20,000–£29,999)										
£4999 and under	1.29	0.49^	2.08**	0.75	1.44	0.35**	1.96**	0.44**	1.34	0.44^
£5000–£9999	0.97	0.54^	1.45	1.02	1.67	0.54**	1.23	0.99	1.73^	0.47^
£10,000–£19,999	0.89	0.78**	1.22	1.65*	0.98	0.84	1.36	0.91	–	–
£30,000–£49,999	–	–	0.82	2.01**	1.34	1.41	1.51*	1.15	–	–

(continued)

Appendix Table 5.5 (continued)

Household characteristics (base category)	1994–1996 (LWD)		1998–2000		2001/2002–2004/2005		2005/2006–2007/2008		2008/2009–2011/2012	
	Burglary	Security	Burglary	Security	Burglary	Security	Burglary	Security	Burglary	Security
	Exponentials of estimated fixed parameters, $Exp(b)$									
£50,000 or more	–	–	0.42	1.73	1.45	2.65^	1.54	1.09	–	–
No income information	0.87	0.58^	1.69	0.94	0.73	0.84	1.68**	0.66^	1.32**	0.68^
Tenure (owner)										
Social rented sector	1.33**	0.39^	1.63**	0.15^	2.25^	0.16^	3.50^	0.11^	3.60^	0.24^
Private rented sector	1.16	0.19^	1.52	0.08^	0.97	0.17^	1.29	0.24^	1.63^	0.23^
Type of accommodation (detached house)										
Semi-detached	0.89	0.71^	0.76	0.59*	0.68	0.64**	1.19	0.82	1.23	0.77*
Terraced	0.88	0.45^	1.03	0.52**	0.84	0.34^	1.36	0.39^	1.72^	0.36^
Flat/maisonette/other	0.66**	0.63^	0.66	0.44*	0.60	0.24^	1.10	0.41^	1.17	0.28^
Hours home left unoccupied (7+ hours)										
Less than 3 hours	–	–	0.57**	1.20	0.91	1.20	–	–	–	–
3–7 hours	–	–	0.60**	1.73**	0.62**	1.04	–	–	–	–
Length of residence[b] (10+ years)										
12 months to 2 years	0.87	1.72^	1.83**	1.04	1.95	0.99	0.95	1.52	–	–
2 to 5 years	1.11	1.35^	–	–	1.64	1.13	1.19	1.03	–	–
5 to 10 years	1.00	1.22*	–	–	–	–	1.08	1.40*	–	–
Number of cars owned/used in last year (two cars)										
No car	1.33**	0.32^	1.37	0.50**	1.28	0.26^	1.00	0.38^	1.52^	0.45^
One car	1.14	0.62^	–	–	1.34	0.59^	0.99	0.86	–	–
Three or more cars	–	–	–	–	–	–	0.85	1.03	–	–
Age of HRP	0.96**	1.10^	0.96	1.23^	1.02	1.11^	0.93^	1.13^	0.99	1.08^
Age of HRP squared	1.000	0.999^	1.000	0.998^	1.000	0.999^	1.001**	0.999^	1.000	0.999^

(continued)

Appendix Table 5.5 (continued)

Area of residence characteristics (base category)	1994–1996 (LWD) Burglary	1994–1996 (LWD) Security	1998–2000 Burglary	1998–2000 Security	2001/2002–2004/2005 Burglary	2001/2002–2004/2005 Security	2005/2006–2007/2008 Burglary	2005/2006–2007/2008 Security	2008/2009–2011/2012 Burglary	2008/2009–2011/2012 Security
	Exponentials of estimated fixed parameters, Exp(b)									
Area type (rural)										
Inner city	2.24^	1.45^	2.66^	2.05*	4.16^	1.76*	1.61*	1.11	2.39^	0.88
Urban	1.49^	1.23**	2.01^	1.55*	1.57**	1.41**	1.34*	1.57^	2.40^	1.38^
Region (South East)										
North East - North in 1994–1996	2.15^	0.8	0.70	0.58	1.56	2.21**	1.24	1.04	1.29	1.32
Yorkshire and Humberside	2.1^	0.89	1.11	0.62	1.94*	0.87	1.39	0.46^	1.41	1.01
North West	2.02^	1.39**	1.06	1.04	1.30	1.64*	1.80**	0.49**	1.49*	0.94
East Midlands	1.34	0.56^	0.61	0.87	1.43	0.94	1.78**	0.49^	0.92	1.49*
West Midlands	1.72^	0.95	1.08	0.69	1.25	0.87	1.10	0.82	1.21	1.39
East - East Anglia in 1994–1996	0.94	0.51^	0.52*	0.93	0.74	1.31	0.96	0.77	1.21	1.14
London	1.87^	2.1^	0.56	2.07	1.25	2.55^	1.57	0.72	1.60**	0.75
South West	1.08	0.65^	1.08	0.67	1.06	1.22	1.01	0.54**	1.16	0.81
Wales	1.22	0.39^	0.46*	0.33^	0.65	0.38^	0.85	0.31^	1.02	0.50^
	Estimated constant term, (b_0)									
Constant	−0.92	−0.79	0.08	−5.57	−2.60	−2.02	−0.13	−2.03	−1.84	−0.89
Constant (baseline)	−1.55	−0.09	−0.83	−0.93	−1.71	−0.01	−0.72	−0.10	−0.86	0.05

Notes: *0.05 < p-value ≤0.10; **0.01 < p-value ≤0.05; ^p-value ≤0.01; − Did not enter the model
[a]Ethnicity of HRP in 2008/2009–2011/2012; [b]1994–1996 refers to length of residence in the same area rather than at the same address

and fourth stages of the modelling strategy. Number of cars the household used in the previous year is a single household characteristic which in the regression is represented via three dummy variables, no car, one car and three or more cars, whilst two-car households are the comparison base. All the categories of this variable were included in the third stage because at least one parameter out of six (three for burglary and three for security) had a p-value less than 0.10. However, since three or more cars had p-value greater than 0.10 in both burglary and security regressions and it is adjacent to the base of two cars, it was omitted. Thus in effect two or more cars become the base of comparison for the effects of number of cars in the household on burglary risk and effective security availability across most periods.

B.2   The Correlation of Burglary Risk and Effective Security Availability Nationally, 1993–2011/2012

Appendix B.2 offers statistical details of the findings discussed in Sect. 5.4.2 with regards to the overall and conditional correlation between burglary and effective security. Unlike the remainder of Chap. 5, it shows the results for all three effective security combinations highlighted in Chap. 4: WD, WDE and WIDE. Appendix Table 5.4 shows the estimated residual correlation between burglary risk and presence of security across the three effective combinations identified in Chap. 4 (WD, WDE and WIDE), over five aggregated CSEW data sets (1994–1996, 1998–2000, 2001/2002–2004/2005, 2005/2006–2007/2008 and 2008/2009–2011/2012) and for two sets of models. In this instance, the correlation coefficient coincides with the estimated covariance between burglary and security, and the respective standard error is given in brackets. Indeed, in joint or bivariate logit models, the respective variances of the two binary dependent variables are one by construct; therefore their covariance is also their correlation. The two sets of models refer to the baseline or empty model, whereby only the random intercepts of the outcomes, burglary risk and presence of effective security, have been estimated in the joint models, and the final two-equation model including all statistically significant covariates for each outcome. The empty model essentially gives the national average correlation of burglary risk and presence of effective security. The final model's correlation gives the residual value after the mediating effects of household and area characteristics and region have been incorporated. In this light it refers to the correlation specific to the RH.[39]

The (conditional) correlation of burglary risk and effective security presence between two randomly selected households with an identical profile is lower than

---

[39] The set of explanatory variables in the final models and as a result the definition of RH differs slightly across periods. This does not compromise the analysis here: the correlations in Appendix Table 5.4 are the same to the ones from preliminary models which include all theoretically relevant household and area characteristics and therefore refer to an identically defined RH.

the national average (see Appendix Table 5.4). However, the household profile can only be delineated with respect to all measurable characteristics in the CSEW and is portrayed in the RH, of which more will be said in the next section. The reduction in the correlation is modest and does not exceed 21 percent (calculated as (50–63)/63 from Appendix Table 5.4). Since context matters (as seen already and in the following sections) we may conclude that it is less than perfectly measured in this work which relies on the CSEW. The relationship between burglary risk and effective security presence depends on household characteristics which are not measured in the CSEW.[40] These may include detailed area socio-economic characteristics, neighbour social networks and house and area of residence physical layout and access to other places (see also Chaps. 2 and 3).

B.3   Estimated Bivariate Logit Regression Models of Burglary Risk and WIDE Security Availability During the Crime Drop

The odds of burglary victimisation are the ratio of the likelihood of being burgled over the complement probability of not being burgled (Long 1997, p. 51). With the exception of age of HRP, the estimated odds of burglary for each household characteristic, area type and region in Appendix Table 5.5 are in comparison and therefore as a ratio to their respective base category (given in brackets in Tables 5.1 and Appendix Tables 5.3 and 5.5 and as a horizontal line at value $y = 1$ in Figs. 5.2, 5.3, 5.4, 5.5, 5.6, 5.7, 5.8 and 5.9). Examples of how to interpret the odds ratios (Long 1997, p. 81) will follow in the last paragraph.

The odds of WIDE security availability is the ratio of having this combination over no security at all. Given the sample employed in this study, the two outcomes are exhaustive and complement each other, although in principle and in reality households have a number of security combinations (see Chap. 4 for a full list). As such WIDE security availability is in principle part of a multinomial logit (at least) three outcome response (contrasting WIDE or any other security combination to no security). However, this statistical specification would have prohibited investigating its correlation with burglary risk in context: '[M]ultinomial response variables cannot be included in multivariate response models, but can be used in univariate response models' (Rasbash et al. 2017, p.228). Therefore the study focused on examining the population group-specific relationship between burglary risk and each effective (WD, WDE and WIDE) security combination in isolation. The results of the estimated bivariate logit regression models of burglary victimisation and WIDE security availability across the five aggregate CSEW data sets from 1994 to 2011/2012 are presented in Appendix Table 5.5.

---

[40] For this reason the estimated correlation is conditional on characteristics included in the models but is 'residual' or 'unexplained' with respect to those that the model (due to data limitations) omits.

The last two rows of Appendix Table 5.5 give the estimated constant terms from the final and baseline model obtained from the fourth and first model fitting stages, respectively (see Appendix B.1 discussion), for each period examined. With the exception of age all variables in the estimated models are qualitative and therefore denoted by dummy variables contrasting each category of the nominal variable to a respective category (see Appendix Table 5.3). For example, ethnicity of the respondent or HRP contrasts each ethnic minority group to White. The effect of all base categories together on the regression outcome, here the likelihood of burglary and effective security availability, is given in the respective intercepts or constant terms (see Appendix B.1). The constant terms are therefore here the *log odds* of experiencing a burglary (attempted or with entry) and WIDE security availability for the RH (penultimate row) and nationally (last row) in each set of CSEW data. For example, in the 1994–1996 CSEW data, the national average log odds of burglary and of WIDE security were −1.55 and −0.09, respectively. For a household with the RH profile generally[41] as given in Sect. 5.4.1, the respective figures were −0.92 and −0.79. Comparing across rows it is worth noting that the RH's burglary risk was higher than the national average except in 2001/2002–2004/2005 and 2008/2009–2011/2012, whilst its WIDE security availability was unequivocally lower.

Burglary risks in this study should only be viewed in relative terms, such as risk of the RH compared to the national average or of population groups under question compared to the RH's risk. Due to the limited sample of the Crime Prevention module and employing all burglary victims' data in this study, our estimates of national or RH-specific burglary risk entailed in the constant terms of the respective regressions in the last two rows of Appendix Table 5.5 are overestimated. A rough calculation[42] of the extent of this overestimation is given in Appendix Table 5.6 which compares burglary risks between the entire CSEW sample and the WIDE samples used here across the five periods. Based on the discrepancy between risks in the two samples over time, the last column of Appendix Table 5.6 gives a deflation factor which can be multiplied with the absolute burglary risk estimates from our models. Interested readers may apply the figures provided in the last column of Appendix Table 5.6 to the estimated risks entailed in the baseline constant terms of Appendix Table 5.5 in order to obtain more realistic estimates of national burglary risks in absolute terms.[43] Simple (without compositional effects) burglary risks of population groups from the entire CSEW samples are provided in Table 5.2 of the main text in Chap. 5.

---

[41] Considering that some population characteristics did not differ from the respective base categories and were therefore omitted in the regressions (see the first two columns of figures in Appendix Table 5.5), the RH here is as defined in the main text of Chap. 5 except earning £20,000 or more, having at least two cars and no routine activity information.

[42] Precise calculation is not possible due to lack of data on security of the entire CSEW samples.

[43] A similar deflation factor of RH-specific burglary risks can be calculated from the total number of households and burglary victims with the RH profile in the entire CSEW samples.

Appendix Table 5.6 Burglary risks in absolute values in the entire and employed (WIDE security focused) CSEW samples over time

| CSEW sweeps | Entire CSEW | | | Employed CSEW sample of households with WIDE or no security | | Deflation factor (C / D) |
	Burglary victims (A)	Sample size (B)	% Burglary risk $(C = (A/B) \times 100)$	Sample size	% Burglary risk (D)	
1994–1996	2140	32,898	6.5	3998	17.5	0.37
1998–2000	1514	34,358	4.4	812	30.4	0.14
2001/2002–2004/2005	2857	107,244	2.7	1434	15.3	0.17
2005/2006–2007/2008	2835	141,982	2.0	1570	32.7	0.06
2008/2009–2011/2012	3555	183,709	1.9	2766	29.8	0.06

Our focus is on burglary and security inequality and how they changed during the crime drop, and therefore absolute estimates are outside the scope of this study and indeed, with regard to security, unattainable due to lack of data. The following paragraphs interpret the relative to the RH estimates of burglary risk and WIDE security across all population groups in the CSEW data.

Apart from the constant terms, all figures in Appendix Table 5.5 present the exponentials of the estimated parameters because, as will be seen shortly, they have a more intuitive interpretation than the parameters themselves (Long 1997, pp. 80–81). The exponentials of the combined estimated parameters for age and age squared of HRP give the (non-linear) change in the odds of burglary (as given in even number columns) and WIDE security availability (shown in odd number columns bar the first) for each year the HRP grows older. The remaining set of figures in Appendix Table 5.5 give the *odds ratios* of experiencing a burglary (attempted or with entry) and WIDE security availability against no security for a household that belongs to the respective population group *in comparison* to the base category for the same characteristic with all other household and area attributes and region being equal. They also provide an indication of their statistical significance. For example, in the 1994–1996 CSEW data, the odds ratio of burglary and of WIDE security of single adult households did not significantly differ from those of a two-adult household. The same can be said for three or more adult households with respect to burglary. However three or more adult households had 46 percent (calculated as $100 \times (0.54-1)$) lower odds of WIDE security availability than two-adult households. Moving along the top rows of figures, the 2001/2002–2004/2005 estimates can be interpreted similarly to the ones from 1994 to 1996 with one exception. Three- or more adult households had, compared to two-adult households, 50 percent (calculated as $100 \times (1.50-1)$) higher odds of burglary (albeit with weak statistical significance, p-value between 0.05 and 0.10). The remaining figures can be interpreted in a similar way bearing in mind the respective indications for their statistical significance.

Prior to our work discussed in Chap. 5, the question of who is at highest risk of burglary and who is least likely to have the most effective security had not been investigated. One of the contributions of this work is using a methodology that accounts for the composition of each population group. Ignoring compositional effects for a moment, the highest burglary risk was faced by the following household types: HRP of Mixed, Chinese or Other ethnicity, single adult households, those with children, lone parents, social renters, households earning under £5000 per year, without a car and living in inner cities of England and Wales in 2008/2009–2011/2012.

With one exception, the above profile remains largely the same when the household profile of maximum burglary risk is examined across all possible contributing factors simultaneously. The exception refers to income; the statistical modelling analysis found that households at £5000–£9999 per year or without income information are at 73 and 32 percent higher burglary risk, respectively, than others during the same period. Although the profile of households mostly at risk of burglary is roughly the same, the estimated effects (odds ratios) of each population socio-economic characteristic on burglary risk disagree between the methodologically rigorous estimates discussed in the previous section (Figs. 5.2, 5.3, 5.4, 5.5, 5.6, 5.7, 5.8 and 5.9 and Appendix Table 5.5) and the bivariate cross-tabulations (Table 5.2).

A few examples of the most startling differences refer to lone parents, inner-city residents and social renters demonstrating in a clear manner the implications of not considering each group's composition in the case of bivariate associations. According to Table 5.2, lone parents have 3.2 times (or 220 percent higher than) the burglary risk of others, whereas from the statistical models, this stands at a much lower 1.8 times (or elevated by 80 percent) the baseline risk (see Sect. 5.5.3 and Appendix Table 5.5). This is likely because the (bivariate analysis calculated) odds ratio entails the two individual effects of 'single adult household' and 'living with children' in the overall lone parent estimate. Indeed, the statistical model produced a similar overall lone parent odds ratio (at 3.45, see penultimate paragraph of Sect. 5.5.3). Similarly, inner-city households' burglary odds ratio is overestimated in the bivariate analysis of Table 5.2. This is because it entails the elevated burglary risk of other (than area type) predominant characteristics of inner-city households, such as single adult ones or HMOs and without car households which are most likely to reside in inner cities. By contrast the odds ratio for social renters based on bivariate analysis underestimated their burglary risk to 2.2 times (or 120 percent more than) that of owner-occupiers – from the statistical models this was 3.6 times (or 260 percent higher than) the RH. The reader would recall that social renters are households on low income but in recent years not worse off (in terms of fuel poverty at least) than private renters and nearly a quarter are lone parents (Wilson and Barton 2017). Therefore, the underestimation possibly confounds the individual effects of income and household composition within tenure. To conclude, the population groups of highest burglary risk in this study are by and large the same regardless of methodology, but the estimate of the effects' magnitude – and therefore prioritisation of preventive resources to those of greatest need – is compromised when using bivariate associations.

Appendix Table 5.7 Household Reference Person (HRP) ethnicity as a percentage of respondents' ethnicity, 2006/2007 CSEW

Ethnicity of Household Reference Person	Ethnicity of respondent				Total
	White	Black	Asian	Mixed, Chinese or Other	
White	99.72	3.68	3.46	12.82	94.21
Black	0.07	95.56	0.27	0.76	1.59
Asian	0.08	0.15	95.48	1.68	2.71
Mixed, Chinese or Other	0.13	0.61	0.80	84.73	1.49
Total	94.12	1.58	2.72	1.58	100

References

Armitage, R., & Monchuk, L. (2011). Sustaining the crime reduction impact of designing out crime: Re-evaluating the secured by design scheme 10 years on. *Security Journal, 24*, 320–343.

Bernasco, W. (2009). Burglary. In M. Tonry (Ed.), *The Oxford handbook of crime and public policy* (pp. 165–190). Oxford: Oxford University Press.

Budd, T. (1999). *Burglary of Domestic Dwellings: Findings from the British Crime Survey* (Home Office Statistical Bulletin 4/99). London: Home Office.

Clarke, R. V. (2012). Opportunity makes the thief. Really? So what? *Crime Science, 1*(3), 1–9.

Cohen, L. E., & Felson, M. (1979). Social change and crime rates trends: A routine activity approach. *American Sociological Review, 44*, 588–608.

Department for Work and Pensions (DWP). (2017). Households below average income: An analysis of the UK income distribution: 1994/95-2015/16. Published 16 March 2017. https://www.gov.uk/government/uploads/system/uploads/attachment_data/file/600091/households-below-average-income-1994-1995-2015-2016.pdf. Accessed 29 Jan 2018.

Ellingworth, D., Hope, T., Osborn, D. R., Trickett, A., & Pease, K. (1997). Prior victimization and crime risk. *International Journal of Risk, Security and Crime Prevention, 2*, 201–214.

Farrell, G., Tseloni, A., Mailley, J., & Tilley, N. (2011). The crime drop and the security hypothesis. *Journal of Research in Crime and Delinquency, 48*(2), 147–175.

Farrell, G., Tilley, N., Tseloni, A. (2014). Why the crime drop? *Why crime rates fall and why they don't, Crime and Justice-A Review of Research, 43*, 421–490. Chicago: University of Chicago Press.

Farrell, G. (2015). Crime concentration theory. *Crime Prevention and Community Safety, 17*(4), 233–248.

Farrell, G., & Brown, R. (2016). On the origins of the crime drop: Vehicle crime and security in the 1980s. *Howard Journal of Criminal Justice, 55*(1–2), 226–237.

Farrell, G., Laycock, G., & Tilley, N. (2015). Debuts and legacies: The crime drop and the role of adolescent-limited and persistent offending. *Crime Science, 4*, 16. https://doi.org/10.1186/s40163-015-0028-3.

Felson, M. (2002). *Crime and everyday life*. Thousand Oaks: SAGE.

Ginsburg, N. (2005). The privatization of council housing. *Critical Social Policy, 25*(1), 115–135. https://doi.org/10.1177/0261018305048970.

Gottfredson, M. R. (1981). On the etiology of criminal victimisation. *Journal of Criminal Law and Criminology, 72*, 714–726.

Green, W. H. (1997). *Econometric analysis*. London: Prentice Hall International.

Hales, J., & Stratford, N. (1997). *1996 British Crime Survey (England and Wales) technical report*. London: Social and Community Planning Research.

Hales, J., & Stratford, N. (1999). *1998 British Crime Survey technical report*. London: Social and Community Planning Research.

Higgins, A., & Jarman, R. (2015). *Safe as houses? Crime and changing tenure patterns*. The Police Foundation: Police Effectiveness in a Changing World Project.

Hunter, J., & Tseloni, A. (2016). Equity, justice and the crime drop: The case of burglary in England and Wales. *Crime Science, 5*(3). https://doi.org/10.1186/s40163-016-0051-z.

Ignatans, D., & Pease, K. (2015). Distributive justice and the crime drop. In M. Andresen & G. Farrell (Eds.), *The criminal act: Festschrift for Marcus Felson* (pp. 77–87). London: Palgrave Macmillan.

Ignatans, D., & Pease, K. (2016). On whom does the burden of crime fall now? Changes over time in counts and concentration. *International Review of Victimology, 22*(1), 55–63.

Johnson, S. J. (2014). How do offenders choose where to offend? Perspectives from animal foraging. *Legal and Criminological Psychology, 19*, 193. https://doi.org/10.1111/lcrp.12061.

Johnston, J. (1984). *Econometrics methods* (3rd ed.). London: McGraw-Hill.

Kennedy, L. W., & Forde, D. R. (1990). Routine activities and crime: An analysis of victimisation in Canada. *Criminology, 28*, 137–152.

Kershaw, C., & Tseloni, A. (2005). Predicting crime rates, fear and disorder based on area information: Evidence from the 2000 British Crime Survey. *International Review of Victimology, 12*, 295–313.

Lee, Y., Eck, J. E., S, O., & Martinez, N. N. (2017). How concentrated is crime at places? A systematic review from 1970 to 2015. *Crime Science, 6*(6), 1–16.

Lewakowski, B. (2012). *Half-locked?: Assessing the distribution of household safety protection in Stockholm*. Student thesis. KTH Architecture and the Built Environment, Department of Urban Planning and Environment. http://urn.kb.se/resolve?urn=urn:nbn:se:kth:diva-97692. Accessed 23 Feb 2018.

Long, S. J. (1997). *Regression models for categorical and limited dependent variables, Advanced quantitative techniques in the social sciences series* (Vol. 7). London: Sage.

Lucy, W., Gilbert, D., & Birkhead, D. (1977). Equity in local service distribution. *Public Administration Review, 37*(6), 687–697.

Metro. (2017). Why you should be worried if you get a call from a 'wrong number'. By Ashitha Nagesh. http://metro.co.uk/2017/05/09/why-you-should-be-worried-if-you-get-a-call-from-a-wrong-number-6625644/. Accessed 10 Jan 2018.

Mooney, G., & Jan, S. (1997). Vertical equity: Weighting outcomes? Or establishing procedures? *Health Policy, 39*(1), 79–87.

Nilsson, A., Estrada, F., & Bäckman, O. (2017). The unequal crime drop: Changes over time in the distribution of crime among individuals from different socioeconomic backgrounds. *European Journal of Criminology, 14*(5), 586–605.

Nottinghamshire Police. (2013). Security Standards of HMO & Rented Properties: Minimum security specifications from Nottinghamshire Police's Pre Crime Unit in partnership with Nottingham City Council Environmental Health Department. https://www.nottinghamcity.gov.uk/housing/private-sector-housing/houses-in-multiple-occupation-hmo/. Accessed 11 Jan 2018.

O, S., Martinez, N. N., Lee, Y., & Eck, J. E. (2017). How concentrated is crime among victims? A systematic review from 1977 to 2014. *Crime Science, 6*(9), 1–16.

Office for National Statistics (ONS) (2012). Ethnicity and national identity in England and Wales: 2011. https://www.ons.gov.uk/peoplepopulationandcommunity/culturalidentity/ethnicity/articles/ethnicityandnationalidentityinenglandandwales/2012-12-11. Accessed 6 Nov 2018.

Official Statistics. (2013). 2011 Rural urban classification. Gov.UK. https://www.gov.uk/government/statistics/2011-rural-urban-classification. Accessed 24 Jan 2018.

ONS (Office for National Statistics). (2017a). Crime in England and Wales: Year ending Dec 2016. Statistical bulletin. 27 April. https://www.ons.gov.uk/peoplepopulationandcommunity/crimeandjustice/bulletins/crimeinenglandandwales/yearendingdec2016. Accessed 4 May 2017.

ONS (Office for National Statistics). (2017b). Statistical bulletin: Nowcasting household income in the UK: Financial year ending 2017. Release date 28 July 2017. https://www.ons.gov.uk/peoplepopulationandcommunity/personalandhouseholdfinances/incomeandwealth/bulletins/nowcastinghouseholdincomeintheuk/financialyearending2017#trends-in-household-incomes. Accessed 13 Feb 2018.

ONS (Office for National Statistics). (2018). Statistical bulletin: Household disposable income and inequality in the UK: Financial year ending 2017. Release date 10 January 2018. https://www.ons.gov.uk/peoplepopulationandcommunity/personalandhouseholdfinances/incomeandwealth/bulletins/householddisposableincomeandinequality/financialyearending2017#median-household-disposable-income-1600-higher-than-pre-downturn-level. Accessed 5 Feb 2018.

Osborn, D. R., & Tseloni, A. (1998). The distribution of household property crimes. *Journal of Quantitative Criminology, 14*, 307–330.

Pateman, T. (2011). Rural and urban areas: Comparing lives using rural/urban classifications. *Regional Trends, 43*(1), 11–86.

Rasbash, J., Charlton, C., Browne, W.J., Healy, M., Cameron, B. (2009). *MLwiN Version 2.10.* Centre for Multilevel Modelling, University of Bristol.

Rasbash, J., Steele, F., Browne, W., Goldstein, H. (2017). *A user's guide to MlwiN version 3.01.* Centre for Multilevel Modelling, University of Bristol.

Rawls, J. (1999). A theory of justice (2nd ed.). Cambridge: Harvard University Press.

Rountree, P. W., & Land, K. C. (1996). Burglary victimisation, perceptions of crime risk, and routine activities: A multilevel analysis across Seattle neighbourhoods and census tracts. *Journal of Research in Crime and Delinquency, 33*, 147–180.

Shaw, C. R., & McKay, H. D. (1942). *Juvenile delinquency and urban areas.* Chicago: Chicago University Press.

Sherman, L. W., Gartin, P. R., & Buerger, M. E. (1989). Hot spots of predatory crime: Routine activities and the criminology of place. *Criminology, 27*(1), 27–56.

Skudder, H., Brunton-Smith, I., Tseloni, A., McInnes, A., Cole, J., Thompson, R., & Druckman, A. (2017). Can burglary prevention be low carbon and effective? Investigating the environmental performance of burglary prevention measures. *Security Journal, 31*, 111. https://doi.org/10.1057/s41284-017-0091-4.

Snijders, T. A. B., & Bosker, R. J. (1999). *Multilevel analysis: An introduction to basic and advanced multilevel modelling.* London: SAGE.

The Guardian. (2012). The great Asian gold theft crisis. Crime. By Emine Saner. https://www.theguardian.com/uk/2012/jan/31/gold-theft-asian-families. Accessed 10 Jan 2018.

The Guardian. (2017). Look at Grenfell and recall when social housing was beloved. Opinion. By Deborah Orr. https://www.theguardian.com/commentisfree/2017/sep/15/social-housing-grenfell-london-modernist-estates. Accessed 12 Jan 2018.

Tilley, N. (2012). Community, security and distributive justice. In V. Ceccato (Ed.), *The urban fabric of crime and fear* (pp. 267–282). New York: Springer.

Tilley, N., & Tseloni, A. (2016). Choosing and using statistical sources in criminology – What can the Crime Survey for England and Wales tell us? *Legal Information Management, 16*(2), 78–90. https://doi.org/10.1017/S1472669616000219.

Tilley, N., Tseloni, A., & Farrell, G. (2011). Income – Disparities of burglary risk and security availability over time. *British Journal of Criminology, 51*(2), 296–313.

Tilley, N., Farrell, G., & Clarke, R. V. (2015). Target suitability and the crime drop. In M. A. Andresen & G. Farrell (Eds.), *The criminal act: The role and influence of routine activities theory* (pp. 59–76). London: Palgrave Macmillan.

Trickett, A., Osborn, D. R., Seymour, J., & Pease, K. (1992). What is different about high crime areas? *British Journal of Criminology, 32*, 81–89.

Tseloni, A. (1995). The modelling of threat incidence: Evidence from the British crime survey. In R. E. Dobash, R. P. Dobash, & L. Noaks (Eds.), *Gender and crime* (pp. 269–294). Cardiff: University of Wales Press.

Tseloni, A. (2006). Multilevel modelling of the number of property crimes: Household and area effects. *Journal of the Royal Statistical Society Series A-Statistics in Society, 169*(Part 2), 205–233.

Tseloni, A. (2011). Household burglary victimisation and protection measures: Who can afford security against burglary and in what context does it matter? Crime Surveys Users Meeting, Royal Statistical Society, London. 13 December. Available online: http://www.ccsr.ac.uk/esds/events/2011-12-13/index.html

Tseloni, A. (2014). Understanding victimization frequency. Chapter 127. In G. Bruinsma & D. Weisburd (Editors in Chief), *Encyclopedia of Criminology and Criminal Justice (ECCJ)* (pp. 5370–5382). New York: Springer.

Tseloni, A., & Pease, K. (2004). Repeat personal victimisation: Random effects, event dependence and unexplained heterogeneity. *British Journal of Criminology, 44*, 931–945.

Tseloni, A., & Pease, K. (2005). Population inequality: The case of repeat victimisation. *International Review of Victimology, 12*, 75–90.

Tseloni, A., & Pease, K. (2015). Area and individual differences in personal crime victimisation incidence: The role of individual, lifestyle /routine activities and contextual predictors. *International Review of Victimology, 21*(1), 3–29.

Tseloni, A., & Pease, K. (2017). So, were you surprised by the BBC/ONS crime risk calculator? *Significance* online London: The Royal Statistical Society. https://www.significancemagazine. com/politics/569-so-were-you-surprised-by-the-bbc-ons-crime-risk-calculator

Tseloni, A., & Thompson, R. (2015). Securing the premises. *Significance, 12*(1), 32–35.

Tseloni, A., & Zarafonitou, C. (2008). Fear of crime and victimisation: A multivariate multilevel analysis of competing measurements. *European Journal of Criminology, 5*(4), 387–409.

Tseloni, A., Osborn, D. R., Trickett, A., & Pease, K. (2002). Modelling property crime using the British Crime Survey: What have we learned? *British Journal of Criminology, 42*, 89–108.

Tseloni, A., Wittebrood, K., Farrell, G., & Pease, K. (2004). Burglary victimisation in the U.S., England and Wales, and the Netherlands: Cross-national comparison of routine activity patterns. *British Journal of Criminology, 44*, 66–91.

Tseloni, A., Mailley, J., Farrell, G., & Tilley, N. (2010). The cross-national crime and repeat victimization trend for main crime categories: Multilevel modelling of the international crime victims survey. *European Journal of Criminology, 7*(5), 375–394.

Tseloni, A., Thompson, R., Grove, L., Tilley, N., & Farrell, G. (2014). The effectiveness of burglary security devices. *Security Journal, 30*(2), 646–664. https://doi.org/10.1057/sj.2014.30.

Tseloni, A., Farrell, G., Thompson, R., Evans, E., & Tilley, N. (2017). Domestic burglary drop and the security hypothesis. *Crime Science., 6*(3). https://doi.org/10.1186/s40163-017-0064-2 Open Access.

Tunstall, R. (2011). Social housing and social exclusion 2000-2011. Discussion paper series (CASEpapers), CASE/153, Centre for Analysis of Social Exclusion July 2011 London School of Economics. http://sticerd.lse.ac.uk/dps/case/cp/CASEpaper153.pdf. Accessed 29 Jan 2018.

Van Dijk, J., & Vollaard, B. (2012). Self-limiting crime waves. In J. van Dijk, A. Tseloni, & G. Farrell (Eds.), *The international crime drop: New directions in research* (pp. 250–267). Hampshire: Palgrave Macmillan.

Vollaard, B., & van Ours, J. C. (2011). Does regulation of built-in security reduce crime? Evidence from a natural experiment. *The Economic Journal, 121*, 485–504.

Weisburd, D. (2015). The law of crime concentration and the criminology of place. *Criminology, 27*(1), 27–56.

Wilcox, P., Lamd, K. C., & Hunt, S. A. (2003). *Criminal circumstance: A dynamic multicontextual criminal opportunity theory.* New York: Aldine de Gruyter.

Wilcox, P., Madensen, T. D., & Tillyer, M. S. (2007). Guardianship in context: Implications for burglary victimisation risk and prevention. *Criminology, 45*(4), 771–803.

Wiles, P., & Costello, A. (2000). *The 'road to nowhere': The evidence for travelling criminals.* London: Home Office.

Wilson, W., & Barton, C. (2017). Allocating social housing (England). Briefing paper number 06397, 9 June 2017. London House of Commons Library.

Winchester, S., & Jackson, H. (1982). *Residential burglary: The limits of prevention.* London: Home Office.

Wolfgang, M. E., Figlio, R. M., & Sellin, T. (1972). *Delinquency in a birth cohort.* London: Chicago University Press.

Chapter 6
An Evaluation of a Research-Informed Target Hardening Initiative

James Hunter and Andromachi Tseloni

Abbreviations

BTF	Burglary Task and Finish group
CSEW	Crime Survey for England and Wales
ESRC-SDAI	Economic and Social Research Council-Secondary Data Analysis Initiative
LSOA	Lower Super Output Area (statistical geographical boundary for UK Census and other data)
NCDP	Nottingham Crime and Drugs Partnership
NCH	Nottingham City Homes
NRV	Near repeat victimisation
OPCC	Office of the Police and Crime Commissioner
PCSOs	Police Community Support Officers
PCU	Pre-crime unit
WIDE	Window locks, internal lights on a timer, double door locks and external lights on a sensor

6.1 Introduction

From the late 1980s until the end of the previous century, extensive burglary reduction initiatives occurred in England and Wales – the history of which is discussed in Chap. 2 of this book. Initiatives, such as the Safer Cities Programme (1988–1995) and the Reducing Burglary Initiative (1998–2001/2002), in effect aimed to target harden the homes of specific population groups (e.g. social renters) and/or particular areas of residence (e.g. inner cities and urban areas). These entailed a mixed bag of home security upgrades and improvements in public footpaths and areas that gave access to

J. Hunter (✉) · A. Tseloni
Quantitative and Spatial Criminology, School of Social Sciences, Nottingham Trent University, Nottingham, UK
e-mail: james.hunter@ntu.ac.uk

© Springer Nature Switzerland AG 2018
A. Tseloni et al., *Reducing Burglary*,
https://doi.org/10.1007/978-3-319-99942-5_6

homes. It is fair to say that no particular security combination underpinned these initiatives. By contrast, the discussion presented here examines the implementation of a WIDE (Window locks, Internal lights on a timer, Double door locks and External lights on a sensor) – informed burglary prevention initiative[1] undertaken from September 2014 to January 2015 in a medium-size English city designed to evaluate the effectiveness of this approach to target hardening in a real-life setting.

For consistency, and in order to utilise standard crime prevention terminology, the demonstration project discussed here is referred to as a *pilot target hardening initiative*. The demonstration project was a *pilot* because it was not generalised but limited to certain addresses within certain selected areas of the city with the aim to maximise the desired objective of burglary reduction and facilitate its evaluation. The selection of target areas and houses for the pilot was determined by the extent of near repeat victimisation (see Chap. 1). *Target hardening* refers to house security upgrades that in principle and according to our research findings would discourage potential burglars (see Chaps. 3 and 4). Finally, the pilot target hardening was a top-down *initiative* rather than the product of natural development, whereby households would have adopted the recommended security of their own initiative. The Nottinghamshire Office of the Police and Crime Commissioner (OPCC) funded the initiative, and this was administered by the local crime and safety partnership, the Nottingham Crime and Drugs Partnership (NCDP). The possibility that individual households may have independently found out about these research findings and/or would have upgraded security of their own volition is not impossible. In fact, quite the opposite will be discussed in the limitation section outlining parallel citywide public awareness initiatives.

Following a brief overview of the theory of repeat and near repeat victimisation and a discussion of methodological issues arising in the evaluation of burglary reduction initiatives in Sect. 6.2, a brief contextual overview of the socio-economic and burglary profile of the city and the NCDP is provided in Sect. 6.3. Section 6.4 includes an outline of the aims and implementation of the demonstration project as well as the collaborative stages of work that went into planning the pilot target hardening initiative. Much of this discussion relies on, and reiterates, the 'Repeat and Near Repeat Burglary Pilot Project Protocol' produced by the NCDP prior to implementation in order to share information and provide an agreed common basis of action across all agencies of the city that were involved in the pilot. No matter how much planning goes into an initiative, human input and changing contexts in reality may deliver something different from that which was originally intentioned. Section 6.4 therefore ends with a discussion of the actual implementation of the pilot target hardening initiative, its cost and immediate effects. Thereafter follows an investigation of whether the pilot target hardening initiative had any measurable outcome on burglary rates – both in the areas where "at risk" households received security upgrades and overall in the city (Sect. 6.5). The chapter ends with a discussion of the evaluation findings and potential caveats that may have affected the outcome of this evaluation (Sect. 6.6).

---

[1] The burglary prevention initiative was developed as a demonstration project that drew on the research findings relating to burglary risk and security devices presented in Chaps. 4 and 5.

6.2 Evaluation of Burglary Reduction Initiatives

6.2.1 Theoretical Underpinnings: Repeat and Near Repeat Victimisation

Crime, even of a less serious nature, can adversely affect victims in multiple ways in relation to psychological and financial costs, perception of security and sense of trust in neighbours and criminal justice agencies (Brunton-Smith and Sturgis 2011; Pease 2009). The impact of victimisation becomes particularly problematic where the same individuals or households experience multiple crime victimisations of the same or different offence types. The phenomenon of repeat victimisation only materialised within criminological research due to an empirical shift towards the results of victimisation surveys and away from police recorded crime data (Tseloni and Rogerson 2018). Pioneering work by Ken Pease and colleagues identified the presence of significant repeat burglary victimisation within local aggregate crime rates (Forrester et al. 1988; Polvi et al. 1991). Subsequent empirical studies based on national crime survey data confirmed this finding (see, e.g. Farrell 1992; Trickett et al. 1992; Osborn et al. 1996; Osborn and Tseloni 1998).

The phenomenon of *crime repetition* that is inherent within repeat victimisation can also manifest itself in other forms. Near repeat victimisation, for example, occurs where the incidence or risk of crime repetition spills over to neighbouring properties following an initial victimisation incident within a neighbourhood. Affluent areas have been shown to suffer higher levels of near repeat burglaries (Bowers and Johnson 2005) where burglaries often occur during daylight during weekdays (Coupe and Blake 2006; see also Townsley et al. 2003). More importantly, analysis of police recorded crime data (Johnson and Bowers 2004a, 2004b; Johnson et al. 2007; Ross and Pease 2007) indicates that the geographical/temporal contamination of neighbouring houses (especially those located on the same street as the original burgled property) is within 400 metres for a period of around 1 month. To put it simply, neighbouring houses are at increased risk of burglary for a month after an initial burglary.

6.2.2 Key Methodological Issues in the Evaluation of Burglary Reduction Initiatives

There is a significant literature concerning the impact and evaluation of burglary reduction initiatives. This has focused upon the evaluation of both generic burglary reduction schemes (e.g. Pease 1991; Foster and Hope 1993; Ekblom et al. 1996) and specific communal and physical forms of target hardening (e.g. Rosenbaum 1987; Tseloni et al. 2017). There are also meta-analyses and systematic reviews that have sought to identify 'what works' in relation to burglary reduction (e.g. Bennett et al. 2006; Sidebottom et al. 2015).

Emerging from this literature are a number of significant methodological issues surrounding the evaluation of burglary (and other forms of crime) reduction initiatives. These methodological problems often arise from weak evaluation designs that are the culmination of poorly designed programmes, questionable approaches to the monitoring and measurement of programme implementation and outcomes, and the failure to consider the possibility of alternative agents of changes in crime levels outside the remit of the crime reduction initiative in question (Lurigio and Rosenbaum 1986). Indeed, set against the 'gold standard' of the Maryland Scale of Scientific Methods for Evaluating Crime Prevention[2], many evaluation designs fail to incorporate a degree of methodological complexity that rises above 'Level 3: A comparison between two or more comparable units of analysis, one with and one without the program' (Sherman et al. 1998, p. 4).

More often than not, however, difficulties surrounding the determination of the precise effectiveness of crime reduction initiatives arise from the lack of practical control afforded to programme evaluators, rather than because of methodological naivety. For example, it is often difficult to control for other forms of police activity in and around initiative test areas (Allatt 1984) – although the presence of additional crime prevention and policing measures may enhance rather than reduce the impact of a specific burglary reduction approach (Millie and Hough 2004). Equally, the desire from policymakers to implement a number of specific target hardening measures (e.g. better street lighting and alley gating) as simultaneous components of a generic burglary reduction initiative can make the methodological diagnosis of the precise impact of the respective policy instruments upon burglary levels problematic (Griswold 1984). Experimental approaches to the evaluation of crime reduction initiatives require the identification of comparative areas for establishing 'test' and 'control' neighbourhoods. Policy interventions outside the remit of the initiative under evaluation, however, can render differences in seemingly identical neighbourhoods in respect of population and economic characteristics (Johnson et al. 2004). In addition, the search for different neighbourhoods to act as comparator areas can be complicated by the tendency for similar localities to be clustered together (i.e. spatial autocorrelation) (Johnson et al. 2004). Establishing the relevant 'before' and 'after' time periods to determine the precise impact of the burglary reduction initiative in question can also prove difficult. This is especially the case given the need to take account of the potential anticipatory impact of burglary reduction measures. For example, publicity concerning the imminent implementation of target hardening measures may bring about a change in offender behaviour prior to the actual

---

[2] The Maryland Scale of Scientific Methods for Evaluating Crime Prevention has five levels of increasing complexity: 'Level One: Correlation between a crime prevention program and a measure of crime or crime risk factors at a single point in time; Level Two: Temporal sequence between the program and the crime or risk outcome clearly observed, or the presence of a comparison group without demonstrated comparability to the treatment group; Level Three: A comparison between two or more comparable units of analysis, one with and one without the program; Level Four: Comparison between multiple units with and without the program, controlling for other factors, or using comparison units that evidence only minor differences; Level Five: Random assignment and analysis of comparable units to program and comparison groups' (Sherman et al. 1998, pp. 4–5).

installation of the policy instruments designed to yield a reduction in victimisation levels (Bowers and Johnson 2003).

Aside from issues surrounding the design and implementation of evaluations, determining the nature and scale of burglary reductions that have occurred can be equally problematic. Bowers et al. (2003) identify the potential existence of impact hotspots within subareas within test and control localities that may prove greater than the more modest decline in burglary rates across the initiative zone as a whole. The context-specific nature of these impact hotspots however may limit the generalisability and policy transfer of the specific burglary reduction initiative to other localities (Santos and Santos 2015). Establishing the statistical significance of reductions in burglary levels also necessitates distinguishing between real changes in crime levels as opposed to random fluctuations (Johnson et al. 2004). Notable changes in burglary levels within test areas can only constitute policy 'success' if the possibility of offence displacement has been ruled out. Bowers et al. (2003) identify three forms of displacement: geographical displacement (i.e. the burglar targets properties within a buffer zone adjacent to the test area), target displacement (i.e. the burglar victimises non-target hardened properties within the test area) or offence displacement (i.e. the offender is dissuaded from committing a burglary but chooses to commit an alternative type of criminal offence). Finally, the ultimate goal of any programme evaluation is to determine not only whether the policy intervention in question worked but also to establish the reason behind the apparent policy 'success'. A significant problem with the evaluation of many burglary (and other forms of crime) reduction initiatives remains the failure on the part of the evaluators to consider the counterfactual possibilities that might actually explain the identified reductions in victimisation levels (Ekblom et al. 1996; Hope 2004; Johnson et al. 2004; Cummings 2006).

6.3 Project Context

6.3.1 The City of Nottingham

The estimated population of Nottingham in 2016 was 325,282 people (NOMIS 2018), making it the 14th largest city in England. Students attending the city's two universities make up around 11.4 percent of the local population (ONS 2018). It has many of the contemporary economic characteristics of urban areas in the Midlands and North of England predicated on traditional industries that have now declined significantly. Currently it is the 10th most deprived local authority area in England according to the 2015 English Indices of Deprivation (DCLG 2015) and was ranked 9th between 2012 and 2016 in terms of workless households across all local authority areas in the United Kingdom (ONS 2017a). The population of the city also features many of the socio-demographic household characteristics associated with higher risk of burglary victimisation outlined in Chap. 5.

6.3.2   Burglary Profile of Nottingham

Between March 2007 and March 2016, police recorded domestic burglary incidents in Nottingham fell from 7641 to 1554 – a decline of 79.7 percent (source of data: ONS 2017b). During the same time, the total number of police recorded offences declined by 54.5 percent. Despite this dramatic fall, domestic burglary incidences remained relatively concentrated within certain parts of the city. Between December 2014 and November 2017, half of all domestic burglaries took place within 29.6 percent of neighbourhoods (Lower Super Output Areas) across the city. The concentration of burglaries is further manifested in that 20 percent of all domestic burglaries occurred in just 7.7 percent of neighbourhoods (source of data: College of Policing 2018).

6.3.3   Nottingham Crime and Drugs Partnership (NCDP)

Community safety partnerships were created by the 1998 Crime and Disorder Act and are designed to bring about the formulation and implementation of a 5-year strategic crime reduction plan developed by local agencies including police forces, local councils, local businesses and organisations drawn from the voluntary and community sectors. The NCDP is the community safety partnership for Nottingham and produces an annual Partnership Plan that identifies strategic targets and initiatives designed to realise the policy objectives set out in the strategic crime reduction plan (further details of the precise nature of NCDP and its activities are set out in Appendix C.1). Discussions about research-informed insights of burglary prevention between Nottinghamshire Police's burglary lead and the academic project's lead had taken place long before the research discussed in this book (Chaps. 4, 5, 6, 8 and 9). NCDP also supported the original academic research through:

(a) Discussions with the principal academic investigator concerning what insights might be useful from a policing burglary prevention perspective and supporting her bid to the Economic and Social Research Council for research funding (see Chaps. 4, 5 and 9)
(b) Sustained collaboration in the research in an advisory capacity and in-kind contribution within the Advisory Committee of the academic research project (see Chap. 9)
(c) Using the research findings to inform burglary prevention initiatives in the city

The burglary prevention pilot project outlined here was therefore the culmination of a sustained collaboration between NCDP and the lead (principal investigator) of the academic research in order to test whether the findings from the national Crime Survey for England and Wales (CSEW) data-based research could underpin effective approaches to burglary reduction at the local level.

6.4   The Nottingham Pilot Burglary Target Hardening Initiative

6.4.1   Project Inception and Operational Framework

The NCDP held a Burglary Summit on 7 August 2013 on behalf of relevant stakeholders, including the Burglary and Security project researchers (outlined in Chaps. 4, 5 and 9), to discuss possible solutions to burglary problems in the city. Partners at this summit generated a number of actions, including target hardening, that were holistic in nature in terms of both joined-up approaches to burglary reduction and the participation of a wide range of organisations. These actions were subsequently taken forward by a multi-agency Burglary Task and Finish group (BTF). As well as including the NCDP (which managed the entire operation and reported back to its own executive group) and academic research partner, the BTF contained representatives from:

- Nottinghamshire Police
- Nottingham City Homes (NCH – the largest provider of social housing in the city)
- Nottingham City Council
- Neighbourhood Watch
- Crimestoppers
- Probation Services
- Professor Ken Pease (acting in a capacity as an unpaid academic advisor)[3]

The BTF began meetings in October 2013, and Nottinghamshire OPCC agreed to make £40,000 total funding available for target hardening homes in the city. NCH were commissioned to carry out this work through their asset management service, on both their own properties and nonsocial housing residences.

6.4.2   Research-Informed Project Aims and Protocol

The concepts of repeat victimisation, and near repeat victimisation, set out in Sect. 6.2.1 underpinned the crime reduction objectives of the pilot target hardening initiative – as well as the selection of the test and control neighbourhoods within Nottingham for the purpose of implementing and evaluating the initiative. The pilot sought to directly test the effectiveness of the identified home security measures in preventing burglary arising from the research discussed in Chaps. 4, 5 and 8 of this book.[4]

---

[3] The primary stakeholders within the BTF which acted as a critical friend to the project were NCH, Nottinghamshire Police, the academic research partner and Professor Ken Pease.

[4] The protocol included the following information: '1.1 The theory of repeat and near repeat victimisation in [relation] to dwelling burglary is well known. Recent on-going and thus preliminary research has also identified the most effective combination of security devices to prevent burglary. The pilot project outlined in this protocol is designed to test the effectiveness of those security devices through a programme of target hardening. The properties selected for target hardening will be determined by the repeat and near repeat victimisation theory within two areas of the city' (NCDP 2014, Sect. 1.1).

In order to ensure that all of the project partners were operating from the same premise, the repeat/near repeat victimisation principles and overarching aims were embedded within guidance issued by NCDP (2014), entitled: 'Repeat and Near Repeat Burglary Pilot Project Protocol'. The introduction to this clearly stated the research evidence base that justified the programme of target hardening. It stated that:

"1. Initial findings of the ESRC-SDAI-funded academic research on 'Which burglary security devices work for whom and in what context' co-advised by the Nottingham CDP indicate that certain measures are an effective deterrent. The four measures which the research shows to be effective in combination are:

- Robust doors with double locks or deadlocks
- Robust windows with secure locks
- External lights on a sensor
- Indoor lights on a timer.

2. Where possible, these measures will be implemented in the Pilot. Earlier published research by the Principal Investigator (currently at Loughborough University) of the ESRC-SDAI project, identified three population groups as being more vulnerable to burglary:

- Households of 3 or more adults
- People living in private rented or social rented housing.
- Lone parent households.

3. Preliminary work of the current ESRC-SDAI project has identified the following population groups at higher risk of burglary and lower availability of effective security:

- Households on a low income
- Households with no car,
- Households in private or social rented accommodation.
- Households of 3 or more adults

4. The pilot will seek to gather data on the composition of the households which take part in the survey. This data will inform the evaluation of the project." (ibid, Section 4).

Points 1 and 3 above to some extent reiterate the research findings that Chaps. 4 and 5 discuss in this book in their preliminary form. Point 2 refers to findings from previous research published by the second author of this chapter (Osborn and Tseloni 1998; Tseloni 2006) which have since been confirmed in subsequent studies (Hunter and Tseloni 2016; Ignatans and Pease 2016) including Chap. 5 of this book.

The protocol also incorporated an introductory definition of repeat and near repeat victimisation from a police operational perspective[5] in order to justify the selection of both test and control target areas in the pilot. 'Previous (near) victimisation has been shown to be the best predictor of future victimisation, for many crime types and in a variety of contexts' (NCDP 2014, Sect. 3.2). The NCDP had gained familiarity with and trusted these findings via (a) supporting the research project application to the ESRC for funding; (b) after successful funding application and the project launch, continuous involvement in the project's Advisory Committee workshops (see also Chap. 9); and (c) the project lead's membership and participation in the NCDP board meetings in her role as an academic expert.

The specific aims of the pilot project set out in the protocol were:

---

[5] Chainey, S. (2012). *Repeat Victimisation*. JDiBrief Series. London: UCL Jill Dando Institute of Security and Crime Science. ISSN: 2050–4853

- To provide target hardening to properties and crime prevention advice to citizens to prevent burglaries
- To test the theory that certain security measures provide a more effective deterrent to burglary
- To prevent burglaries from occurring by target hardening neighbouring properties based on the near repeat theory of victimisation

6.4.3   Selection of Participating Areas

The pilot was designed to empirically test the effectiveness of the four security devices outlined above in two wards of the city. The selection of these wards was based on analysis of repeat and near repeat victimisation (NRV) undertaken by the NCDP. In 2013, 1.59 percent of households in Nottingham were victims of burglary, and 0.05 percent experienced repeat, i.e. two or more, burglaries. Repeat burglaries accounted for 3.29 percent of all burglaries. The city also had high rates of near repeat burglaries that occurred near the time and in close proximity (within a 200 m vicinity) of an initially reported burglary. Between January and December 2013, of the 1952 burglaries in the city, 395 (20 percent) happened within 200 m and 0–7 days from an initial reported burglary, giving a 34 percent chance of a near repeat within this period. More specifically, 5.3 percent of all burglaries in the city were near repeats *within a day* within 200 m of an initially reported burglary. The NCDP predicted that citywide super cocooning could contribute to an almost 23 percent reduction in the number of burglary reports for the 2014 calendar year. This involved swift (in principle within a week) target hardening, and advice, to all those households situated near a burglary victim ($5 \times 5 \times 5 \times 5$ to the front, back and either side of the initial victim, respectively). Additional crime reduction benefits were to be gained if the deployment of this approach took into account the different levels of NRV across wards.

Table 6.1 provides the rate of repeat and near repeat burglaries across all Nottingham wards. To retain anonymity, the actual names of wards (numbered from I to XX) and number of burglaries are omitted.

The city wards selected for the pilot had to meet the following criteria:

(a) High rates of near repeat burglaries within 7 days and 200 m of an initially reported burglary.
(b) They were not to come from areas with the highest volume of burglary in the city. The rationale for avoiding areas with high burglary rates was that there were already other interventions taking place within those wards that would make it more difficult to evaluate the effectiveness of this study.
(c) Areas with high levels of student housing were avoided since it was deemed socially and economically inequitable that properties owned by students' landlords would be eligible for free upgrades if they fell in the test area.
(d) Areas of high levels of social housing were also avoided since NCH were carrying out a series of upgrades in their own properties that had started prior and independently to the academic research the pilot was testing (Jones et al. 2016).

Table 6.1 The percentage of repeat and near repeat burglaries within 200 metres of initial burglary during the year prior to the initiative, January to December 2013 (Source: NCDP 2014)

Wards	% Repeats	Near repeat victimisation within 200 metres of initial burglary	
		% Within 24 h	% Within 7 days
I	0.05	11.3	38.7
II[a]	0.04	4.5	30.1
III	0.04	14.4	30.1
IV	0.05	3.2	22.6
V	0.04	4.8	21.4
VI	0.04	3.8	16.6
VII	0.03	6.0	15.4
VIII	0.03	3.4	13.7
IX[a]	0.02	3.6	13.5
X	0.00	7.5	13.4
XI	0.01	4.5	12.7
XII	0.00	2.4	12.0
XIII	0.03	4.3	11.2
XIV	0.02	1.1	10.9
XV	0.01	3.6	9.6
XVI	0.04	3.7	7.4
XVII	0.00	1.6	6.5
XVIII	0.00	0.0	5.0
XIX	0.10	2.7	2.7
XX	0.06	2.5	2.5

[a]Wards selected for the purposes of designated test and control areas

(e) Finally, wards designated as conservation areas were avoided due to potential complications surrounding the installation of new windows and doors.

Balancing these respective criteria was challenging, and there were lengthy deliberations in order to select the most appropriate wards for the pilot.[6] Two areas, Wards II and IX from Table 6.1 above, were eventually selected with respective rates of near repeat burglaries of 30.1 and 13.5 percent. Thus, the design allowed for assessing the effectiveness of cocooning in areas with different levels of the NRV problem since Ward II had more than double the Ward IX rate of NRV.

Two Lower Super Output Areas (LSOAs)[7], which will be referred to as neighbourhoods in the following discussion, of similar population size and household

---

[6]At an earlier stage, two wards with similar near-repeat burglaries (Ward IX at 13.5 percent and Ward XI at 12.7 percent) within the set time and space configurations were selected. Due to the high presence of council housing, Ward XI was subsequently replaced with Ward II.

[7]Lower Super Output Areas are a geographical statistical area employed by the Office for National Statistics in the UK primarily for the Census and the country's sampling frame of a number of social surveys. They are analogous to Census tracks for other countries. LSOAs underpin the collation and analysis of data relating to a wide range of social problems and issues including crime in England. They have minimum and maximum population thresholds of between 1000 and 3000 people in order to enable valid comparison of data across similar sized neighbourhoods.

Table 6.2 Population and household size and burglary rate per 100 households of Lower Super Output Areas participating in the pilot target hardening initiative (Source: NCDP 2014)

	Number of individuals living in households[a]	Number of households with at least one usual resident[a]	Average number of individuals per household	3 Years, 2011–2013, average burglary rate (per 100 households)
Ward II Test	1800	700	2.4	6.11
Ward II Control	1600	700	2.3	5.92
Ward IX Test	1600	800	2.0	4.64
Ward IX Control	1600	800	2.1	4.81

Note: [a]The numbers have been rounded to the closest 100th value for area confidentiality

size and with comparable burglary rates were selected within each of these wards in order to act as test (Ward II Test and Ward IX Test) and control areas (Ward II Control and Ward IX Control). The respective test and control neighbourhoods were not adjacent to each other. 'This is because any displacement which may occur from the test area would be more likely to go into the adjacent control [neighbourhood] and this would undermine the design of the pilot. This is to get as near to perfect as test conditions as is possible when conducting a study out in the community' (NCDP 2014, Sect. 7.2). Table 6.2 shows the population/household size and burglary rates for the two test and two control neighbourhoods in the 3 years preceding the pilot.

6.4.4   Pilot Process: Planning, Implementation, Security Cost and Evaluation

Those houses selected within the two test neighbourhoods were offered target hardening in order to upgrade the presence of security measures to the level of those prescribed by WIDE. This activity was triggered by a burglary in a dwelling taking place in one of the identified neighbourhoods. This event would initiate a Home Security Assessment visit conducted by a member of Nottinghamshire Police's pre-crime unit (PCU). The assessment would take place in the burgled property, and the same assessment would be offered to neighbouring properties (five in front, five either side and five behind: $5 \times 5 \times 5 \times 5$). A working group made up of NCDP, Nottinghamshire Police, NCH and the academic research partner institution agreed this process jointly. Diagram 6.1 sets out a systematic flow diagram of the process to be followed in both the test and control areas by all the agencies involved. A marker was placed on the designated neighbourhoods for the project within the police control room, whereby a reported burglary was given a tag that automatically notified the PCU that a burglary had taken place in one of the designated neighbourhoods. Appendix Table 6.5 gives the steps that were to be taken following a burglary tag.

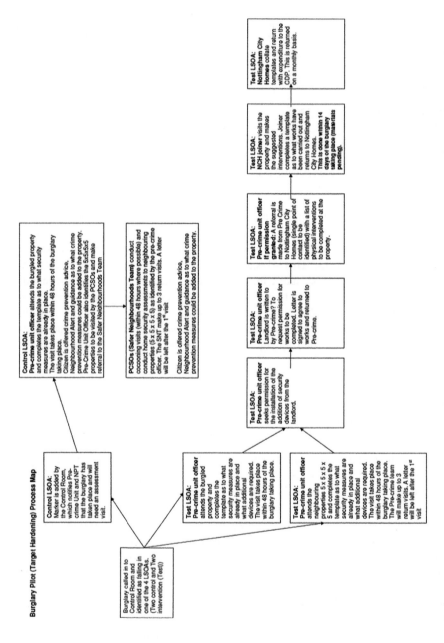

Diagram 6.1 Flow chart detailing the steps of the pilot target hardening process in the control and test areas. (Source: NCDP 2014)

The PCU visited the burgled property and conducted a Security Survey (see forms in Appendix C.2 and Appendix Tables 6.6, 6.7a and 6.7b). The Security Survey forms and, in the cases of rented property or absent occupiers, the pilot information letter were prepared in advance. Small packs including crime prevention advice and a property marking pen and guidance[8] which in the protocol were mentioned in relation to the control areas were also prepared in advance. Whilst the assessment looked at all aspects of the property, it had a specific focus on recommending an upgrade to the four identified security measures. Officer rotas were created to task two officers to visit properties wherever possible within 48 h of a burglary. When the occupier was not at home at the time of the visit, the case was kept open for up to 7 days from when the pilot information letter was posted. Additionally, these visits were intended to capture the household composition details required for the pilot evaluation (via the Data Capture Form presented in Appendix C, Appendix Table 6.6.)

In the test areas, the recommended security improvements identified in the Security Survey were offered to the property owner. As part of the pilot study, NCH offered to undertake the recommended security upgrades free of charge. Upon agreement the owner signed to give their permission for the works to be undertaken, and the referral (see Appendix Table 6.7a) was made to NCH. Where the house was privately rented, a letter was sent to the landlord giving details of the pilot, the recommendations and a form to be signed giving permission for the works to be completed. Signed forms by owner-occupiers and landlords of privately rented properties in the test areas were returned to the NCDP, which generated the referral to NCH.

Any new doors and windows or window locks required were ordered, whereas solar lights for external use and timers for internal lights were already in stock (and at a very low cost). All replacement doors and windows were composite together with Secured by Design window locks (see Chap. 3). The solar lights had LED lamps with no cable connection to eliminate the need for electrical certificates/ upgrades, a detection range of 7 m and operating time of 80–90 min with a fully charged battery. The work to be undertaken was priced, and the budget profile updated. Following a visit to the relevant properties, and completion of the required works, NCH completed a form (shown in Appendix C, Appendix Table 6.7b) detailing the dates for the different works undertaken. The completed form was then returned to the NCDP for monitoring.

For the duration of the pilot project, each morning an analyst from Nottinghamshire Police identified burglaries occurring in the test and control areas with pre-mapped polygons for highlighting the respective LSOA's borders (see Diagram 6.2). Based on this, an aerial map was utilised to identify those properties due to receive a

---

[8] Property marking, which had been commonly provided in the city since 2006, was deemed a successful preventive measure (NCDP 2015a). According to the protocol, it was due to be given out to burgled properties and their cocooning in the control areas together with advice about Neighbourhood Alert (NCDP 2014). In reality however the property marking pen and guidance were given to any properties which did not already have it regardless of whether they fell in the test or control areas. According to the pilot's Activity Log, nearly 60 percent of the burgled and cocoon properties in the test areas received the property marking pen and guidance (see later Table 6.3).

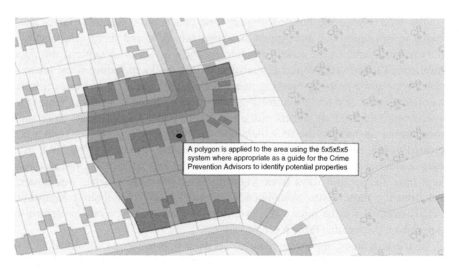

A polygon is applied to the area using the 5x5x5x5 system where appropriate as a guide for the Crime Prevention Advisors to identify potential properties

Diagram 6.2 Example of aerial map for identification of the cocoon area around a burgled property. (Source: NCDP 2015b)

cocoon (5 × 5 × 5 × 5) Home Security Assessment visit and, if the properties fell in each test area (Ward II Test and Ward IX Test), further security surveys and target hardening.

Following the visit to the burgled property, the PCU also conducted Security Surveys at neighbouring properties (5 × 5 × 5 × 5 as detailed above). The same process was followed as for the burgled property, and target hardening was offered (subject to the property owner granting permission) and a referral generated to NCH. The full list of properties given the opportunity to access the cocooning was sent to NCH at the point they were identified. NCH then had a checklist that was ticked off when a full referral came through to them. It was acknowledged that it might be difficult to make contact with the occupier of the property to conduct the assessment; therefore a letter was left after the initial visit which explained the pilot and asked the householder to contact the PCU to book an appointment to receive a Security Survey. In all previous cases, wherever possible, target hardening was delivered within 14 days of the original burglary taking place.

Following a burglary in the control area, the PCU completed the standard Nottinghamshire Police Home Security Assessment[9] at the burgled property within 14 days, if possible. The recommendations of the Home Security Assessment were shared with the occupier of the property as advice on how they could make their home safer, but they were not offered the target hardening security upgrades. The occupier was also offered advice about Neighbourhood Alert. Once the PCU conducted the Home Security Assessment in the burgled property, they then determined which properties would receive the cocooning visits (5 × 5 × 5 × 5). The PCU con-

---

[9] The Home Security Assessment form differs from the Security Survey conducted in the test areas and is included in Appendix C.

tacted the Safer Neighbourhoods Team for the relevant neighbourhood and gave them a list of properties to receive cocooning. The relevant Police Community Support Officers (PCSOs) then conducted the visits using the $5 \times 5 \times 5 \times 5$ model as specified by the PCU.[10] The cocooning included the Home Security Assessment, and the occupier was given advice as to alterations they could make to their property themselves which could make it more secure, but they were not given the target hardening offer. As with the burgled property, the occupier was offered advice about Neighbourhood Alert.[11] The PCSOs were workloaded to make up to three return visits to a property to conduct the Home Security Assessment. Thus, even in the control areas where the property owners were not offered target hardening, the level of advice and support they were offered exceeded that of burglary victims residing within neighbourhoods outside of the pilot areas.

There was a slow start to the project initiative arising from the fortunate occurrence of no burglaries within the city during the first 2 weeks of the pilot. In the early stages, referrals were slow (due to a disbelief of free security upgrades by owners of burgled dwellings and their neighbouring properties within the test areas) which initially made it seem that the project would last a lot longer than envisaged (since the Nottinghamshire OPCC funding was not used up at the anticipated rate). In order to encourage households to participate, the following steps were taken at the beginning: there was some flexibility on the 2-week deadline as the overall objective was to get as many houses protected with the four devices as possible. Since people were not always at home when the PCU first attended the property, the Neighbourhood Policing Team started doing follow-up visits in the evenings. Once take-up started to increase, however, it seemed that the budget (which was monitored weekly) would run out before enough homes had been protected to generate a sufficiently large enough sample size for the pilot evaluation. For this reason, the budget was extended by an additional £10,000 in November 2014, raising the overall target hardening budget to £50,000 of which £49,097 was eventually spent.

Property landlords/owners of rented homes were often not easily contactable in respect of granting permission for the required works, and when permission was granted, they were subsequently not easy to contact in order to secure access to undertake the required work. NCH also encountered some issues with planning work and paperwork. The solar lights provided for external lighting had limited time to power up, especially in the winter, and there were delays with regard to replacement door and windows manufacturing. Despite these issues, the pilot had immediate positive effects. It provided physical security in high near repeat burglary areas, helping owners who could not afford to replace windows and doors that did not lock or close properly. The solar lights were quickly installed and thus offered instant security. As with any crime prevention initiatives, the increased activity by the PCU may have enhanced public reassurance in the areas (Linning et al. 2017). All householders who took up the target hardening offer commented positively

---

[10] It should be noted that in the test areas, the Home Security Assessment of the neighbouring properties of a burgled dwelling was undertaken by the PCU and not the Safer Neighbourhoods Team PCSOs (as was the case in the control areas).

[11] Advice about Neighbourhood Alert does not seem to have been offered in the test areas according to the protocol.

about the workmanship and that they felt more secure once the works had been completed. Finally, the pilot provided an opportunity for close coordination of activities across NCDP, Nottinghamshire Police and NCH in a manner that portrayed excellent partnership working.

6.5   Evaluation

6.5.1   Pilot Data

The NCDP kept three sets of detailed activity files concerning properties in the pilot that contained the following information:

(a) An Activity Log containing details of all communication with households, including acceptance of timers to operate with internal lights, property marking and completion of the Security Survey (Appendix Tables 6.7a and 6.7b) during the first contact or leaving a letter to property owners or residents who were out at the time
(b) A file of any WIDE[12]-related target hardening upgrades, including front or rear doors, windows, solar lights and window locks, individual cost per device installed per property and dates of contact and completion
(c) Copies of the Data Capture Forms and Security Surveys (similar to Appendix Tables 6.6, 6.7a and 6.7b) with handwritten details of the properties and the residing households

From the above three, the first two are examined here to gauge whether the target hardening initiative worked in preventing burglaries in the test areas more than in the control areas and rest of the city.[13,14]

Table 6.3 presents data from the Activity Log of the 256 pilot properties, the vast majority of which (239 or 93.4 percent) were located within the test areas for reasons possibly relating to Safer Neighbourhoods Team PCSO workloads. As seen in Sect. 6.4.4, the PCU did not undertake the Security Surveys of the cocooning properties surrounding burgled victims in the control areas; these were done by the PCSOs of the Safer Neighbourhoods Teams of the relevant neighbourhoods. It seems therefore that the Activity Log was kept by the PCU which attempted to con-

---

[12] The 'I' from WIDE has been crossed out here since internal lights operating on a timer were not part of the initiative in the test neighbourhoods but across both control and test neighbourhoods.

[13] The last one requires labour intensive work to prepare it for analysis, consequently the household-related contextual information of the pilot remains for future study.

[14] Additional analyses which distinguish areas with high burglary rates (defined as above two standard deviations of the mean citywide burglary rate), those bordering the test areas and the rest of the city have been completed, but they are not presented here for economy. These analyses showed considerable diffusion of benefits from the pilot target hardening initiative in the test to the neighbouring areas.

Table 6.3 Summary of the Activity Log for the burglary pilot project

	Number of properties		
	Test areas (%)	Control areas (%)	Total
Security Survey			
Accepted	68 (28.5) including three indicating no required upgrades	4 (23.5)	72 (28.1)
Refused	12 (5.0)	1 (5.9)	13 (5.1)
No required upgrades	9 (3.8)	4 (23.5)	13 (5.1)
No immediate contact – letter left	119 (49.8)	7 (41.2)	126 (49.2)
Missing data	31 (13.0)	1 (5.9)	32 (12.5)
Property marking			
Accepted	143 (59.8)	5 (29.4)	148 (57.8)
Refused	15 (6.3)	3 (17.6)	18 (7.0)
Missing data	81 (33.9)	9 (52.9)	90 (35.2)
Timers			
Accepted	52 (21.8)	4 (23.5)	56 (21.9)
Refused	104 (43.5)	4 (23.5)	108 (42.2)
Missing data	83 (34.7)	9 (52.9)	92 (35.9)
Number of properties in Activity Log	239	17	256
Upgrades (some element or all) for achieving at least WIDE security[a]	71 (29.7)	–	–

Note: [a]The file documenting the WIDE-related target hardening upgrades and costs per property does not include internal lights on a timer. Timers are however mentioned in the Activity Log, a file which includes a small number of properties in control areas

tact owners of all burgled properties in both test and control areas and was responsible for cocooning properties in the test areas (see Diagram 6.1).

The PCSOs could not establish contact with residents of about half of these properties (see last column and fourth row of figures; Table 6.3). The Security Survey was completed for just over a quarter of these (see last column and first row of figures; Table 6.3) or just over half the ones which were immediately contacted. The missing data refer to cases where the Activity Log was blank, containing, respectively, no information about the Security Survey, property marking and timers.

Readers may have noticed that whilst property marking was only mentioned in relation to control areas when discussing the planning stages of the pilot (Sect. 6.4.4), it does not appear in the flow chart (Diagram 6.1). In addition, it was not included in the original research evidence the pilot target hardening initiative aimed to test (Tseloni et al. 2014; Sect. 6.4.2 of this chapter). During the actual implementation, the occupier in both the test and control areas was offered property marking, which has been given out to all Nottingham residents since around 2006 (NCDP 2015a). According to the Activity Log, 166 properties (or 65 percent of the total number in the log from both test and control areas where they had possibly not been in receipt of it already) were offered property marking. Of these, 148 properties (89.2 percent out of 166 or 57.8 percent of all properties in the Activity Log) subsequently received it with the majority being in the test areas.

Timers are another security device included in the Activity Log and therefore offered to occupiers in both test and control areas notwithstanding the pilot's design (and absent in the flow chart; Diagram 6.1). In particular, the timers were offered to 164 properties (or just under two thirds of the properties in the log) of which 56 (just over a third, 34.1 percent of those offered or 21.9 percent of all properties in the Activity Log) accepted them at comparable ratios between test and control areas. Overall nearly a third of the Activity Log had no information about property marking and timers but fewer missing data with regard to immediate contact with the occupiers and Security Survey completion.

The last row of Table 6.3 denotes the number of properties that received the offered security upgrades with the number taken from the second file of any WIDE-related target hardening which was mentioned in the beginning of this section. Overall, 71 homes were target hardened in about equal numbers across the two wards (36 in Ward II Test and 35 in Ward IX Test area).[15] A further 168 households within these test localities failed to respond to the PCU letter or refused the target hardening offer. The poor take-up from homeowners was possibly due to a 'too good to be true' ethos and suspicion regarding works offered. About a third of burglaries that elicited a response were because of entry through open windows or doors (i.e. insecurity type burglaries, which present a challenge for target hardening approaches). It is worth highlighting here that it was not possible to discern whether homes had additional security features to those prescribed within WIDE.[16] This is also true for security outside the set examined in the CSEW, including property marking which was recorded only for households it was offered to but there were no records of which households already had it.

The pilot target hardening initiative described in this chapter is far from constituting a randomised control experiment. For example, timers were offered to few and accepted by four control areas homes (see Table 6.3) – thus the I in WIDE is not strictly speaking part of the initiative in the test areas, and it will be referred to as W(I)DE in the following discussion. In addition, property marking which has not been documented in (and therefore examined with) national data from the CSEW was also part of the package offered to both test and control areas homes in Nottingham.

---

[15] The difference between the 71 target hardened properties in the test areas and the number of properties in the Activity Log that had the Security Survey and required upgrades (65, calculated as 68 minus 3 from the first row and column of figures in Table 6.3) refers to properties of which the owners responded to the letter left and accepted the offered upgrades at a subsequent to first contact point in time.

[16] For a full list of security devices identified within the national data in the CSEW, see Chap. 4.

6.5.2 Evaluation Data: Did It Work?

The pilot ran for 4 months in total, from the end of September 2014 (due to slow start-up) to the end of January 2015. In order to evaluate whether the W(I)DE target hardening worked in reducing burglaries, the NCDP provided the research academic team with citywide police recorded anonymised monthly burglary data from September 2013 to January 2016, covering a year before and a year after the initiative. Although the two wards that were selected to participate in the pilot had comparable household profiles, the precise number of resident households in each test and control LSOA differs. The burglary figures were therefore transformed into burglaries per 1000 households to render the police recorded burglary incidents comparable across neighbourhoods (LSOAs) for the entire city. Furthermore, the 3-month moving average was calculated following classic mathematical transformation and standard police analyst process to smooth monthly crime trends.[17] The resulting 3-month moving average burglary rate per 1000 households in the test, control, and the rest of the city is given in Fig. 6.1.

The W(I)DE pilot target hardening initiative reduced burglaries in the test area right from about 2 months into the pilot (December 2014). The success of the W(I) DE pilot target hardening initiative in the test areas is perhaps emphasised by the

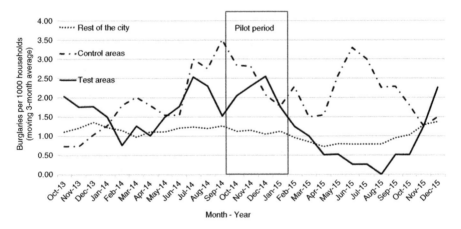

Fig. 6.1 Burglaries per 1000 households in test, control and other areas, September 2013–January 2016 (moving 3-month average)

---

[17] As a result of this process, the end months of the original burglary data, September 2013 and January 2016, were subsumed within those of the following and previous month, respectively. Therefore the data presented here concern 3-month moving average burglary rates per 1000 households from October 2013 to December 2015. For example, the burglary rate for October 2013 was calculated as the mean value (or the sum divided by three) of the burglary rates for September, October and November 2013; that of November 2013 as the mean value of the burglary rates for October, November and December 2013; and so on and so forth.

Table 6.4a Total number of burglaries per 1000 households before (October 2013–September 2014) and after (January–December 2015) the pilot and percent change in test, control and other areas excluding the pilot period

	October 2013–September 2014[a]	January–December 2015[b]	% Change
Other areas	14.19	11.50	−18.96
Control areas	21.78	25.08	15.15
Test areas	19.77	10.12	−48.81
Net % reduction in burglaries between test and control areas			−63.95

Notes: [a]The 3-month moving average burglary rate for September 2014 includes burglaries in August 2014 which is outside the pilot period
[b]The 3-month moving average burglary rate for January 2015 includes burglaries in February 2015 which is outside the pilot period. In effect, the available data did not allow for calculating 12-month burglaries

Table 6.4b Total number of burglaries per 1000 households during the calendar years before and after the mid-point of the pilot and percent change in test, control and other areas including the pilot period

	January–December 2014	January–December 2015	% Change
Other areas	13.83	11.50	−16.86
Control areas	27.00	25.08	−7.11
Test areas	21.09	10.12	−52.00
Net % reduction in burglaries between test and control areas			−44.90

fact that burglary in the control areas remained high or increased during and for the year following the pilot. However, a year later in November 2015, burglary rates in the test areas started surpassing those in the control areas. It is important to remember that the goal of target hardening in the test areas was to prevent repeat and near repeat burglaries. The later increase in burglary in the test areas may therefore simply be due to a different set of properties being victimised (rather than the target hardened properties). Table 6.4 (panels a and b) presents the changes in 3-month moving average burglary rates a year before and a year after the pilot.

If we focus upon the outcomes arising before and after the initiative (but excluding the pilot period), then the net percentage reduction in burglaries between the test and control areas was just under 64 percent (Table 6.4a). The scale of the net reduction in burglaries between the test and control areas drops to just under 45 percent once the pilot period is included within the analysis and data are compared across the two calendar years (Table 6.4b). As with any evaluation, the cut-off points that defined the 'before' and 'after' timelines is important – and this is demonstrated by the different comparator burglary trends in the control areas which increased when the pilot period was excluded (+15.15 percent change [Table 6.4a]) and dropped slightly when it was included (−7.11 percent change [Table 6.4b]). The difference may indicate some diffusion of benefits from the pilot target hardening initiative to the control areas, especially as crime prevention advice was given to burgled properties and their surrounding neighbours in these areas.

From a cost-benefit perspective, the pilot target hardening initiative had a clear benefit. Whilst the final average cost of target hardening was £691.50 per property, the actual amount spent ranged from £56 for properties requiring only a solar external light (which covered a considerable number of target properties) to about £4000 for houses which, in addition to external lights, required replacement doors and a number of windows. Excluding the two properties which required many replacement windows, the average cost of the security upgrades was £603.25 per property. This is well below the cost of a single burglary incident, estimated at £2300 nearly 20 years ago (Brand and Price 2000), and potentially costing more than this sum in 2014, when the pilot target hardening initiative took place.

6.6 Discussion and Conclusion

As identified at the start (Sect. 6.2), evaluating the impact of any crime reduction initiative within specific localities through a 'before and after' lens that compares changes in crime levels within test and control areas is fraught with methodological problems. The evaluation design is predicated upon the ability of the researchers to identify the precise, and independent, impact of the initiative in question (Bowers et al. 2003; Johnson and Bowers 2004b) – and to successfully control for any other potential factors that may have contributed to any identified reduction in crime levels (Allatt 1984). In the case of the Nottingham target hardening pilot initiative to reduce burglary, the results presented above are designed to yield initial findings and reflections on the implementation of W(I)DE measures within two high repeat and near repeat burglary hotspots. Built into the evaluation design is an assumption that the number of police recorded burglaries identified within the test and control areas is the product of common previous victimisation and sociodemographic household characteristics that might shape any under-reporting of burglary incidents within each area. Equally, to the knowledge of the researchers, there were no other crime reduction interventions, or changes to police patrol routines, that may have directly or indirectly impacted upon burglary levels within the test and control areas.

Potential contamination of both the control and other neighbourhoods outside the test areas may, however, have been an issue. There was an extensive programme of awareness raising by Nottinghamshire Police and the NCDP around the issue of target hardening measures designed to reduce burglary through an extensive advertising campaign that featured across the entire bus network within Nottingham during and after the pilot period. Although the adverts in question did not specifically outline all of the WIDE measures identified as being effective by the researchers, they did highlight the need to lock doors and windows and heavily focused on the potential of lighting as a means of deterring would be burglars. Furthermore, the more detailed crime prevention information provided on the Nottinghamshire Police website that backed up the advertising campaign did outline a range of Secured by Design measures that included all of the components of WIDE. Thus it is possible that householders living within test and control areas (but who were not victims of a burglary nor included in any cocooning activities) may have taken it upon themselves to implement (at their own cost) the crime reduction measures advocated by WIDE. This is not quite in the same ilk as the potential anticipatory impact phenomena identified by Bowers and Johnson (2003) – but it is within the same ballpark.

To this end, the empirical analysis and discussion above represent an initial evaluation of a burglary reduction initiative driven by the findings relating to enhanced security presented earlier within the book. What can be determined is that following the implementation of the missing WIDE security measures within some victimised properties (and the immediate surrounding dwellings), burglary rates dropped significantly within the test areas within the city. In design, the implementation and evaluation of the Nottingham Pilot Burglary Reduction Initiative sought to meet the accepted minimum Level 3 standard within the Maryland Scale of Scientific Methods for Evaluating Crime Prevention. As identified above, 'comparison between multiple units with and without the program, controlling for other factors, or using comparison units that evidence only minor differences' (Sherman et al. 1998, p. 4) proved to be beyond both the scale and financial remit of the initiative and the control of the evaluators. Furthermore, in essence, the findings to date relate to the micro impact of enhanced security within the specific context of burglary hotspots within a small proportion of neighbourhoods within the city. Until the scope of the implementation of this approach to target hardening can be extended within and beyond the original test sites, the findings presented here constitute 'what's promising' rather than 'what (definitively) works'.

Appendix C

C.1   The Nottingham Crime and Drugs Partnership (NCDP)

The following text has been provided by the NCDP and briefly delineates the partnership at the time of writing (2017) which is almost 3 years after the pilot.

C.1.1   Statement

The CDP is the local Community Safety Partnership. The Crime and Disorder Act 1998 established Community Safety Partnerships, placing a statutory duty on public authorities (referred to as Responsible Authorities[18]) to co-operate in order to formulate and implement a strategy for the reduction of crime and disorder and for

---

[18] The Responsible Authorities are the Local Authority, Nottinghamshire Police, Nottinghamshire Probation, Nottinghamshire Fire and Rescue Service and Clinical Commissioning Group.

combating substance misuse in the area. The CDP produces an annual Partnership Plan which performs this function and gives strategic direction to the partnership.

C.1.2   For Recognition

The CDP is supported by a small partnership team of Community Safety Officers, commissioners and analysts who produce the necessary documents to enable the partnership to discharge its statutory obligations.

C.1.3   History

The Crime and Drugs Partnership was formed in 2005 after the merger of the Nottingham Drug and Alcohol Action Team (DAAT) and the Nottingham Crime and Disorder Reduction Partnership. The merger recognised the inextricable link between crime and drug use (particularly heroin and crack cocaine). The CDP, by virtue of the fact that it incorporates the DAAT, is distinct from other CSPs in the country, and this unique model has provided excellent opportunities for partnership working.

Crime and drug-related offending in Nottingham has dropped significantly over recent years. Since 2002 crime in the city has reduced by over 60% (representing over 40,000 less crimes).

The nature of crime and substance misuse has changed significantly since the formation of the partnership in 2005. As crime has fallen in Nottingham, the profile has changed considerably so that offending is now more evenly distributed across the city and across a wider range of offence types.

Partners have responded with the development of case working approaches to manage repeat offenders and victims such as the Complex People's Panel, Multi-Agency Risk Assessment Conferences (addressing high-risk domestic violence), Young People's Panel and the Priority Families programme.

As partnership funding has reduced for the delivery of local community safety initiatives and individual partners have subsumed the work into their core business, the CDP has significantly reduced its investment in the direct provision of community safety measures. Consequently the function of the partnership has evolved to give greater focus to coordination and facilitation in locality working and more effective use of partner's core resources in people, systems and process improvements.

C.2   Selected Protocol and Home Security Assessment Templates

Appendix Table 6.5   Burglary Pilot (Target Hardening) Process

Burglary called in to control room in one of the four neighbourhoods (LSOAs) (one control and one test area in each ward)		
Marker is added by the control room, which notifies pre-crime unit and Safer Neighbourhoods Team that the burglary has taken place and will need an assessment visit		
Pre-crime unit officer attends the property and completes the template as to what security measures are already in place The visit takes place within 48 h of the burglary taking place		
Does the property fall into one of the control areas?		
If yes	If no (test area)	
Citizen is offered crime prevention advice, Neighbourhood Alert and guidance as to what crime prevention measures could be added to the property	Permission is sought to target harden the property from the landlord	Pre-crime unit visit neighbouring properties 5 x 5 x 5 x 5 to conduct Home Security Assessments. Then the process is followed as for the burgled property
PCSOs conduct cocooning visits and conduct Home Security Assessments to neighbouring properties $5 \times 5 \times 5 \times 5$ Citizen is offered crime prevention advice, Neighbourhood Alert and guidance as to what crime prevention measures could be added to the property	Landlord is written to by pre-crime to request permission for works to be completed	
	Letter is signed to agree to works done and returned to pre-crime unit	
	If permission granted:	
	A referral is made from pre-crime unit to Nottingham City Homes (single point of contact to be identified) with a list of physical interventions to be completed at the property	
	Nottingham City Homes joiner visits the property and makes the suggested interventions. Joiner completes a template as to what works have been carried out and returns to Nottingham City Homes **This is done within 14 days of the burglary taking place**	
	Nottingham City Homes collate templates and return with expenditure to the Nottingham Crime and Drugs Partnership. This is returned on a monthly basis	

Appendix Table 6.6 Data Capture Form

Partnership burglary pilot	
The Crime and Drugs Partnership are co-ordinating a pilot project to assess the effectiveness of different home security measures. To this end, we are conducting surveys of properties within the pilot area which have suffered a burglary or are neighbouring a burgled property. This is to determine what security measures are currently in place and to make recommendations as to how the property could be made more secure	
Date assessment completed:	
Time assessment completed:	
Name and position of person completing assessment:	
Address of property:	
Tenure of property: (owner occupier/private rented/social housing tenant)	
Type of accommodation: (detached/semi-detached/terraced/flat (floor))	.
Landlords details (name/address/phone)	
Does landlord consent to works being undertaken?	
Composition of household:	
Number of adults	
Number of children	
Ethnic origin of household representative	
Number of cars owned by the household	
Is this a Neighbourhood Watch area?	
If yes, does the household participate?	
How many hours per day is the property usually left empty (please circle):	0 h
	Less than 3 h
	Less than 7 h
	7 or more hours
The information you give here will be used by the Nottingham Crime and Drugs Partnership, Nottingham City Homes and Loughborough University and will help us to evaluate the effectiveness of the project. Your data will be stored securely in accordance with the requirements of the Data Protection Act and will not be shared with any third parties other than those listed above	
Please sign to confirm that you consent to this:	

Appendix Table 6.7a Security Survey

Location	Specify requirements and include any additional recommendations (e.g. reinforcing internal doors)	Cost
Front exit door, i.e. new PAS 24 2012 door		
Rear exit door, i.e. new PAS 24 2012 door		
Security of windows to ground floor, i.e. new window or extra locks		
Security of windows to upper floors, i.e. new window or extra locks		
Lighting	**External on sensor?** *Please be clear where to erect lights and how many*	
	Internal on timer?	
Alarm system[a]	*For guidance only. Alarms are not offered as part of this project*	
Outbuildings [a]	*For guidance only. Alarms are not offered as part of this project*	
Fencing[a]	*For guidance only. Alarms are not offered as part of this project*	
Any other recommendations – please specify[a]		

[a]These rows were included in the protocol but after were omitted from the forms which were actually used during the implementation of the pilot project following academic advice to avoid confusion and unnecessary effort

Appendix Table 6.7b Details of security works carried out

Referral number:	
Address of property:	
Date and time completed:	
Name of person completing form:	

Location	Specify requirements and include any additional recommendations (e.g. reinforcing internal doors)	Cost	Date added
Front exit door, i.e. new PAS 24 2012 door			
Rear exit door, i.e. new PAS 24 2012 door			
Security of windows to ground floor, i.e. new window or extra locks			
Security of windows to upper floors, i.e. new window or extra locks			
Lighting	External		
	Internal		
Any other works carried out – please specify			
Total cost of works carried out:			

This form to be returned to Nottingham Crime and Drugs Partnership

References

Allatt, P. (1984). Residential security: Containment and displacement of burglary. *Howard Journal, 23*(2), 99–116.

Bennett, T., Holloway, K., & Farrington, D. (2006). Does neighbourhood watch reduce crime? A systematic review and meta-analysis. *Journal of Experimental Criminology, 2*, 437–458.

Bowers, K., & Johnson, S. (2003). *The role of publicity in crime prevention: Findings from the Reducing Burglary Initiative.* London: Home Office.

Bowers, K., & Johnson, S. (2005). Domestic burglary repeats and space-time clusters: The dimensions of risk. *European Journal of Criminology, 2*(1), 67–92.

Bowers, K., Johnson, S., & Hirschfield, A. (2003). *Pushing back the boundaries: New techniques for assessing the impact of burglary schemes.* London: Home Office.

Brand, S., & Price, R. (2000). *The economic and social costs of crime* (Home Office Research Study 217). Economics and Resource Analysis Research, Development and Statistics Directorate. London: Home Office.

Brunton-Smith, I., & Sturgis, P. (2011). Do neighbourhoods generate fear of crime? An empirical test using the British Crime Survey. *Criminology, 49*(2), 331–369.

College of Policing. (2018). Crime and policing in England, Wales and Northern Ireland – Police. UK. https://data.police.uk/data.

Coupe, T., & Blake, L. (2006). Daylight and darkness targeting strategies and the risks of being seen at residential burglaries. *Criminology, 44*(2), 431–464.

Cummings, R. (2006). 'What If': The counterfactual in program evaluation. *Evaluation Journal of Australasia, 6*(2), 6–15.

Department for Communities and Local Government (DCLG). (2015). *The English indices of deprivation 2015.* London: DCLG.

Ekblom, P., Law, H., Sutton, M. (1996). *Safer cities and domestic burglary* (Home Office Research Study 164). London: Home Office.

Farrell, G. (1992). Multiple victimisation: Its extent and significance. *International Review of Victimology, 2*, 85–102.

Forrester, D., Chatterton, M., & Pease, K. (1988). *The Kirkholt burglary prevention project, Rochdale.* London: Home Office.

Foster, J., & Hope, T. (1993). *Housing, community and crime: The impact of the priority estates project* (Home Office Research Study 131). London: Home Office.

Griswold, D. (1984). Crime prevention and commercial burglary: A time series analysis. *Journal of Criminal Justice, 12*, 493–501.

Hope, T. (2004). Pretend it works: Evidence and governance in the evaluation of the reducing burglary initiative. *Criminal Justice, 4*(3), 287–308.

Hunter, J., & Tseloni, A. (2016). Equity, justice and the crime drop: The case of burglary in England and Wales. *Crime Science, 5*(3), 1–13.

Ignatans, D., & Pease, K. (2016). On whom does the burden of crime fall now? Changes over time in counts and concentration. *International Review of Victimology, 22*(1), 55–63.

Johnson, S., & Bowers, K. (2004a). The stability of space-time clusters of burglary. *British Journal of Criminology, 44*(1), 55–65.

Johnson, S., & Bowers, K. (2004b). The burglary as clue to the future. The beginnings of prospective hot-spotting. *European Journal of Criminology, 1*(2), 237–255.

Johnson, S., Bowers, K., Jordan, P., Mallender, J., Davidson, N., & Hirschfield, A. (2004). Evaluating crime prevention scheme success: Estimating 'outcomes' or how many crimes were prevented. *Evaluation, 10*(3), 327–348.

Johnson, S., Bernasco, W., Bowers, K., Elffers, H., Ratcliffe, J., Rengert, G., & Townsley, M. (2007). Near repeats: a cross national assessment of residential burglary. *Journal of Quantitative Criminology, 23*, 201–219.

Jones, A., Valero-Silva, N., & Lucas, D. (2016). *The effects of 'Secure Warm Modern' homes in Nottingham: Decent Homes impact study.* Nottingham: Nottingham City Homes http://www.nottinghamcityhomes.org.uk/EasySiteWeb/GatewayLink.aspx?alId=2472. Accessed 12 June 2018.

Linning, S. Eck, J., & Bowers, K. (2017, November 15–18). The temporal effects surrounding place-based crime prevention interventions. Paper presented at the American Society of Criminology 73rd annual meeting, Philadelphia.

Lurigio, A., & Rosenbaum, D. (1986). Evaluation research in community crime prevention: A critical look at the field. In D. Rosenbaum (Ed.), *Community crime prevention: does it work?* (pp. 19–44). Beverley Hills: Sage.

Millie, A., & Hough, M. (2004). *Assessing the impact of the Reducing Burglary Initiative in Southern England and Wales*. London: Home Office – Second Edition.

NOMIS. (2018). Population estimates – Local authority by single year of age 2016. NOMIS Official Labour Market Statistics/Office for National Statistics. https://www.nomisweb.co.uk.

Nottingham Crime and Drugs Partnership (NCDP). (2014). Repeat and Near Repeat Burglary Pilot Project Protocol (Restricted report). Nottingham: Crime and Drugs Partnership.

Nottingham Crime and Drugs Partnership (NCDP). (2015a). Strategic assessment 2015/2016. http://www.nottinghamcdp.com/wp-content/uploads/2017/06/REVISED-FINAL-Strategic-Assessment-2015-16.pdf. Accessed 12 June 2018.

Nottingham Crime and Drugs Partnership (NCDP). (2015b). Repeat and near repeat burglary pilot project: Operation Paddlewood. Burglary and security conference presentation, Galleries of Justice, Nottingham, 21 January 2015.

Office for National Statistics (ONS). (2017a). *Workless households for regions across the UK: 2016*. London: Office for National Statistics.

Office for National Statistics (ONS). (2017b). *Recorded crime data by community safety partnership area 2017*. London: Office for National Statistics.

Office for National Statistics (ONS). (2018). *Mid-year population estimates 2016*. London: Office for National Statistics.

Osborn, D., & Tseloni, A. (1998). The distribution of household property crimes. *Journal of Quantitative Criminology, 14*, 307–330.

Osborn, D., Ellingworth, D., Hope, T., & Trickett, A. (1996). Are repeatedly victimised households different? *Journal of Quantitative Criminology, 12*, 223–245.

Pease, K. (1991). The Kirkholt Project: Preventing Burglary on a British Public Housing Estate. *Security Journal, 2*, 73–77.

Pease, K. (2009). The carbon cost of crime and its implications. An ACPO Secured by Design research project. http://www.securedbydesign.com/wp-content/uploads/2014/02/The-Carbon-Cost-of-Crime.pdf. Accessed 17 Dec 2015.

Polvi, N., Looman, T., Humphries, C., & Pease, K. (1991). The time course of repeat burglary victimisation. *British Journal of Criminology, 31*, 411–414.

Rosenbaum, D. (1987). The theory and research behind neighbourhood watch: Is it a sound fear and crime reduction strategy. *Crime & Delinquency, 33*(1), 103–134.

Ross, N., & Pease, K. (2007). Community policing and prediction. In T. Williamson (Ed.), *Knowledge-based policing* (pp. 305–321). Chichester: Wiley.

Santos, R., & Santos, R. (2015). Practice-based research: Ex post facto evaluation of evidence-based police practices implemented in residential micro-time hot spots. *Evaluation Review, 39*(5), 451–479.

Sherman, L., Gottfredson, D., Mackenzie, D., Eck, J., Reuter, P., & Bushway, S. (1998). *Preventing crime: What works, what doesn't, what's promising*. Washington, D.C.: US Department of Justice, Office of Justice Programs.

Sidebottom, A., Tompson, L., Thornton, A., Bullock, K., Tilley, N., Bowers, K., & Johnson, S. (2015). *Gating alleys to reduce crime: A meta-analysis and realist synthesis. What works: Crime reduction systematic review series*. London: What Works Centre for Crime Reduction http://whatworks.college.police.uk/About/Documents/Alley_gating.pdf. Accessed 11 Apr 2018.

Townsley, M., Homel, R., & Chaseling, J. (2003). Infectious burglaries: A test of the near repeat hypothesis. *British Journal of Criminology, 43*, 615–633.

Trickett, A., Osborn, D. R., Seymour, J., & Pease, K. (1992). What is different about high crime areas? *British Journal of Criminology, 32*, 81–89.

Tseloni, A. (2006). Multilevel modelling of the number of property crimes: Household and area effects. *Journal of the Royal Statistical Society: Series A (Statistics in Society), 169*(2), 205–233.

Tseloni, A., & Rogerson, M. (2018). Estrategias para la prevención de la revictimización. En M. Tenca, y E. Mendez Ortiz (Coordinadores) *Manual de Prevención del Delito y Seguridad Ciudadana* (pp. 251-276). Buenos Aires: Ediciones Didot [Tseloni, A., & Rogerson, M. (2018). Strategies for preventing repeat victimisation. In M. Tenca & E. Mendez Ortiz (Eds.) *Handbook of crime prevention and citizen security* (pp. 251–276). Buenos Aires: Ediciones Didot].

Tseloni, A., Thompson, R., Grove, L., Tilley, N., & Farrell, G. (2014). The effectiveness of burglary security devices. *Security Journal, 30*(2), 646–664. DOI: 10.1057/sj.2014.30.

References

Chapter 7
The Role of Security Devices Against Burglaries: Findings from the French Victimisation Survey

Amandine Sourd and Vincent Delbecque

Abbreviations

CAPI	Computer-assisted personal interviewing
CASI	Computer-assisted self-interviewing
CESDIP	Centre d'études sociologiques sur le droit et les institutions pénales
CSEW	Crime Survey for England and Wales
CVS	Cadre de Vie et Sécurité
INSEE	The National Institute of Statistics and Economic Studies
ONDRP	French National Observatory of Crime and Criminal Justice
SSMsi	French Ministerial Statistical Department for Internal Security

7.1 Introduction

There are many benefits in the development of international work in criminology in terms of complementary, discussion, comparison of results and deepening knowledge. Since the beginning of the 1980s, the issue of burglaries and their determinants has given rise to an abundance of academic literature, in particular in Great Britain, the USA and the Netherlands. In prior research, particular attention has been paid to assess causing factors of burglaries in order to better prevent them. On a conceptual level, the routine activity theory (Cohen and Felson 1979) and the lifestyle theory (Hindelang et al. 1978) have been significantly called on to give a reference framework to these empirical analyses.

Several issues and perspectives on the subject of burglaries have been developed during the last decades. The analysis of trends in such property crime is the subject of regular publications, by using victimisation surveys or police-recorded data (see, e.g. Office for National Statistics 2016; National Observatory of Crime and Criminal Justice (ONDRP) 2015; Bureau of Justice Statistics 2016; Australian Institute of

A. Sourd · V. Delbecque (✉)
National Observatory of Crime and Criminal Justice, Paris, France
e-mail: vincent.delbecque@gmail.fr

© Springer Nature Switzerland AG 2018
A. Tseloni et al., *Reducing Burglary*,
https://doi.org/10.1007/978-3-319-99942-5_7

Criminology 2016; Morgan and Clare 2007). The trends have also been studied as part of a crime-drop context, which has emerged since the mid-1990s (Aebi and Linde 2010; van Dijk et al. 2012).

Spatial and environmental analysis is a significant technique for addressing the burglary phenomenon. The work which developed these analyses has stressed the significance of environmental factors as well as the concentration of burglaries in criminal hotspots linked to ecological factors (see, e.g. Lynch and Cantor 1992; Miethe and David 1993; Rountree and Land 2000; Rountree et al. 1994; Ceccato et al. 2002; Bernasco and Nieuwbeerta 2005) and distance patterns of perpetrators towards the crime scene referencing the 'journey to crime' notion (Brantingham and Brantingham 1975; Capone and Nichols 1975; Gabor and Gottheil 1984; Rhodes and Conly 1981).

Although a large proportion of the literature on burglaries deals with the issue from the perspective of victimisation, a few works have explored the phenomenon from the angle of the perpetrators, their motivations, their choices and their views on the evolution of the phenomenon (see, e.g. Cromwell et al. 1991; Nee and Meenaghan 2006; Kuhns et al. 2012; Brown 2015). In particular, these works enable the opportunist nature of burglary to be highlighted, as described by Cohen and Felson (1979) in the routine activity theory.

As individual features for protection against burglaries, security devices have been the focus of particular attention. Several pieces of work carried out in this field come to the conclusion that housing units equipped with security devices are burgled less often (Pease and Gill 2011; van Dijk 2008; Mayhew et al. 1993; Bettaieb and Delbecque 2016). When the number and the combination of devices are taken into account, the findings in terms of effectiveness are more precise. Based on Murphy and Eder (2010), the Crime Survey for England and Wales (CSEW) groups security devices into four categories: no security, less than basic, basic and enhanced. Housing units with no security are found to be more at risk than those with basic or enhanced security (Tilley 2009; Flatley et al. 2010). Finally, it is shown that security devices are more efficient when used in combination rather than individually (Tseloni et al. 2014).

This chapter follows this line and aims at deepening the assessment of security devices efficiency. Based on the literature previously mentioned, our analysis aims at deepening the assessment of security devices efficiency and to refine the knowledge on burglaries. Firstly, this study is based on French victimisation data which, until now, has little been exploited for research purposes (Bettaieb and Delbecque 2016; Perron-Bailly 2013). Secondly, we are focusing on security devices as protection features for housing units against burglaries. These devices are analysed both individually and in combination (Tseloni et al. 2014). The main contribution of this work is the approach to burglary, not as a single and uniform event but as a process which can be sequenced.

Indeed, in the study of burglaries, their causes and characteristics, the implicit assumption is generally made that a burglary is a binary event during which a housing unit is either a victim or not. Furthermore, even though a few works acknowledge attempted burglaries, these are generally not included in the analysis. Yet, on the contrary, we assume that it is essential to take into account attempts, not only as a failed burglary but also as evidence of the targeting of the housing unit by a perpetrator. When a distinction is made between an attempted break-in (the perpetrator targeted the accommodation and tried to break-in), a forced entry (the perpetrator did succeed in breaking in) and a theft (the perpetrator managed to steal items following the entry), different characteristics are revealed in terms of security equipment (van Kesteren et al. 2000; Tseloni et al. 2014) and the perpetrators' degree of preparation (Hough 1987).

In this line, the present study offers a detailed analysis of the effectiveness of households' security features against burglaries. Here burglary is studied as a three-step sequence – the targeting, the forced entry and the theft – and not as a homogenous whole. In France, burglary has no legal definition per se. However, the Penal Code outlines specific cases of thefts when they occur in a dwelling (Penal Code, Art.311-4 al.6) and when they are following a damage to the dwelling (Penal Code, Art.311-4 al.8). Besides, in common French language, burglary is defined as a theft in a property where the perpetrator has broken in, entered by climbing or picking a lock by any means.[1] The same definition is used by police services to describe a burglary. In this way, this breach is based on a series of actions carried out by the perpetrator and not one single action. Relying on the routine activity theory, we assume that most perpetrators choose their target rationally based on the information they have at their disposal concerning the target, the place and the time of the burglary (Bernasco and Luykx 2003). This information is available before they commit the deed; however, they will need to re-evaluate it at the time of the burglary (during attempted break-ins or thefts). We follow the hypothesis proposed by Bernasco and Luykx (2003, p. 985) according to which 'burglars' target selection is a sequential decision process' during which factors related to the housing's environment, the potential gain and the risk of failure are constantly being re-evaluated. This re-evaluation is based on whether the information is easily accessible or not. Thus we assume that some of the information is directly accessible by the perpetrator. This is the first information to be taken into account and relates to the environment of the housing and its external appearance. On the other hand, the housing units' specific factors are more difficult for the perpetrator to perceive as they are less accessible or visible.

The concept of information held by the perpetrator has in particular been dealt with through the prism of households which have experienced repeat burglaries. Indeed, it has been shown that information regarding burgled housing was spread verbally (Bernasco and Luykx 2003). Furthermore, these repeat victimisations took place in a relatively short space of time (Lammers et al. 2015). Cases of repeat victimisations change the quality and the quantity of information held by the perpetrator. Following the first burglary, both environmental and housing unit-specific factors are known by the perpetrator, thus enabling him/her to better evaluate the situation during a subsequent burglary of the same place. We therefore investigate further the case of repeat victimisation and more specifically the effect of security devices in those specific cases.

---

[1] Larousse: http://www.larousse.fr/dictionnaires/francais/cambriolage/12485

The chapter is organised as follows. Section 7.2 presents in detail the data sources and the specific modelling of the different stages of the burglary process; Sect. 7.3 describes the results of the experiments, whilst Sect. 7.4 discusses these results in regards to both the findings and the limitations.

7.2 Source, Contextual Data and Modelling

7.2.1 Source

For this study, we use data from the French victimisation survey. France has had several victimisation survey since the 1980s, when the first surveys were conducted by the *Centre d'études sociologiques sur le droit et les institutions pénales* (CESDIP). Different victimisation surveys have been conducted both at the national and local level, as well as on specific topics such as violence against women. The current survey, named *Cadre de Vie et Sécurité (CVS)*, has been conducted every year since 2007, designed and monitored by the French National Institute of Statistics and Economic Studies (Insee) and the French National Observatory of Crime and Criminal Justice (ONDRP). The survey is conducted annually in Metropolitan France (Mainland France including Corsica). Sweeps in 2011 and 2015 have also been conducted in the French West Indies, Guyana and Reunion Island. Data is collected through computer-assisted personal interviewing (CAPI) and computer-assisted self-interviewing (CASI) for sensitive questions. All respondents are asked questions on victimisation and perceptions of insecurity.

The French victimisation survey CVS has a random sample selection of around 22,000 households with a response rate between 75 and 80 percent every year. This leads to an average sample of around 15,000 respondents aged 14 and over for certain parts of the survey and of around 12,000 respondents aged between 18 and 75 for other parts. Households are randomly selected from a master sample that is derived from the population census. All of the data is weighted and therefore representative of the households and population of Metropolitan France.

A specific feature of the *Cadre de Vie et Sécurité (CVS)* is that the survey is divided into three questionnaires:

- A questionnaire on household victimisation and the perception of insecurity, answered by the head of household (aged 14 and over)
- A questionnaire on personal victimisation and the perception of insecurity, answered by a randomly selected member of the household (aged 14 and over)
- A questionnaire on personally sensitive victimisation (sexual violence, violence within the household), answered by the previous respondent (aged between 18 and 75)

Respondents are asked a wide range of questions on victimisation experience (i.e. household, vehicles, personal and victimisation within the household). Detailed

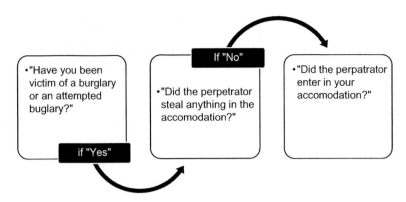

Diagram 7.1 Question sequencing on burglary victimisation. (Source: French CVS survey Insee-ONDRP-SSMsi, 2007–2015). Area covered: permanent residences, households in Metropolitan France

questions are also asked on the specific circumstances of the victimisation event(s) (i.e. time, place, cost, consequences, author/s, modus operandi, report to the authorities, issues). These questions are related to events that happened during the last 2 years.

For this chapter, we combine every sweep of CVS between 2007 and 2015, leading to a total sample of almost 150,000 households, representative of the 24.5 million households living in Metropolitan France during that period.

The dependent variable is derived from the questionnaire on households' victimisations. A set of questions relates to the experience of burglary. The introduction question for this victimisation is formulated as follows: 'During the last year or the year before, have you been victim of a burglary or an attempted burglary of your primary residence? Burglary occurs when a perpetrator breaks into your accommodation or an attached structure, including cases where no item is stolen'. If the answer is positive, then the respondent is asked more specific questions on the time of the victimisation, the sequence, the perpetrator (if seen), the reporting to the police and the follow-up on the procedure. It is precisely asked whether the perpetrator stole anything in the accommodation and, if not, whether the perpetrator did enter the dwelling or not. The sequencing of the questioning is displayed in Diagram 7.1.

7.2.2 Defining the Three Stages of the Burglary

In order to match the theoretical framework presented in the Introduction section, we operationalise each stage of the burglary process from the CVS questionnaire detailed above. Diagram 7.2 displays the processing of the burglary through three sequential steps. In this process, aiming at committing a burglary, the perpetrator first targets a housing unit, then tries to enter the dwelling with forced entry and, once he/she entered the accommodation, completes the theft.

Diagram 7.2 The process
of burglary. (Source:
ONDRP)

Although the questioning is not ordered the same way as the sequencing of the burglary process, we can measure the number of victims at each of the three stages: the targeting, the forced entry and the theft.

1. The targeting step is assessed through the first question displayed in Diagram 7.1. All housing units which have answered positively to this question have, at least, been victim of an attempted break-in. That means that their dwelling has been targeted by a burglar.
2. The forced entry step refers to the second and the third question displayed in Diagram 7.1. A positive answer to both questions indicates that the accommodation has been targeted (question 1) and that, whether any item has been stolen or not, the perpetrator entered the dwelling.
3. Finally, the theft step assessment relies on the second question displayed in Diagram 7.1. Housing units answering 'yes' to this question have been victim of a theft following the forced entry.

Victims at each step are then subsets of the same sample. Victims of theft are also victims at the forced entry step, and victims of forced entry are also victims at the targeting step. The questioning on the victimisation covers the 2 years prior to the sweep (i.e. the 2015 sweep covers the victimisations that have occurred in 2013 and 2014). Victims are also asked to indicate the number of victimisations during the covered period. The related results enable us to identify repeat victims (those with more than one victimisation) (Diagram 7.3).

The targeted housing units are all those which have been a victim of an attempted break-in or breaking and entering, followed by a theft or not. On average, over the 2007–2015 sweeps, 3.3 percent of housing units have been victims at this first step during the 2 years prior to the sweep (i.e. 900,000 households) (Diagram 7.2). On average, 2 percent of housing units are victims at the second stage of the burglary process, namely, when the perpetrator enters with force in the housing unit (on average 545,000 households each sweep). This step only concerns the households targeted during the first step. Finally, households studied at the theft stage are those that declared that the perpetrator actually entered the housing unit. At this point, we note a high proportion of failure during burglaries since only 1.7 percent of house-

Diagram 7.3 Sequential victimisation rate of burglary. (Source: French CVS survey Insee-ONDRP-SSMsi, 2007–2015). Area covered: permanent residences, households in Metropolitan France

holds report a theft. In other words, in almost 50 percent of cases, the targeting was not followed by theft. Indeed, as can be noted, there is a significant difference between the number of households which were victims of targeting and those where the perpetrator actually entered the housing unit with force and stole items. The failure rate of burglaries is then close to 50 percent. However, once the perpetrator has entered the housing unit, the odds of the burglary being completed are more than eight in ten. Subsequently the decisive stage of the burglary appears to be whether the perpetrator enters the housing unit or not.

The failure rate appears relatively higher in the British survey. The latest figures from the Crime Survey for England and Wales (CSEW)[2] show a burglary prevalence rate (all burglaries) of 1.6 percent between April 2015 and March 2016. This rate is made up of 0.9 percent instances of entering and 0.6 percent thefts. The attempt is only followed by the perpetrator entering the housing unit in 53 percent of cases and by theft in 34 percent of cases, i.e. the estimations regarding failure are significantly higher than those obtained in France from the CVS survey.

As part of our study, we distinguish between houses and apartments as victimisation rates, and the rates and types of security devices differ for these two subsamples. The victimisation rates at the different stages of burglaries for houses are close to those estimated for all housing units (respectively, 3.5, 2.2 and 1.9 percent) due to the main contributions to the results as a whole. The proportion of burglaries followed by thefts is higher if the target is a house: 63 percent of perpetrators entered the housing unit, and 54 percent completed the theft. Fewer households living in an apartment declared themselves to be victims. Indeed, only 2.9 percent of these households reported that they had been victims of a burglary, or an attempt, during the 2 years prior to the questioning. In addition, it is more likely that the burglary would fail. Forty-six percent of households living in an apartment which were targeted were not victims of forced entry, and 54 percent did not report any item stolen.

---

[2] The data from the survey in England and Wales is available on the website of Office for National Statistics.

7.2.3   Security Features and Information Regarding the Presence of Someone in the Housing Unit

Information regarding the installation of security features within housing units is available for all of the households' surveyed. For the households which were victims, whether these features were installed before or after the burglary (or attempt) is noted. We only take into account devices which had been installed before the burglary, if it took place.

Four questions corresponding to security devices are included in the survey covering: alarms, security doors,[3] cameras and digital lock. Information is also available regarding features which add to the security of housing: presence of a caretaker or owning a dog.

There are differences in equipment depending on the nature of the housing unit. Indeed, 80 percent of households living in an apartment have digital locks versus only 14 percent of those living in a house (Fig. 7.1). In addition, houses are equipped with an alarm more often than apartments (respectively, 13 and 2 percent). Between 2007 and 2015, the number of households owning cameras increased (+3.4 points). This increase is greater for apartments (3 percent in 2007 versus 7 percent in 2015). However, it should be noted that the survey does not give any information regarding the positioning of the camera. Yet, in the case of apartment blocks, the camera could be a device installed on the ground floor of the building and not in the housing unit.

Information regarding the presence of someone at home at the time of the burglary is only available, by definition, for households who are victims. Accordingly, it is not used in the estimation of the targeting model. It is only incorporated in the break-in and theft models which are only estimated on the households which are victims.

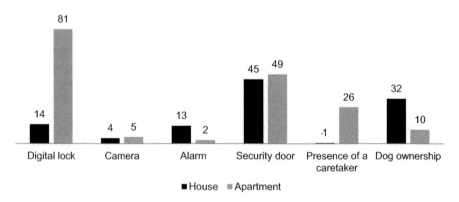

Fig. 7.1 Security features according to housing type (%). (Source: French CVS survey Insee-ONDRP-SSMsi, 2007–2015). Area covered: permanent residences, households in Metropolitan France

---

[3] A security door is defined as a door that is reinforced with internal steel plates or steel bars and can be added with multiple locks.

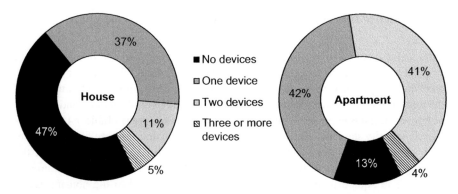

Fig. 7.2 Number of security devices according to housing type. (Source: French CVS survey Insee-ONDRP-SSMsi, 2007–2015). Area covered: permanent residences, households in Metropolitan France. Note: The security devices are alarms, security doors, digital lock and cameras

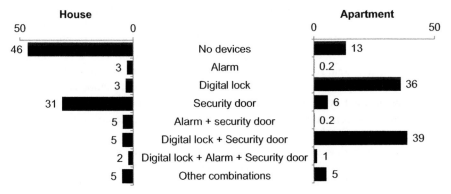

Fig. 7.3 Combination of security devices according to housing type (%). (Source: French CVS survey Insee-ONDRP-SSMsi, 2007–2015). Area covered: permanent residences, households in Metropolitan France

In addition, whether or not one or several security devices are owned also depends on the housing type. Households living in an apartment are more likely to own at least one security device, only 13 percent do not report any equipment listed above versus 43 percent of households living in a house. Likewise, apartments are more likely to be equipped with at least two security devices (45 percent versus 16 percent of houses) (Fig. 7.2).

Even though the devices are sometimes used in combination, in most cases houses are only equipped with a single device: a security door (31 percent). However, the situation is different for those living in an apartment. Even though 36 percent of these households only have digital lock, 39 percent have a digital lock and a security door (Fig. 7.3).

7.2.4 Environmental Factors and Lifestyle

Risk factors related to the housing unit, its environment and the household were stressed in the literature as explanatory burglary risk factors. This chapter looks into characteristics related to the household itself such as the employment situation of the household's reference person, his/her profession, level of income, and the type of household (single person, couple, with or without children) (Table 7.1). In addition to security devices, the characteristics of the housing unit are also called on, such as the housing type (detached house, attached house, apartment, etc.), the housing occupancy status (owner/tenant) and the surface area of the accommodation (size in square metres). Finally, several factors related to the housing unit's environment are extracted from the data source, namely, the type of neighbourhood, how close it is to the town centre and whether there is vandalism, deterioration or burglaries in the immediate surroundings.

The variables are presented separately whether they refer to the environment or the housing unit and, by extension, by degree of visibility to the perpetrator (Table 7.1). The following are listed as environmental factors: the geographic situation of the housing unit, the demographic environment and whether there is vandalism, deterioration or burglaries in the neighbourhood. Housing-specific factors, which are potentially difficult for the perpetrators to assess, refer to the housing type, its surface area, security devices, wealth (income) and the presence of someone in the housing unit or a dog. The remaining factors, which cannot necessarily be taken into account by the perpetrator, are nonetheless called on as control variables for the correct specification of the models. They include age, employment situation and marital status.

From a methodological point of view, the predominant group (most frequent characteristic) of variables used are not always the same according to the housing type (Appendix D). For example, the majority of households living in a house are found in residential areas mainly made up of detached houses (68 percent) whereas

Table 7.1 List of environmental, housing unit's specific and control factors

Environmental	Housing unit specific	Control
Rural/urban	Surface area of the housing	Age
Housing unit's environment	Camera	Gender
Population size	Alarm	Employment situation
Awareness of burglaries	Security door	Marital status
Acts of vandalism in the neighbourhood	Digital lock	Region
Deterioration of the neighbourhood	Caretaker	
	Income	
	Presence (if burgled)	
	Dog ownership	

Source: French CVS survey Insee-ONDRP-SSMsi, 2007–2015
Area covered: permanent residences, households in Metropolitan France

the households living in apartments are most often in areas of apartment blocks (59 percent). Likewise, a high proportion of houses are in rural municipalities (39 percent), whilst the majority of apartments are in towns with over 100,000 inhabitants (74 percent). The size of housing units also differs, apartments are generally smaller (between 40 and 70 m^2) than houses (between 100 and 150 m^2). The contextual data presented reveals significant differences between the security equipment of houses and apartments.

7.2.5   Modelling

In the remainder of this chapter, we aim to determine the risk of being a victim at each of the three steps of the burglary sequence as described above. We thus measure the effect of each of the factors presented in Table 7.1 on the probability of being targeted, victim of a forced entry and victim of a theft. This exercise should enable the effects in this respect to be checked at each stage of the burglary process and more particularly an estimation of the effect of the environmental and housing-specific factors by measuring specifically the impact of security devices as a form of dissuasion and protection. First and foremost we evaluate the probability that a household is targeted, whatever stage the burglary gets to. Accordingly, we will be able to estimate the quantitative effect of each of the variables forming part of this analysis on this probability. The same exercise is then carried out on housing units which have been targeted to measure the probability of being a victim of a forced entry. Finally, we measure the probability of being a victim of a theft following a forced entry depending on the set of explanatory factors including security features.

The explanatory factors forming part of this analysis (Appendix D) are the same for each of the three models in order to be able to compare their importance and significance at each stage (Diagram 7.1). As the descriptive elements brought out the differences, in particular in terms of security devices between houses and apartments, the estimations are carried out separately for the two housing types.

The factors' effects are estimated by fitting logistic regression models at each stage and the findings expressed as odds ratios. Logistic regressions are run using the *Logistic procedure* in the *SAS 9.4* software. We applied the *stepwise selection option* for independent variables. Using this option allows the independent variables to enter the model when significant at the 0.3 level and to stay in the model when significant at the 0.1 level. Consequently, only variables which are significant at the 0.1 level are displayed in Figs. 7.4, 7.5, 7.6, 7.7, 7.8, 7.9, 7.10, 7.11, 7.12, 7.13, 7.14 and 7.15. The findings are presented in graphs in the following part with a view to making them easier to read.

7.3 Results

7.3.1 The Role of Security Devices

We are considering, as a first step, the effect of security devices one by one regardless of the number of devices installed. The models also include the factors specific to households, housing units and their environment. Housing unit-specific factors relate to the particular characteristics of the households and the reference person of the household, and environmental factors relate to the situation and the environment of the housing (geographic situation, deterioration or not of the neighbourhood, etc.) (Table 7.1).

7.3.1.1 Targeting

The choice of housing unit as a target for burglary is particularly influenced by environmental factors. Whatever the housing type (house or apartment), the households which are aware of burglaries in their accommodation's surroundings are three times more likely to be targeted than a household which is not aware.[4] Likewise, frequent observation of acts of vandalism in the neighbourhood also have an increasing impact on the probability of being targeted (+60 percent for houses and +100 percent for apartments). Houses and apartments with large surface areas (more than 100 m^2) are also more likely to be targeted. Indeed, houses with a surface area between 40 and 100 m^2 are around 20 percent less likely to be targeted by a burglar than houses between 100 and 150 m^2.

Focusing more particularly on houses, those in large urban areas are more at risk of being targeted by burglars. Housing units in an urban area of over 100,000 inhabitants are almost twice as likely to be targeted compared to those in a rural municipality (Fig. 7.4). Income, a proxy for wealth, also has an impact on the likelihood of victimisation. Indeed, high-income households living in a house have a 27 percent higher risk of being a perpetrator's target.

We found that the effects of security devices are different for houses and apartments. Indeed, for houses, an alarm has a dissuasive effect (if weak) right from the stage of the burglary when the housing unit is targeted (−13 percent risk). Digital locks also seem to have a dissuasive effect (−11 percent). Surveillance cameras also have a protective effect when it comes to the targeting of housing units (−25 percent). At this step, we do not find any significant effect of security doors on the probability of being targeted.

---

[4]The timing between burglary and awareness of burglary in the surroundings is not specified in the CVS questionnaire so that we cannot avoid reverse causality between the two indicators. Moreover, as the data has no time dimension, Granger causality tests cannot be run. In order to check for potential endogeneity bias or reverse causality, we ran a set of different regression models excluding the burglary awareness variable and compared the results with the full model. Taking out the variable reduces the quality but does not change the relevancy of the specification. However, when taking out the variable from the model, we estimate a higher effect of other environmental variables, such as acts of vandalism and deterioration of the neighbourhood. This indicates that these three variables do proxy for the same concept, namely, local crime and delinquency. Results of the tests are available upon request.

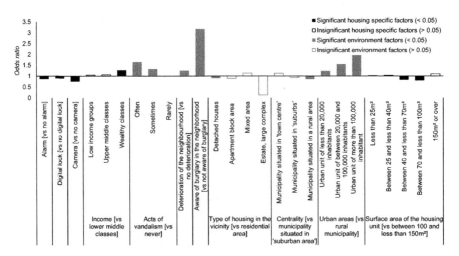

Fig. 7.4 Effect of variables on the probability of houses being targeted. (Source: French CVS survey Insee-ONDRP-SSMsi, 2007–2015). Area covered: permanent residences, households in Metropolitan France. Note to the reader: only variables which are significant at the 0.1 level are displayed in the figure

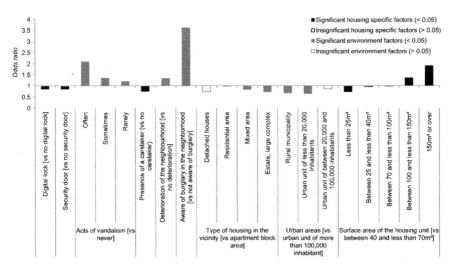

Fig. 7.5 Effect of variables on the probability of apartments being targeted. (Source: French CVS survey Insee-ONDRP-SSMsi, 2007–2015). Area covered: permanent residences, households in Metropolitan France

With regard to apartments, households living in small towns are less at risk of being burgled. Households living in apartments in a rural municipality reduce their risk of being a victim by a third and by 36 percent for households living in areas with less than 20,000 inhabitants, compared to those living in an urban area with more than 100,000 inhabitants (Fig. 7.5). Unlike large apartments, small apartments are less likely to be targeted by a burglary (−27 percent for households living in less than 25 m²). In the same way as for houses, security features have a dissuasive effect right from the targeting stage, but this effect is limited in comparison to factors linked to the environment of the housing. This includes the impact of digital locks (−15 percent) and also of a security door (−14 percent). A caretaker also has a dissuasive effect on the targeting of the housing unit – apartment blocks with a caretaker are almost 25 percent less at risk. We do not estimate any significant dissuasive effect of cameras and alarms at the targeting step for apartments.

7.3.1.2   Forced Entry

For both houses and apartments, alarms and security doors have a protective effect at the time of the forced entry. Alarms decrease the risk for houses by 34 percent and for apartments by 47 percent. Security doors reduce the risk by 27 and 24 percent, respectively. In addition, one of the effects shared by households is the presence of someone at home at the time of the break-in. However, this effect is higher for houses. Indeed, it reduces the risk by more than a half (−55 percent) versus by less than a quarter for apartments (−22 percent).

With regard, more particularly, to the security features which have a protective effect in houses, owning a dog is significant, whereas it is estimated as non-significant at the targeting step. Households owning a dog are almost 27 percent less likely to have the burglar enter the housing unit (Fig. 7.6). Conversely, as observed with the targeting model, it is more likely that the burglar will succeed in breaking into the house if a wealthy household lives there (+16 percent). In addition, unlike apartments, awareness of acts of vandalism still has an effect in the model for houses. In other words, if a household has observed acts of vandalism in its neighbourhood, it is less likely that a perpetrator will enter the housing unit.

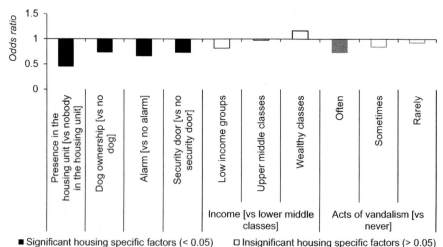

Fig. 7.6 Effect of variables on the probability of forced entry in houses. (Source: French CVS survey, Insee-ONDRP-SSMsi, 2007–2015). Area covered: permanent residences, households in Metropolitan France

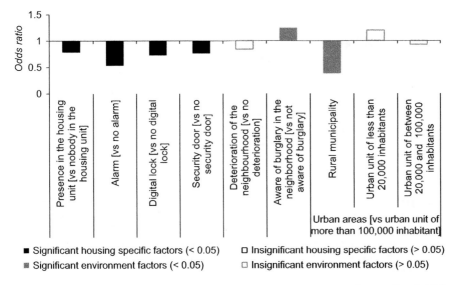

Fig. 7.7 Effect of variables on the probability of forced entry in apartments. (Source: French CVS survey, Insee-ONDRP-SSMsi, 2007–2015). Area covered: permanent residences, households in Metropolitan France

With regard to apartments, we found that a third security device, digital locks, has a protective effect.[5] Digital locks reduced the risk for this housing type by 28 percent. In the same way as for houses, environmental factors still play a significant role in this model. Awareness of burglaries in the neighbourhood will increase the risk of forced entry (+24 percent) (Fig. 7.7). In addition, for households living in apartments, being in a rural municipality reduces the risk that being targeted is followed by a forced entry (−62 percent).

7.3.1.3 Theft

At the theft stage, only the presence of someone in the housing unit at the time of the burglary remains a significant protective factor for houses and apartments. This effect is protective in both cases, respectively, −43 and −44 percent. In the theft model for houses, awareness of burglaries in the surroundings of the housing unit also increases the likelihood that the theft will be successful (+47 percent) (Fig. 7.8). This was not found in the model for forced entry. In houses, the only device which has a protective effect against theft is an alarm (−58 percent). A house equipped

[5] The data from the CVS survey does not enable us to have information regarding where the digital locks are located in the housing unit (at the entrance of the apartment block or in the housing unit itself).

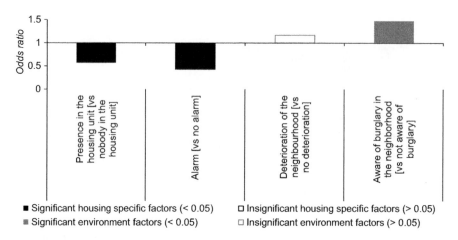

Significant housing specific factors (< 0.05)   Insignificant housing specific factors (> 0.05)
Significant environment factors (< 0.05)   Insignificant environment factors (> 0.05)

Fig. 7.8 Effect of the factors on the probability of thefts in houses. (Source: French CVS survey, Insee-ONDRP-SSMsi, 2007–2015). Area covered: permanent residences, households in Metropolitan France

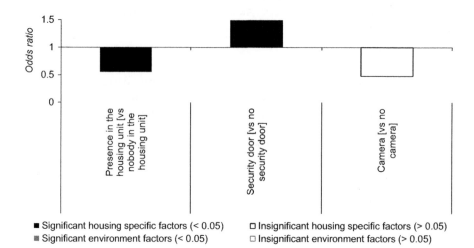

Significant housing specific factors (< 0.05)   Insignificant housing specific factors (> 0.05)
Significant environment factors (< 0.05)   Insignificant environment factors (> 0.05)

Fig. 7.9 Effect of the factors on the probability of thefts in apartments. (Source: French CVS survey, Insee-ONDRP-SSMsi, 2007–2015). Area covered: permanent residences, households in Metropolitan France

with at least one alarm is more protected against theft. With regard to apartments, the environmental factors no longer have any influence on the probability of theft. However, there is a paradoxical effect of one of the security devices – the security door. Here, a security door increases the likelihood of theft (+49 percent) which may suggest a better-prepared perpetrator to tackle a device which did not fulfil its protective role at the time of the break-in (Fig. 7.9).

7.3.2 Analysis of Combinations of Devices

After demonstrating the effectiveness of security devices, we are interested here in evaluating the effect of different combinations of devices. To achieve this, we are focusing on the most common device combinations (Fig. 7.3) and estimating their impact on the risk of burglary at each of the three steps.

The previous findings showed the key role of security devices during the forced entry and theft stages. As such, we will only present the findings relative to those stages for houses and apartments separately. Nonetheless, the findings obtained relating to the targeting of housing by perpetrators are comparable to those presented in the previous section. The introduction of combinations of devices does not affect the predominance of environmental factors on the likelihood of being a victim. With respect to security devices, their impact is quite limited in terms of the risk of being the target of a burglary. For both houses and apartments, when only one device is installed, we estimate little or no effect on the risk of being targeted. The combination of a digital lock, an alarm and a security door provides greater security, but this effect remains low in relation to the effect of environmental characteristics. Although the effect is weaker, combinations including a camera[6] also dissuade burglars.

7.3.2.1 Entering the Housing Unit

We previously identified the more significant role of security features at the forced entry stage, with environmental factors having, during this stage, a much lesser effect. Here we are focusing on the protective effect of different combinations of devices against the risk of a perpetrator entering a housing unit following a targeting.

A common feature appears for houses and apartments when it comes to protective features. For the two types of housing, the presence of someone at home at the time of the entry reduces the likelihood that the perpetrator will enter the housing by half for houses and by a quarter for apartments. Besides this shared characteristic, two distinct patterns emerge for houses and for apartments.

Alarms are an effective protection for houses. Whether they are the only device or combined with others (security door and digital lock), alarms are the most effective security feature when it comes to a perpetrator entering the housing unit. Indeed, the alarm may not be visible from the outside and may not be taken into account when the housing unit is chosen. It may have a 'surprise effect' during the forced entry, in the same way as someone being home at that time. Alarms reduce the likelihood that the targeting is followed by the perpetrator entering in houses by 50 percent if the alarm is the only device and by 66 percent when an alarm is combined with a digital lock and a security door (Fig. 7.10). Beside the

---

[6] Combinations brought together under 'other combinations'.

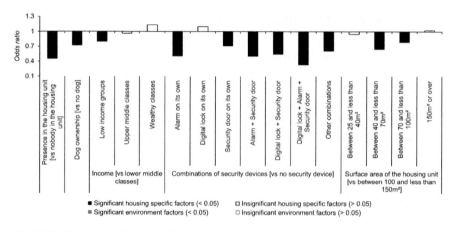

Fig. 7.10 Effect of combinations of security devices on the probability of forced entry in houses. (Source: French CVS survey, Insee-ONDRP-SSMsi, 2007–2015). Area covered: permanent residences, households in Metropolitan France. Note: The 'other combinations' field includes the devices mentioned in less than 1% of the responses. More particularly, the combinations included in the 'other' category are those with cameras and the combination of digital lock and alarm

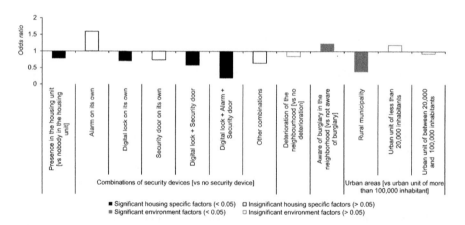

Fig. 7.11 Effect of combinations of security devices on the probability of forced entry in apartments. (Source: French CVS survey, Insee-ONDRP-SSMsi, 2007–2015). Area covered: permanent residences, households in Metropolitan France

effects of security devices, it is estimated that, all other things being equal, perpetrators are more likely to enter large housing units than housing units with a smaller surface area.

Digital locks have a moderate protective effect when it comes to apartments. It reduces the likelihood that the perpetrator will enter by around 25 percent (Fig. 7.11).

The most effective combination is when a digital lock is combined with a security door and an alarm. In this case, the likelihood that the targeting is followed by the perpetrator entering the housing unit is reduced by nearly 85 percent. When compared with the previous findings, this estimation indicates that an alarm on its own is less effective for apartments than it may be for houses. This result, however, though significant may be read with caution since the proportion of apartments equipped with an alarm is very small.

7.3.2.2   Theft

The findings relating to the last stage of the process are weaker than the previous findings, due to the small sample size studied. In this context, combinations of security devices may not appear to be significant. This is particularly the case for apartments. The majority of the estimated parameters of the model are not significant. In other words, the numerical findings do not enable a conclusion to be drawn on the impact of security devices.

However, for houses, it is noted that alarms have a significant protective effect as does the combination of an alarm and a security door (Fig. 7.12). In a counterintuitive way in theory, the combination of a digital lock and a security door has a positive effect on the probability of theft taking place. However, at the theft stage,

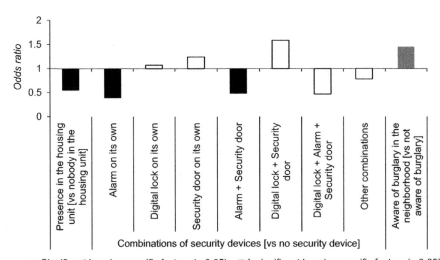

■ Significant housing specific factors (< 0.05)   □ Insignificant housing specific factors (> 0.05)
▨ Significant environment factors (< 0.05)   □ Insignificant environment factors (> 0.05)

Fig. 7.12 Effect of combinations of security devices on the probability of theft for houses. (Source: French CVS survey, Insee-ONDRP-SSMsi, 2007–2015). Area covered: permanent residences, households in Metropolitan France

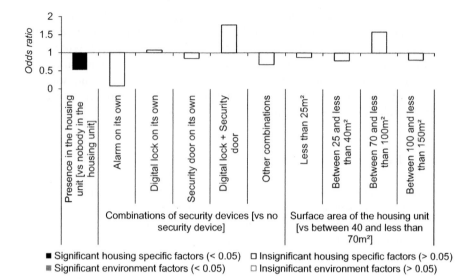

Fig. 7.13 Effect of combinations of security devices on the probability of theft for apartments. (Source: French CVS survey, Insee-ONDRP-SSMsi, 2007–2015). Area covered: permanent residences, households in Metropolitan France

these two devices have already played their part. We have seen at the previous stage that they do actually have a protective effect when the perpetrator tries to enter in the housing unit.

For apartments, none of the combinations are seen as having a significant effect. The only factor which seems to be significantly protective for apartments is the presence of someone at home at the time of the burglary (Fig. 7.13). This effect is also highlighted for houses. In both cases, the presence of someone at home almost halves the probability that the perpetrator will commit a theft following the break-in.

7.3.3 The Specific Case of Repeat Victimisations

Here we focus on the specific case of households which have been victims of burglaries multiple times, in other words those which have been victims more than once during the last 2 years. As part of the French victimisation survey, the respondents are asked to indicate, if necessary, the number of victimisations experienced during each of the last 2 years. In this way, we are able to identify the households which have been repeat victims of burglaries.

We wish to give specific attention to this within this sub-section since it has been identified that repeat victimisations have specific features (Bernasco and Luykx 2003). According to the literature on the subject, repeat victims seem to correspond to a different victimisation pattern than other victims; it is also of interest to find out whether the factors previously identified are still relevant when taking into account the repeat victimisation of households. In order to study this phenomenon, we have therefore chosen, amongst the households which were victims at least once during the year before the sweep, those which were victims more than once during the 2 years.[7]

If a household has already been the victim of a burglary, we estimate that it is six times more likely to be targeted for another burglary. When this population is divided according to the housing type (house or apartment), it is noted that having already been a victim has different effects when it comes to the perpetrator entering the housing unit. If a house has already been burgled, this has a negative effect on the success of a forced entry. The likelihood that the burglar will successfully enter the housing unit is reduced by 40 percent if the household has already been a victim of a burglary during the previous year or the same year (Fig. 7.14). One hypothesis to interpret these findings is that the households were able to protect themselves more effectively following the first burglary. Indeed, 25 percent of households had at least one security device installed after a burglary. Most of these devices are alarms (17 percent) and security doors (10 percent).

A study on burglary in New Zealand also highlighted this phenomenon: 'Burglary victims had lower levels of household security at the time of burglary than other households had at the time of interview, but tended to increase their use of security after a burglary, to a higher level of security on average than other households' (Triggs 2005, p. 54).

Unlike houses, a perpetrator is more likely to enter an apartment if it has already been burgled. The risk increases by 61 percent for apartments which have already been burgled (Fig. 7.15). One possible interpretation of this finding is that these households are less likely to install a device after a burglary than those living in houses (12 percent versus 25 percent). In addition, the particularity of housing units in an apartment block means that some of the security is linked to the protection of the entrance to this block. Security devices are not therefore necessarily installed by the households but as part of co-ownership agreements.

However, the interpretation of these effects remains quite limited due to the small sample size. In addition, the findings of the CVS survey do not enable all of the information regarding the time frame of the installation of security devices to be collected. It is therefore impossible to know whether the households which are repeat victims installed devices as a result of the first burglary, beforehand or subse-

---

[7] Unlike the previous model, the victim households studied are only those which were victims during the year preceding the study, which reduces our sample size.

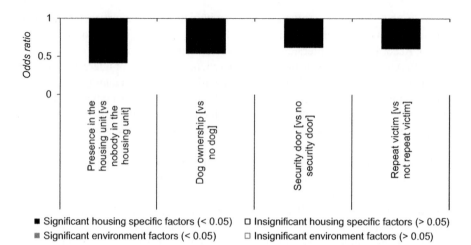

Fig. 7.14 Effect of repeat victimisation on the probability of forced entry to houses. (Source: French CVS survey, Insee-ONDRP-SSMsi, 2007–2015). Area covered: permanent residences, households in Metropolitan France

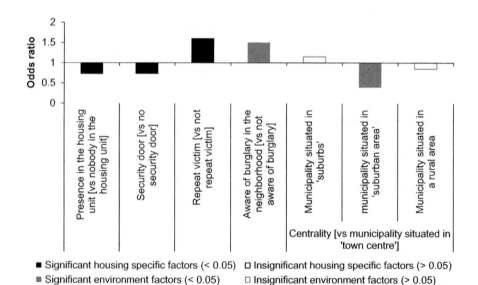

Fig. 7.15 Effect of repeat victimisation on the probability of forced entry to apartments. (Source: French CVS survey, Insee-ONDRP-SSMsi, 2007–2015). Area covered: permanent residences, households in Metropolitan France

quently. Accordingly, the interpretations made of the last set of findings must be considered as points for discussion rather than as conclusions.

7.4 Discussion

The analysis of burglaries as a process involving several stages enables distinct roles to be brought to light with regard to the different factors regarding the environment of the housing units and the housing units themselves. Furthermore, this work enables the way in which burglars incorporate information concerning their target to be better understood as the same factors do not have the same impact at each stage of the process. We indeed see that the choice of target for the burglary largely depends on factors linked to the environment of the housing unit. Thus, local crime and delinquency and burglaries in the neighbourhood are aggravating factors which increase the probability of being a victim. These first results must however be taken cautiously since we cannot determine the direction of the causality between victimisation and perception. Households in a highly populated area also have an additional risk. Therefore it seems that housing units are not targeted at random and that they are targeted in relation to their environment. Certain factors specific to the housing units, such as the surface area, also have a significant impact on the risk of houses and apartments being targeted for a burglary. Over and above what has already been discussed, the distinction between environmental factors and factors specific to the housing units seems important. By environmental factors we mean those that the burglar can take into account before his/her attempt as they can be evaluated before he/she even enters the housing unit. Thus, we find that the majority of security devices have a low dissuasive effect at this stage of the burglary, with the exception of surveillance cameras. Indeed, in the case of houses, these cameras can be visible from the outside and thus dissuade perpetrators from choosing such housing units as a target.[8] Alarms and security doors only have a very low dissuasive effect on the risk of being targeted. As for apartment blocks, the presence of a caretaker, who may also be visible, effectively dissuades perpetrators to target the dwelling.

The distinction between housing-specific factors and environmental factors tends to be corroborated by the analysis of determinants regarding the perpetrator entering the housing unit. During this stage, the target has already been chosen, and the perpetrator tries to enter the housing unit with force. At this stage, it is the housing unit-specific factors, such as owning a dog, someone being at home or an alarm, which have a protective effect. In parallel, most of the environmental factors no longer have a significant effect at this stage, with the exception of awareness of burglaries in the neighbourhood. The protective features, unlike during the stage

---

[8] There is no legal obligation in France to inform of the presence of CCTV in private premises.

when the target is chosen, are the features which cannot be seen from outside the home and may be 'unexpected' for the perpetrator. They thus play a protective rather than dissuasive role. The presence of security features in the residences will play a role in burglary attempts. Indeed, as noted by Tseloni et al. (2014), we also found that security devices are protective features against forced entry. During the last stage of the burglary process, the perpetrator has succeeded in entering the housing unit. Therefore any dissuasive and protective features have not fulfilled their function. At this stage, only the presence of someone in the home (and, for houses, the presence of an alarm) is considered to effectively reduce the risk of theft.

The detailed analysis of the combinations of security devices reveals more detailed findings about the effectiveness of protective features. In particular, with regard to houses, an alarm remains the most protective device against the perpetrator entering the housing unit. This device is all the more effective when it is combined with a security door and with a digital lock. When these three features are combined, the probability that the perpetrator will actually enter the housing unit decreases by two thirds. For apartments, this combination is also the most protective against perpetrators entering. However, for this housing type, an alarm on its own does not have a significant effect in the findings we obtain.

Upon reading all of the findings of this study, we observe that the environment of the housing unit has a very strong impact on the probability of being targeted during a burglary. Security devices also have a dissuasive effect, although this effect is limited. This is particularly the case for surveillance cameras for houses and digital locks for apartments. The protective role of these security devices is greater when the perpetrator attempts to enter the housing unit. During this stage, alarms have the most protective effect, in particular if they are combined with other devices. In parallel, and independent of security devices, the presence of a person or dog in the home significantly reduces the probability of the perpetrator entering the housing unit and committing a theft.

Through these findings we show that the information which perpetrators base their choice for a burglary on is not complete at the time they select their target. The fact that housing-specific factors have little effect on the models for selecting the housing unit shows that these factors are not particularly influential and that perpetrators neglect to properly take security devices into account. The choice of target is made on the basis of information related to the environment. At the following stage, when the perpetrator attempts to enter with force, he/she may discover new information. This is the case in particular for certain security features, such as an alarm or the presence of someone at home. This information may have been unavailable when the perpetrator chose his/her target and is revealed when he/she attempts to enter the housing unit. Indeed, some alarms may be visible from the outside and have a deterrent effect when targeting the housing. However, the questioning does not give such precision in the French CVS. Once inside the housing unit, some information and the presence of certain factors may still be unknown to the perpetrator and be brought to his/her attention during the burglary. Thus, the delayed triggering of an alarm or the presence of someone at home is going to change the information the perpetrators have at their disposal and cause them to re-evaluate the situation they find themselves in.

The findings presented in this study are based on a survey system which, in the way it has been constructed, cannot be exhaustive regarding the security features surrounding the housing units. Even though the devices presented here are the most common, they do not specify, for example, whether the video surveillance devices are 'connected'. Neither is account taken of other so-called exogenous security features which are based on the presence of residents in the neighbourhood and on how isolated the places are (cul-de-sacs, ground floors, crossroads, etc.) as well as on ad hoc schemes implemented at local level (Neighbourhood Watch, municipal video surveillance and specific police surveillance devices). The Crime Survey for England and Wales enabled factors specific to the neighbourhoods and capable of having a protective role against burglaries to be brought to light such as Neighbourhood Watch.

Other limitations related to the questioning in the French CVS remain. The first relates to the fact that the census of digital locks and video surveillance cameras is carried out using a single question. As a result, if these two devices were installed on different dates, it would not be possible to trace the history of each of them or their time frame in relation to the potential burglary using the survey. The second relates to the person responsible for the installation. The respondent is asked whether the household is responsible for the installation and, only if the response is positive, when this installation took place. If the response is negative, we assume that the device was installed before the current occupants moved into the housing unit. However, in some cases, it is possible that the device was installed after the current occupants moved in but that this was carried out by a third person and not by the occupants themselves (e.g. the landlord in the case of households which are renting). The third limitation relates to the time frame of the installation in relation to the burglary and specially in the case of repeat victimisations. The question on time frame of installation is formulated as follows: 'Are you responsible for the installation of this device?' if yes, 'Was this following a burglary at your primary residence?' In the case of repeat victimisations, and when no details are given, the declared installation may have taken place following a burglary, yet other burglaries may have occurred after this installation. Thus if someone responds negatively to this question, he/she could wrongly be considered to have not been equipped at the time of the burglary. Future work focusing both on survey data and on police data could take the analysis of security further with a greater number and more types of devices.

Appendix D

Appendix Table 7.2 References for all of the variables used

Factor type	Variables	Predominant group according to housing type		
		House	Apartment	Together
Environmental	Awareness of burglary	No (59%)	No (86%)	No (69%)
	Acts of vandalism in the neighbourhood	Never (61%)	Never (47%)	Never (56%)
	Type of housing in the vicinity	Residential areas mainly of detached houses (68%)	Apartment block area (59%)	Residential areas mainly of detached houses (43%)
	Housing type	Detached house (65%)	Apartment block w/ten housing units or more (68%)	Detached house (39%)
	Deterioration of the neighbourhood	No criticism (50%)	No criticism (55%)	No criticism (52%)
	Centrality[a]	Suburban (31%)	Town centre (52%)	Suburbs (31%)
	Region of residence	Paris Basin (21%)	Paris region (32%)	Paris region (18%)
	Location in relation to urban areas	Rural municipality (39%)	Urban unit of more than 100,000 inhab. (74%)	Urban unit of more than 100,000 inhab. (46%)
Housing	Digital lock	No digital lock (86%)	Digital lock (81%)	No digital lock (60%)
	Camera	No camera (96%)	No camera (95%)	No camera (96%)
	Alarm	No alarm (87%)	No alarm (98%)	No alarm (91%)
	Security door	No security door (55%)	No security door (51%)	No security door (53%)
	Presence of a caretaker	No caretaker (99%)	No caretaker (76%)	No caretaker (90%)
	Dog ownership	No dog (68%)	No dog (90%)	No dog (77%)
	Presence in the unit (if burgled)	NoNo (99%)	No (99%)	No (99%)
	Surface area (housing unit)	Between 100 and 150m² (41%)	Between 40 and 70m² (42%)	Between 70 and 100m² (35%)

(continued)

Appendix Table 7.2 (continued)

Factor type	Variables	Predominant group according to housing type		
		House	Apartment	Together
Household	Household type	Couple w/ children (34%)	One person (48%)	One person (35%)
	Marital status	In a couple, married (54%)	Single (29%)	In a couple, married (43%)
	Socio-professional category	Blue-collar worker (25%)	Employee (28%)	Blue-collar worker (25%)
	Age group	65 years old and over (32%)	65 years old and over (23%)	65 years old and over (28%)
	Work situation	In employment (55%)	In employment (57%)	In employment (56%)
	Income	Lower middle classes (31%)	Low income group (35%)	Lower middle classes (30%)

Source: French CVS survey, Insee-ONDRP-SSMsi, 2007–2015
Area covered: Permanent residences, households in Metropolitan France
[a]Location of housing unit in relation to the town centre

References

Aebi, M., & Linde, A. (2010). Is there a crime drop in Europe? *European Journal on Criminal Policy and Research, 16*(4), 251–277.

Australian Institute of Criminology. (2016). *Australian crime: Facts & figures 2014*. Canberra: Australian institute of Criminology.

Bernasco, W., & Luykx, F. (2003). Effects of attractiveness, opportunity and accessibility to burglars on residential burglary rates of urban neighborhoods. *Criminology, 41*(3), 981–1002.

Bernasco, W., & Nieuwbeerta, P. (2005). How do residential burglars select target areas? A new approach to the analysis of criminal location choice. *British Journal of Criminology, 45*(3), 296–315.

Bettaieb, I., & Delbecque, V. (2016). *Mesure de l'exposition aux cambriolages*. Paris: INHESJ.

Brantingham, P., & Brantingham, P. (1975). The spatial patterning of burglary. *The Howard Journal of Crime and Justice, 14*(2), 11–23.

Brown, R. (2015). *Explaining the property crime drop*. Canberra: Australian institute of criminology.

Bureau of Justice Statistics. (2016). *Criminal victimization, 2015*. Washington D.C. Bureau of Justice Statistics.

Capone, D., & Nichols, W. (1975). Crime and distance: An analysis of offender behavior in space. *Proceedings of the Association of American Geographers, 7*, 45–49.

Ceccato, V., Haining, R., & Signoretta, P. (2002). Exploring offence statistics in Stockholm City using spatial analysis tools. *Annals of the Association of American Geographers, 92*(1), 29–51.

Cohen, L., & Felson, M. (1979). Social change and crime rate trends: A routine activity. *American Sociological Review, 44*, 588–608.

Cromwell, P., Olson, J., & Avary, D. W. (1991). *Breaking and entering. An ethnographic analysis of burglary*. Newbury Park: Sage.

Flatley, J., Kershaw, C., Smith, K., Chaplin, R., & Moon, D. (2010). *Crime in England and Wales: Findings from the British crime survey and police recorded crime*. London: Home Office.

Gabor, T., & Gottheil, E. (1984). Offender characteristics and spatial mobility: An empirical study and some policy implications. *Canadian Journal of Criminology, 26*, 267–281.

Hindelang, M., Gottfredson, M., & Garofalo, J. (1978). *Victims of personal crime: An empirical foundation for a theory of personal victimization.* Cambridge: Ballinger.

Hough, M. (1987). Offenders' choice of target: findings from victim surveys. *Journal of Quantitative Criminology, 3*(4), 355–369.

Kuhns, J., Blevins, K., & Lee, S. (2012). *Understanding decisions to burglarize from the offender's perspective.* Charlotte: UNC.

Lammers, M., Menting, B., Ruiter, S., & Bernasco, W. (2015). Biting once, twice: The influence of prior on subsequent crime location choice. *Criminology, 53*(3), 309–329.

Lynch, J., & Cantor, D. (1992). Ecological and behavioral influence on property victimization at home: Implications for opportunity theory. *Journal of Research in Crime and Delinquancy, 29*(3), 335–362.

Mayhew, P., Aye Maung, N., & Mirrlees-Black, C. (1993). *The 1992 British crime survey.* London: HMSO.

Miethe, T., & David, M. (1993). Contextual effects in models of criminal victimization. *Social Forces, 71*(3), 741–759.

Morgan, F., & Clare, J. (2007). *Household Burglary Trends in Western Australia.* State Government of Western Australia.

Murphy, R., & Eder, S. (2010). Acquisitive and other property crime. In *Crime in England and Wales 2009/10: Findings from the British crime survey and police recorded crime* (Home Office Statistical Bulletin 12/10 ed., pp. 79–107). London: Home Office.

Nee, C., & Meenaghan, A. (2006). Expert decision-making in burglars. *British Journal of Criminology, 46*, 935–949.

Office for National Statistics. (2016). *Crime in England and Wales: Year ending Sept 2016. Statistical bulletin.* London: Office for National Statistics.

ONDRP. (2015). *La criminalité en France. Rapport annuel de l'Observatoire national de la délinquance et des réponses pénales 2015.* Paris: CNRS Editions.

Pease, K., & Gill, M. (2011, September). Home security and place design: Some evidence and its policy implications. Retrieved February 8, 2017, from http://www.securedbydesign.com/wp-content/uploads/2014/02/home-security-and-place-design.pdf

Perron-Bailly, E. (2013). *Caractéristiques des cambriolages et des tentatives de cambriolages de la résidence principale décrites par les ménages s'étant déclarés victimes sur deux anslors des enquêtes "Cadre de vie et sécurité" de 2011 à 2013.* Paris: INHESJ.

Rhodes, W., & Conly, C. (1981). Crime and mobility: An empirical study. In P. B. Brantingham (Ed.), *Environmental criminology* (pp. 167–188). Prospect Heights: Waveland Press.

Rountree, P., & Land, K. (2000). The generalizability of multilevel models of burglary victimization: A cross-city comparison. *Social Science Research, 29*, 284–305.

Rountree, P., Land, K., & Miethe, T. (1994). Macro-micro integration in the study of victimization: A hierarchical logistic model analysis across Seattle neighborhoods. *Criminology, 32*(3), 387–413.

Tilley, N. (2009). *Crime prevention.* Cullompton, Devon: Willan Publishing.

Triggs, S. (2005). *Surveys of household burglary Part one (2002): Four police areas and national data compared.* Canberra: Ministry of justice.

Tseloni, A., Thompson, R., Grove, L., Tilley, N., & Farrell, G. (2014). The effectiveness of burglary security devices. *Security Journal, 30*(2), 646–664. https://doi.org/10.1057/sj.2014.30

van Dijk, P. (2008). *The world of crime, breaking the silence on problems of security, justice and development across the world.* London: Sage Publications.

van Dijk, J., Tseloni, A., & Farrell, G. (2012). *The international crime drop – New directions in research* (van Dijk, TSeloni, Farrell ed.). Basingstoke: Palgrave Macmillan.

van Kesteren, J., Mayhew, P., & Nieuwbeerta, P. (2000). *Criminal victimisation in seventeen industrialised countries: Key findings from the 2000 international criminal victimization survey.* The Hague: Ministry of Justice.

Chapter 8
The Role of Security in Causing Drops in Domestic Burglary

Nick Tilley, Graham Farrell, Andromachi Tseloni, and Rebecca Thompson

Abbreviations

BCS	British Crime Survey
CSEW	Crime Survey for England and Wales
DAPPER	Default, aesthetically pleasing, powerful, principled, effortless, rewarding
ICVS	International Crime Victims Survey
SPF	Security Protection Factor
WIDE	Window locks, internal lights on a timer, door double or deadlocks, external lights on a sensor

8.1 Introduction

In the years following the Second World War, crime levels seemed to rise inexorably in most Western societies and did so in spite of improvements in income, housing, education, welfare and employment. In 1957 the then British Prime Minister Harold Macmillan said in a speech in Bedford that 'most of our people have never had it so good'. What went for improving social conditions in Britain went for most industrialised societies. Common sense links between adverse social conditions and criminality seemed to be contradicted. They might lead one to suppose that crime should

N. Tilley (✉)
Jill Dando Institute, Department of Security and Crime Science, University College London, London, UK
e-mail: n.tilley@ucl.ac.uk

G. Farrell
School of Law, University of Leeds, Leeds, UK

A. Tseloni · R. Thompson
Quantitative and Spatial Criminology, School of Social Sciences, Nottingham Trent University, Nottingham, UK

© Springer Nature Switzerland AG 2018
A. Tseloni et al., *Reducing Burglary*,
https://doi.org/10.1007/978-3-319-99942-5_8

have been falling. Instead, increased wealth and welfare spending went alongside rises in crime. This posed a major challenge for criminology.

Cohen and Felson (1979) provided a powerful explanation for the post-war crime rises in the USA in terms of 'routine activities', although their ideas are relevant to the crime rises that occurred in the UK also. Cohen and Felson's starting point was disarmingly simple: three crucial conditions are needed for a crime to take place – they referred originally to direct contact predatory offences although with only slight modifications the same conditions are needed for most other crimes also. The three conditions are a potential offender (sometimes referred to as a 'likely' offender), a suitable target (which could be a person or thing) and the absence of a capable guardian. If these three converge in space and time, then a crime can occur. If any is absent, a crime cannot occur: no target and/or no likely offender and/or the presence of a capable guardian means that there will be no crime. The co-presence of a suitable target and absence of capable guardian comprise a crime opportunity. The arrival of a likely offender provides someone liable to take advantage of the opportunity.

The supply, distribution and movement of suitable targets, likely offenders and capable guardians create crime patterns. In the years following the Second World War, there were massive increases in the supply of suitable targets: for example, cars, televisions and audio equipment. Likely offenders, especially adolescent males, were less tied to the home, for example, because of increased wealth, better transport opportunities and reduced involvement in domestic chores, and were thereby more liable to encounter suitable targets. Added to this, there was some reduction in the capable guardianship of dwellings as women increasingly joined the labour market. Crime rose accordingly. It was not necessary to postulate a growth in disposition to commit crime to make sense of increasing levels of crime.

Specific crime trends could be explained by focusing on specific changes in everyday life relevant to them. Increases in the number of car thefts could be explained in terms of the growth in the supply of cars coupled with their suitability as targets for theft (there were many cars, they were easy to steal – getting into them, hotwiring them and driving them away was not difficult; moreover, they were fun to drive and from the 1960s they contained valuable radio cassettes). Furthermore, many easy-to-steal cars were parked in locations with little guardianship. Increases in shop theft could be explained in terms of target suitability (a growing supply of attractive goods on sale in readily accessible stores) and in terms of the weak guardianship furnished by self-service, rather than over-the-counter, retailers who were trying both to reduce the costs of selling goods with minimal staff numbers and also to tempt customers with appealing goods they could see and touch (Tilley 2010). With regard to domestic burglary, reductions in the guardianship of homes and increases in the supply of suitable targets for theft within them shaped the suitability of dwellings as places from which to steal. One indicator of the usefulness of routine activities theory is the high rate at which the key paper outlining the theory is cited: more than 7000 times by mid-2017, according to Google Scholar.

Until the peak in crime levels observed in many countries in the 1990s, the trend in rising crime seemed to be unstoppable. The public, politicians, the news media and the criminological community all took it for granted that crime would tend to rise year on year. Then, remarkably, crime levels began to fall, first in the USA and then in many other industrialised countries. This fall has now been widely documented (Blumstein and Wallman 2000; Van Dijk et al. 2012; Tonry 2014).

Police-recorded crime statistics are not dependable for most crime types, given the fact that many crimes are not reported, and of those reported, many are not recorded. Moreover, these reporting and recording practices vary widely by country and are apt to change over time (see Chap. 1). Recognising the problems with recorded crime data, large-scale victimisation surveys, starting in the USA in 1973 but later run in many other countries, were developed to try to obtain more robust measures of crime and of trends in crime (Tilley and Tseloni 2016). These are invaluable both for analysing trends by country and for testing some of the hypotheses relating to the causes of the crime drop. However, due to variations in methodology and of crime categories, comparing patterns across countries can be risky. In an effort to overcome these problems, the International Crime Victims Survey (ICVS) was instigated in 1989, and there have been four major sweeps since then (1992, 1995–1997, 1999–2000, 2004–2005). The ICVS uses a standard survey instrument across different countries and across sweeps and comprises the most robust source of information on international crime trends and on variations in crime rate by country (see Van Dijk et al. 2007, 2012). Unfortunately, different sets of countries have taken part in each sweep and the sample sizes tend to be quite small (around 2000 per country). This limits what can be said with confidence about trends and relative rates of different types of crime. Nevertheless, the ICVS data comprise the best available for gauging international crime trends. Analyses of ICVS data accord well with findings of many much larger national victimisation surveys: there has indeed been a substantial and widespread drop in rates of high-volume crimes following a period of increase (Tseloni et al. 2010).

This chapter examines the role of security in generating falls in domestic burglary. It begins by briefly outlining some general theories that have been advanced to explain the international crime drop, the basic requirements that must be met by any satisfactory theory and the reason why security improvements comprise the most plausible explanation advanced so far. It then goes on to outline the security hypothesis in more detail and to show how it applies specifically to reductions in domestic burglary. Next, it spells out the data signatures that would be expected were the theory to be adequate and then indicates how the theory fares when confronted by victimisation survey data from multiple sweeps of the Crime Survey for England and Wales (CSEW) going back to 1981. The chapter acknowledges that not all security measures are effective; indeed, it highlights that burglar alarms seem to have lost the crime-reducing efficacy they once enjoyed. It also acknowledges that some security measures have serious downsides and emphasises the importance of designing security measures that are both effective and 'elegant'.

8.2   A Comprehensive Theory of the Crime Drop

8.2.1   Seventeen Propositions and Four Tests

Many of the initial efforts to explain the crime drop focused on violence in the USA (see, e.g. Blumstein and Wallman 2000). They tended to invoke particular developments there that had not occurred elsewhere. These were valiant attempts that lost plausibility as similar falls began to be observed in other countries and in particular as the similarity in trends in Canada and the USA became evident (Ouimet 2002).

Farrell et al. (2014) identified four tests that any explanation for the crime drop should satisfy if it is to be a serious contender:

1. *The cross-national test*: any satisfactory explanation must apply across countries as the crime drop is international. This rules out explanations that turn on idiosyncratic developments in specific countries.
2. *The prior increase test*: any satisfactory explanation must be consistent with the patterns of crime increase prior to the crime drop. This rules out explanations that invoke causes that would lead to expectations that crime would not have increased up to the point at which downward trends began.
3. *The e-crimes and phone theft test*: any satisfactory explanation has to be consistent with the patterns of crime increase that are observed alongside the major falls, and these rises are especially notable in relation to mobile telephony and cybercrime. This rules out explanations that imply that crime would necessarily fall across the board.
4. *The variable trajectories test*: any satisfactory explanation has to square with variations in trajectory across crime types and across jurisdictions. This rules out explanations that assume that all crime drops at the same time within any jurisdiction or that crimes will drop at the same time in all jurisdictions.

Following on from Farrell (2013), Farrell et al. (2014) list 17 explanations and indicate how they fare in relation to these four tests, as summarised in Table 8.1. It can be seen that the security hypothesis is the only one that satisfies all tests. This does not necessarily make it correct, though there is a range of additional supporting evidence to be considered. In addition, we suggest that at this stage the likelihood that an as-yet-undiscovered explanation will emerge that is consistent with the tests and refutes the security hypothesis whilst remaining consistent with the evidence is remote.

8.2.2   The Security Hypothesis

The broad starting point for the security hypothesis is clear and simple (Farrell et al. 2008, 2011a): Following widespread post-war rises in crime, steep falls have been produced in the main by reductions in opportunity as a result of deliberate increases

Table 8.1 Hypotheses to explain the crime drop and their a priori plausibility as indicated by four crucial tests

	Crime drop hypothesis	Test 1	Test 2	Test 3	Test 4
1	General economic improvement reduced crime	✓	✗	✗	✗
2	More concealed weapons increased deterrence	✗	✓	✗	✗
3	Increased use of death penalty induced greater deterrence	✗	✓	✗	✗
4	Gun control reduced crime due to gun control laws	✗	✓	✗	✗
5	Increased imprisonment reduced crime via incapacitation and deterrence	✗	✗	✗	✗
6	Better preventive policing reduced crime	✗	✓	✗	✗
7	Police staff increased so crime fell	✗	✓	✗	✗
8	Abortions in the 1970s meant less at-risk adolescents in the 1990s	✗	✓	✗	✗
9	Immigrants commit less crime and promote social control in inner cities	✓	✗	✗	✗
10	Strong economy shifts consumers away from stolen second-hand goods	✓	✗	✗	✗
11	Decline in hard drugs markets reduced related violence and property crime	✗	✓	✗	✗
12	Lead damaged children's brains in the 1950s, causing crime wave from the 1960s when they reached adolescence, then cleaner air from the 1970s caused crime drop of the 1990s	✓	✓	✗	✗
13	Aging population means proportionally fewer young people offenders and victims, so crime rates fall	✓	✓	✗	✗
14	Institutional control weakened in the 1960s causing crime increase, then strengthened in the 1990s causing crime drop	✓	✓	✗	✗
15	Improved quality and quantity of security reduced crime opportunities	✓	✓	✓	✓
16	Attractive displacement of offenders to e-crimes and changed lifestyles of victims	✓	✓	✗	✗
17	Portable phones spread rapidly in the 1990s and provide guardianship	✓	✓	✓	✗

in the use of security and improvements in the quality of security devices. These security increases and improvements are widespread (Clarke and Newman 2006; Van Dijk et al. 2007). They have directly reduced criminal opportunities and hence the numbers of high-volume crimes, notably burglaries and thefts of and from vehicles, both in the UK and in many other countries (Farrell et al. 2011a; Tseloni et al. 2017).

Indirect crime prevention effects are also produced as a result of crime opportunities reduced through increases in the uses of security measures and improvements in their quality. If high-volume 'debut' crimes are directly reduced as a consequence of security developments, typical gateways to further involvement in criminal activity involving a wider array of offences are also narrowed, thereby leading to reduced criminality and hence fewer crimes (Owen and Cooper 2013; Svensson 2002). The crimes avoided by reducing the onset of criminal careers may, of course, be of many types. Hence, in particular with regard to vehicle theft, an unintended beneficial knock-on effect of security improvements is that fewer young people have been inducted into criminality with consequential reductions in the array of crimes that they would otherwise commit (Farrell et al. 2015).

A further side effect of reducing vehicle theft following security improvements is that the supply of a major tool or resource for the commission of other crimes is reduced, and subsequent crimes that arise as a consequence of the initial crime are shrunk. Stolen cars are used, for example, in burglaries, commercial robberies and drive-by shootings. The idea that the prevention of one kind of crime, notably car theft, prevents others is known as the keystone hypothesis: the analogy is with the removal of the keystone of an arch after which the rest of the arch collapses (Farrell et al. 2011a).

The spread of security has been widely noted by criminologists. The commentary has been mostly critical (e.g. Davis 2000; Zedner 2003; Neocleous 2007). Much of the negative commentary has focused on private security guards where numbers have grown rapidly and where the activities of some providers have been brutal and themselves criminal. But criticisms have also been levelled, for example, at gated communities and their divisiveness; the exclusion of some from semi-public places such as shopping malls; the creation of ugly fortress societies; the side effects of increasingly manifest security on fear of crime; the deliberate fostering of fear in the interests of selling security products; the risks of displacement of offences from those who can afford security to those who cannot do so; external costs in terms of false alarms or measures designed to keep young people out of the way by producing ear-splitting noise that affects them but cannot be heard by the old; and the failure to address underlying causes of crime and criminality that lie in the fabric of society. In the social science literature, security and security increases have, thus, been commonly recognised but almost universally rued.

Proponents of the security hypothesis have likewise been struck by the growth in security but have been more interested in hard evidence of its consequences for crime levels than those subjecting it to negative commentary. They have acknowledged that the growth in some forms of security is open to criticism but conclude that the evidence suggests that the widely observed growth in security has led to crime drops and that the practical policy issue lies in determining what comprises 'bad' rather than 'good' security and how the latter can be made to thrive and the former wither. We return to this issue towards the end of the chapter.

The strongest evidence in favour of the security hypothesis currently relates to vehicle thefts, in particular theft of cars (Farrell et al. 2011b). The security built into modern cars generally lacks the downsides emphasised by critics. The security of cars is inconspicuous. The default is that the doors lock, the alarm is armed, the immobiliser is activated, the windows are shut and the wing mirrors are folded in as the driver walks away. The car is left in a secure state without effort or altering the appearance – the security measures do not detract from the appearance of the vehicle. The components of the audio system and GPS sat-nav are distributed in ways that prevent them being pulled out for resale. The aerial is an immovable ridge atop the car, the wheel nuts are locked and the opener to the fuel tank is accessible only from inside the car – all making stealing components from the car tricky. On driving away, the doors lock automatically to thwart roadside intruders and warnings sound or cameras show the driver when he/she risks causing damage to another when parking. And so on. These forms of security have driven falls in vehicle crime. They also lack those downsides that are homed in on by critics denouncing securitisation *tout court*.

This is not to say that there are no forms of security that suffer negative attributes, only that security per se does not necessarily have those attributes. In some cases, of course, there may be a trade-off between the use of security measures to reduce crime risk and negative aspects of security. Discussion of how trade-offs should be made lies beyond the compass of this volume (but see Ekblom 2011). All that we would say is that those designing and regulating the provision of security are in a good position to devise new security measures that are effective whilst minimising or eliminating negative characteristics.

8.3   Testing the Security Hypothesis for the Burglary Drop: A Data Signatures Approach

This chapter focuses on the security hypothesis as it relates specifically to domestic burglary. Its main concern is to adduce evidence that speaks to the conjecture that increases in and improvements of the security of dwellings have played a large part in reducing domestic burglary rates. The evidence drawn on comes from successive sweeps of the Crime Survey for England and Wales (CSEW, previously the British Crime Survey or BCS). The nature, strengths and limitations of the CSEW, notably for understanding burglary and trends in burglary, are discussed in Chap. 1 and will not be repeated here. Instead we focus on how the data were used and on findings that speak to the security hypothesis.

Our approach to analysis was to ask what specific findings from the data available from the CSEW would be expected were the security hypothesis to be correct. Unsurprisingly, the data collected from the CSEW do not provide the data needed to test all aspects of the overall hypothesis, as will become clearer later in this chapter. Given Peter Medawar's precept that science is 'the art of the possible' (Medawar 1967), we concentrate first on what we could do by way of security hypothesis testing with the data available, before moving to extensions to the hypothesis for which we have not yet found adequate data to conduct satisfactory tests.

If the security hypothesis relating to the drop in domestic burglary were correct, we would expect the following 'signatures' to be found in CSEW/BCS data[1]:

---

[1] See Farrell et al. (2016) for an explanation of our use of 'data signatures' as our method for testing the security hypothesis using CSEW/BCS data.

1. There would be an overall increase in the level of security of dwellings: other things being equal, more security is expected to reduce levels of burglary.
2. There would be a reduction in the proportion of dwellings unprotected by security measures: other things being equal, reducing the supply of unprotected dwellings reduces the supply of those most suitable for theft and hence levels of burglary.
3. Dwellings with more security would generally be less vulnerable to burglary than those with less security: more security leads to less vulnerability and hence lower levels of burglary.
4. The use of more effective security devices and combinations will grow more than the use of less effective security devices and combinations: the wider use of more effective security devices and combinations will be more effective in reducing burglary.
5. The protection conferred by security devices would increase over time: the quality of security devices tends to increase over time leading to increased effectiveness of those that are in place.
6. There will be no downward trend in burglary amongst properties with no security: for dwellings with no security, there is no protection from burglary.
7. There would be a greater fall in burglary with forced entry where the offender has to overcome security devices, than in unforced entry where this is not necessary.

Analysis of the succession of CSEW sweeps broadly finds that all seven of these expected signatures are found in practice. Although the best available, the data are far from perfect for our purposes. In particular (as shown in Appendix A in Chap. 4), questions about security devices fitted have changed over time. This has two important consequences that the reader needs to understand: first, we only know about the trends in security devices in relation to which questions were asked and we cannot, therefore, estimate trends for all devices; second, the meaning of 'no security' changes before and after 1998 given that no security refers to the absence only of those devices about which questions were asked (see also Chap. 4, Appendix A and Appendix Table 4.7). With these qualifications in mind, let us turn to findings.

8.4   Security-Led Burglary Drop in England and Wales

8.4.1   Signature 1: There Would Be an Overall Increase in the Level of Security of Dwellings

Figure 8.1, which presents the trends in the installation of security devices from 1992 to 2011/2012 as found in the CSEW, shows that there was a substantial increase in the use of window locks and double door locks/deadlocks as well as increases in the use of burglar alarms and outdoor lights, with some increase in the use of indoor lights (on a timer) followed by a later fall. In contrast there were falls in the use of security chains/bolts and bars (perhaps reflecting concerns about fire safety) from 1998 and in window bars/grilles from 1994. On aesthetic grounds, as discussed below, these falls in security device usage may be welcome in view of the overall fall in levels of burglary, especially if they are also not very effective.

8.4.2   Signature 2: There Would Be a Reduction in the Proportion of Dwellings Unprotected by Security Measures

The solid line in Fig. 8.1 refers to the trend in dwellings where there were no devices of the kind asked about in the respective sweeps. This shows an overall falling trend, as expected, although the figures for the earlier years may be inflated because of the lower number of security devices about which questions were asked before 1998. Even with a consistently narrow list of security devices across the 1992–1996 periods, the fall in households with no security was 40 percent in just a few years (Tseloni et al. 2017).

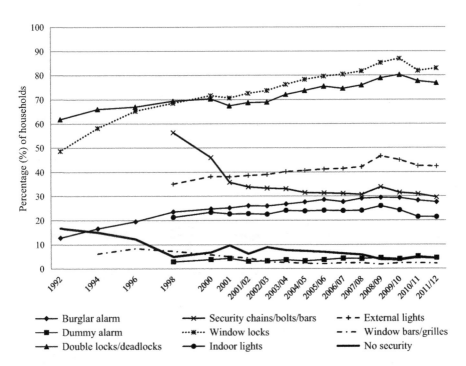

Fig. 8.1 Trends in the proportion of dwellings with security devices installed or 'no security' in England and Wales, CSEW/BCS 1992–2011/2012. Note: Adapted from Tseloni et al. (2017), p. 6

8.4.3   Signature 3: Dwellings with More Security Would Generally Be Less Vulnerable to Burglary than Those with Less Security

The effectiveness of security device combinations and the methods used for estimating this, is dealt with in Chap. 4. The interested reader is referred to this discussion if they want to know how the calculations were made (see also Tseloni et al. 2014). Suffice it here to say that the Security Protection Factor (SPF) measures the risk of burglary in comparison to dwellings with no security. Statistical power for the analysis was increased by merging data for multiple years (e.g. 2008/2009 to 2011/2012).

Figure 8.2 shows that in general larger combinations of security devices confer more protection against burglary than do single devices or small combinations, at least up to four of them. It should be noted here that the high effectiveness of combinations with six devices relies on only two sets, of which one is an outlier.

Figure 8.3 shows variations in the effectiveness of a selected number of different security combinations. Combinations generally produce higher collective SPFs than the sum of the SPFs for the individual devices comprising the combination, with the exception of burglar alarms, which will be discussed later in this chapter. What Fig. 8.3 shows is that certain combinations are especially efficacious, notably those

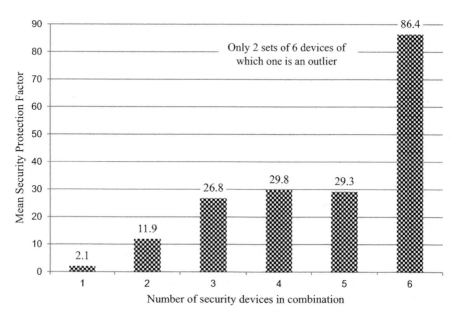

Fig. 8.2 Average protective effects against burglary with entry across numbers of security devices in combination, CSEW 2008/2009–2011/2012. Note: Calculated from Tseloni et al. (2014), Table 4

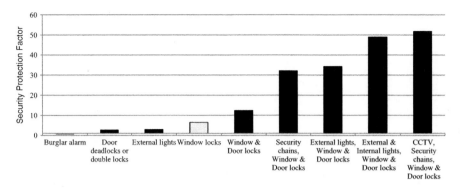

Fig. 8.3 Security Protection Factors (SPFs) for selected home security against burglary with entry (significant at 5% level unless shaded in grey) based on the 2008/2009–2011/2012, CSEW (Source: Tseloni and Thompson (2015), p. 34)

that include window locks, internal lights, door double or deadlocks and external lights. These add up to the 'WIDE' acronym referred to elsewhere in this book.

8.4.4   Signature 4: The Use of More Effective Security Devices and Combinations Will Grow More than the Use of Less Effective Security Devices and Combinations

Figure 8.4 shows the trend in the most effective security combinations from 1992 to 2011/2012, during which the rate of domestic burglary fell. These include dead or double locks to doors alongside window locks, the use of both of which increased considerably in England and Wales. This comprises a highly efficacious set of security devices (see Fig. 8.3 and also Chap. 4). The dotted line represents trends of consistent security definitions over time, i.e. before and after the 1998 CSEW. The bottom line shows the trend in the availability of only one device in households which, as seen earlier, confers minimal protection (Fig. 8.2). Overall the trends in Fig. 8.4 suggest that the steep increase in effective combinations and decrease in reliance on single security devices occurred as the steep fall in domestic burglary was taking place.

8.4.5   Signature 5: The Protection Conferred by the Presence of Security Devices Would Increase over Time

Figures 8.5, 8.6 and 8.7 show over time changes in the effectiveness of security combinations (Fig. 8.5 presents SPFs for two or three devices where Fig. 8.6 shows four or more) and of security devices when used on their own (Fig. 8.7), as measured through the SPF. Increases in apparent effectiveness patterns are most marked

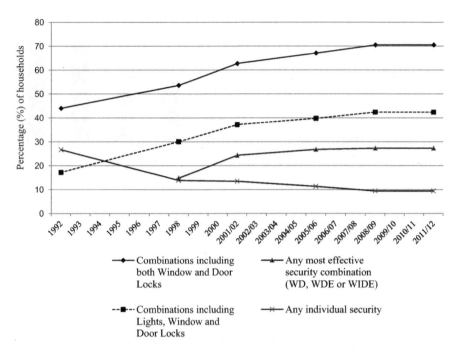

Fig. 8.4 Trends in most effective security combinations and single security, 1992–2011/2012, CSEW/BCS. Note: The 1992–1996 CSEW data about lights is assumed to correspond to external lights, internal lights or both in the post-1996, CSEW sweeps (Source: Tseloni et al. (2017), p. 8)

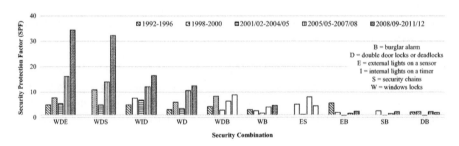

Fig. 8.5 Over time Security Protection Factors against burglary with entry for security device combinations (pairs or triplets) 1992–2011/2012 (significant in at least burglary with entry or attempted burglary) (*p*-value <0.05 unless shaded in white). Note: For the 1992–1996 CSEW sweeps, external and internal lights are confounded hence values for E and I equal L. In addition, security chains and dummy alarms were not recorded in this period, hence they are missing

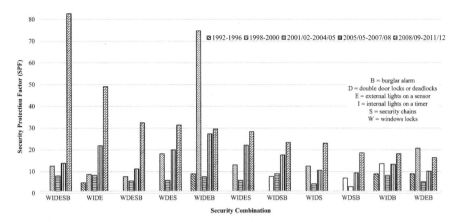

Fig. 8.6 Over time Security Protection Factors against burglary with entry for security device combinations (four or more) 1992–2011/2012 (capped at 80) (significant in at least burglary with entry or attempted burglary) (*p*-value <0.05 unless shaded in white). Note: For the 1992–1996 CSEW sweeps, external and internal lights are confounded hence values for E and I equal L. In addition, security chains and dummy alarms were not recorded in this period, hence they are missing

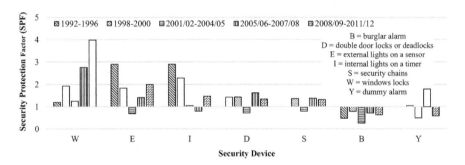

Fig. 8.7 Over time Security Protection Factors against burglary with entry for single security devices 1992–2011/2012 (significant in at least burglary with entry or attempted burglary) (*p*-value <0.05 unless shaded in white). Note: For the 1992–1996 CSEW sweeps, external and internal lights are confounded hence values for E and I equal L. In addition, security chains and dummy alarms were not recorded in this period, hence they are missing

for larger combinations, especially those using all or most WIDE measures. Increases in effectiveness are less noticeable with smaller combinations, and especially in relation to single measures operating on their own. Burglar alarms on their own and in combination are a special case to which we return later. Overall the trend is towards increasing efficacy of security devices, especially of WIDE-related combinations.

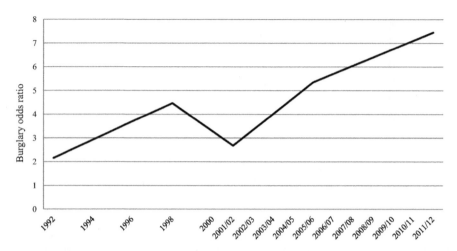

Fig. 8.8 Trends in burglary risk of dwellings with no security relative to national risk in England and Wales, CSEW/BCS 1992–1996 to 2008/2009–2011/2012. Note: Adapted from Tseloni et al. (2017), p. 7

8.4.6 Signature 6: There Will Be No Downward Trend in Burglary Amongst Properties with No Security

Figure 8.8 shows that those properties with no security, as measured in respective sweeps of the CSEW, not only failed to experience the overall fall in burglary but experienced a substantial rise in relative burglary risk between 1992 and 2011/2012. Chapters 4 (Table 4.5) and 5 (Figs. 5.2, 5.3, 5.4, 5.5, 5.6, 5.7, 5.8, 5.9 and Appendix Table 5.5) of this book show the household types which tend not to have security.

8.4.7 Signature 7: There Would Be a Greater Fall in Burglary with Forced Entry Where the Offender Has to Overcome Security Devices, than in Unforced Entry Where This Is Not Necessary

Figure 8.9 decomposes burglary trends into those where entry was forced, unforced and others (e.g. through distraction). The lowest (most darkly shaded) part of the figure shows the trend for burglary where entry was forced, where the burglar had to break in forcibly to gain entry. It is here that the major drop is found, from 1992 to 1993 onwards. The top (most lightly shaded) part of the figure shows the trend in unforced burglary. This typically occurs where the house is left insecure and the burglar can simply walk or climb in. The change here is moderate. Physical security

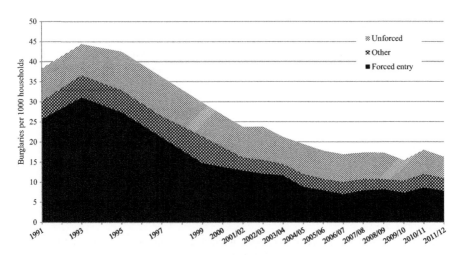

Fig. 8.9 Trends in burglary mode of entry, CSEW/BCS 1992–2011/2012 (Source: Tseloni et al. (2017), p. 10)

measures providing a barrier to entry are most likely to thwart efforts to commit burglary. Physical security measures are less likely to inhibit burglaries in properties that do not require forced entry.

8.5 Discussion

None of these patterns on its own clinches the case that increases in the amount of security or in the quality of devices have played a large part in producing the substantial fall in domestic burglary in England and Wales. Still less do they speak directly to the causes of falls in domestic burglary in other countries. Collectively, however, they accord with those patterns that would be expected were security increases and improvements to have been causally important in producing the drop. The improbability that security did not play an important part in producing the fall in burglary increases as the number of different signatures that accord with it grows. Security increases and improvements also speak clearly to obvious and plausible mechanisms through which a fall in rates of domestic burglary was produced. Security measures increase real and perceived effort and risk from breaking and entering into dwellings. Strong locks to windows and doors, when applied, make it more difficult to get into properties unlawfully. Internal and external lights increase the risk that any effort by those trying to break in will be observed and subsequently held to account for committing a burglary.

The findings described in this chapter are, we think, sufficiently persuasive that it is now up to sceptics to produce evidence that the observed patterns internal to the fall in burglary could have been produced in some other way. The conjecture that

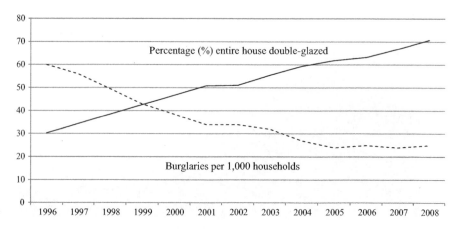

Fig. 8.10 Trends in percentage of dwellings with double glazing and burglaries per 1000 dwellings, 1996–2008 (Source: Farrell et al. (2016), p. 8)

security improvements have been crucial to the drop in domestic burglary in other jurisdictions would require that similar detailed analysis be undertaken there, with results that accord with those reported here. What seems to be common internationally is the growth in security, and our working conjecture is that it has been pivotal in producing those widespread crime drops that have been observed.

In regard to security increases, the discussion so far has focused on evidence relating to England and Wales as collected through the CSEW. However, we need to acknowledge that the data are imperfect for testing our hypothesis. They were collected for other purposes and we have drawn on them opportunistically, as the best available source of data relevant to testing our theory.

One gap in the CSEW data relates to double glazing.[2] Over the period of the crime drop, the proportion of dwellings in England and Wales with double glazing grew dramatically. Full double glazing, covering all windows and doors, builds in increased security, even though crime prevention is not its main purpose. Double glazing was installed in existing properties in the interests of energy conservation and protection from the cold, but will have simultaneously upgraded security. Moreover, new housing stock with double glazing will have incorporated better security than in much older, unimproved stock. Figure 8.10 shows the trend in the proportion of dwellings with full double glazing (solid line), setting it against the changing burglary rate (dotted line). The inverse relation suggests there may be a causal relationship, but we lack data that would allow us to determine whether the relationship is indeed a causal one.

A second gap relates to improvements in the robustness of front doors. Figure 8.11 shows the change in the percentage of dwellings in the UK rented from local authorities, so-called council houses (see Chaps. 2 and 5). Some of this change reflects

---

[2] The CSEW has incorporated this question in recent sweeps.

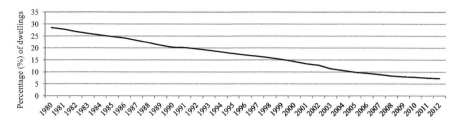

Fig. 8.11 Trends in percentage of dwellings rented from councils, 1980–2012 (Source: Live tables on dwelling stock (including vacants), Table 101, DCLG. Dwelling stock by tenure, 1980 to 2012)

right to buy, which was introduced in 1980. The front doors of many council houses were notorious for the ease with which their bottom panels could be kicked in by burglars (Tilley et al. 1991 report this for the St Ann's Estate in Nottingham). A typical early move by those who bought council houses and flats was to change the front door, making the properties more secure. Moreover, even when right to buy was not exercised, front doors were replaced in many places to improve security, as occurred in St Ann's in the early 1990s. There were several general initiatives in Britain focused on particular estates, and many of these included improvements to the security of dwellings, making them less vulnerable to burglary in ways that are not fully captured in the CSEW (see Chap. 2).

A third gap relates to the quality of the security measures that are in place. The CSEW asks about the presence of security measures of various types. But security type varies in quality, which is liable to influence its effectiveness. Over time one might expect the quality of devices generally to improve, although as we shall see in the next section this is not necessarily always the case!

8.6   The Curious Case of Burglar Alarms

Figure 8.12 presents findings relating to the marginal effects of adding a burglar alarm to other security measures and combinations across two time periods. We calculated the marginal effects by comparing the SPF for different combinations including and not including burglar alarms amongst them. A score of more than 1 indicates that the risk of burglary is to that extent reduced, so a 2 would mean that the risk of burglary was halved by the addition of a burglar alarm to those labelled in the x-axis. A score of less than 1 indicates that the risk of burglary is increased. Thus, a score of 0.5 would indicate that the risk of burglary was doubled. Panel A (top) shows the findings for 1992–1996 and Panel B (bottom) those for 2008/2009–2011/2012. The different time periods reflect those over which common questions about security devices were asked (remembering that one set of questions relates to security devices in place at the time of the (first) burglary for respondents reporting they had suffered a burglary with entry and the others those who had not suffered a

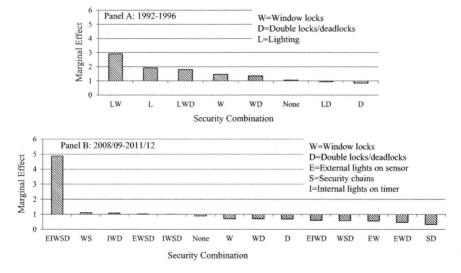

Fig. 8.12 Marginal effects of burglar alarms on burglary with entry, CSEW/BCS 1992–1996 and 2008/2009–2011/2012 (Source: Tilley et al. (2015a), pp. 11 and 12)

burglary). Moreover, the data from several waves had to be merged to create a large enough sample for the analyses undertaken.

The surprising and striking finding shown across the two panels is that the presence of burglar alarms in the earlier period is pretty consistently associated with reduced risks of burglary with entry, whilst the opposite is found for the later period when the presence of a burglar alarm is associated with increased risk. Why might this be the case and how does it fit with the security hypothesis? Figure 8.1 shows that the proportion of properties with burglar alarms doubled from around 15 percent to around 30 percent between 1992 and 2009–2010. One plausible possibility is that many burglar alarms were installed as a form of grudge spending (Goold et al. 2010), often at the behest of insurance companies. If this is the case, then it seems likely that those incurring the cost will tend to opt for systems that are inexpensive, nominally fit the requirement and are of poor quality. Moreover, the rate of false alarms is very high, at well over 90 percent, and attention to alarms by the public and police has therefore become muted at best (Cahalane 2001; Sampson 2011; LeBeau and Vincent 1997). Under these circumstances alarms will cease to elicit direct responses that significantly increase real or perceived risks to burglars.

Burglars may interpret alarms in two quite different ways: (a) as an indication that risks to them are thereby made higher and therefore that they should avoid burglary or (b) as an indication that they are there to protect property that is worth stealing (Wright and Decker 1994; Nee and Taylor 2000). Hence what may have happened is that the first reading (a) tended to prevail in 1992–1996, when there were far fewer burglar alarms (but which elicited a response), but that the second reading (b) tended to prevail in the second period, when burglar alarms had become much more widely installed (but ceased to elicit the same level of response). A second hypothesis that has been suggested to us is that burglar alarms may be fitted, but they are not necessarily armed either when no one is home or when residents retire to bed. This would mean that the potential of alarms to increase real risk to burglars when dwellings are otherwise vulnerable is compromised. A third hypothesis is that offenders were initially deterred by what at the time appeared to be high-tech burglar alarms, but subsequently realised they were unlikely to increase their risk of arrest. The same learning process has been suggested for car thieves who in the 1980s and 1990s were significantly deterred by vehicle alarms but then learned there was little additional risk (Farrell and Brown 2016). A number of alternative hypotheses are discussed in detail in Tilley et al. (2015a).

The findings for burglar alarms do not dent the overall thesis that increases in security have played a major part in driving down rates of domestic burglary, but they do indicate that security measures are heterogeneous. Some are better than others and some measures, such as burglar alarms, may vary in effectiveness over time depending on the context.

8.7   The Importance of Design and Detailed Understanding

It is important to understand which measures work on their own and in combination and in which circumstances to determine what security devices to recommend for householders, developers, insurance companies and those managing public and private property portfolios. From research reported in this book, the WIDE combination appears to be an economical default, whilst Skudder et al. (2017) have also found it carbon cost-effective. Moreover, there is good reason to welcome the findings relating to burglar alarms. We have advocated the development of elegant security, whose attributes we summarise under a DAPPER mnemonic (Tilley et al. 2015b; Farrell and Tilley 2017). D refers to 'default' (being in place as the normal state of affairs), A to 'aesthetically pleasing' (either neutral or unobjectionable), P to powerful (effective), P to principled (equitable and jeopardising no one's rights), E to effortless (activated without bother) and R rewarding (cost-effective). Alarms fail the DAPPER test on several scores. Alarms normally have to be set (they are not the default in houses as they are in cars). Alarms are not aesthetically pleasing (they intentionally emit nuisance nasty noises). Alarms are not powerful (on balance they increase rather than decrease the risk of burglary with entry). Alarms require action on entering and leaving property (they are not effortless). Alarms are not cost-effective (they are quite expensive and do not reduce burglary risk). Of course, it is possible to conceive of alarms that are DAPPER. Indeed, early ones were more DAPPER than later ones in some respects. For example, they were inconspicuous and sounded in local offices of alarm companies rather than publicly and the alarm company staff would quickly go and check whether a burglary was in progress (Tilley et al. 2015a). However, such a system is expensive and to that extent available only to the relatively affluent. Moreover, so-called silent alarms that are sometimes fitted to dwellings at (short-term) high risk of repeat incidents and which feed through to the police who respond quickly function differently from alarms fitted as a precautionary measure intended to deter offenders rather than help catch them (Chenery et al. 1997).

Figure 8.1 shows a tendency for most types of security to dwellings to increase, but with interesting exceptions. In particular the proportion of dwellings with security chains, bolts and bars has dropped as has that of properties with window bars/grilles. In DAPPER terms these are welcome changes. They mark a reduction in the stock of houses bearing non-DAPPER security attributes. Window bars/grilles in particular are ugly and betoken high levels of risk to residents and visitors alike. In the context of measures that have been effective in reducing levels of burglary, their reduced use is to be welcomed. In policy terms, it would be desirable for more DAPPER devices to drive out those that are less DAPPER.

8.8   Conclusion

The drop in domestic burglary in England and Wales has been dramatic. From a high of close to an estimated 2.5 million incidents in 1993, there was a fall of more than two thirds to just below 800,000 by 2013/2014. The evidence offered in this chapter corroborates the security hypothesis as it speaks to the direct effects of changes in the volume and efficacy of security measures in reducing the level of domestic burglary. Moreover, our findings are liable to understate the direct contribution of security improvements to the drop in domestic burglary, because of the lack of data capturing many forms of security to dwellings that have apparently improved. This chapter has not, however, presented data that speak to the indirect effects of security improvements on domestic burglary. Whilst it is plausible that security improvements in cars have meant that would-be burglars do not make use of previously easily stolen cars in the commission of their crimes, we have no direct evidence on this or on whether it has contributed to the overall falls in burglary. Equally, although elsewhere we have adduced evidence that changes in age-related criminality are as expected from the hypothesised role that security improvements have made to the onset of criminal careers (Farrell et al. 2014, 2015), we lack evidence that speaks directly to any effect on domestic burglary in particular. These are issues that warrant further empirical research.

References

Blumstein, A., & Wallman, J. (2000). *The crime drop in America*. Cambridge: Cambridge University Press.

Cahalane, M. (2001). Reducing false alarms has a price – So does response: Is the real price worth paying? *Security Journal, 14*, 31–53.

Chenery, S., Holt, J., & Pease, K. (1997). *Biting back II: Reducing repeat victimisation in Huddersfield, Crime detection and prevention series, Paper 83*. London: Home Office.

Clarke, R. V., & Newman, G. (2006). *Outsmarting the terrorists*. Westport: Praeger.

Cohen, L. E., & Felson, M. (1979). Social change and crime rate trends: A routine activities approach. *American Sociological Review, 44*, 588–608.

Davis, M. (2000). *City of Quartz*. London: Verso.

Ekblom, P. (2011). *Crime prevention, security and community safety using the 5Is framework*. Basingstoke: Palgrave Macmillan.

Farrell, G. (2013). Five tests for a theory of the crime drop. *Crime Science, 2*(5), 1–8.

Farrell, G., & Brown, R. (2016). On the origins of the crime drop: Vehicle crime and security in the 1980s. *The Howard Journal of Criminal Justice, 55*(1–2), 226–237.

Farrell, G., & Tilley, N. (2017). Technology for crime prevention: A supply side analysis. In B. Leclerc & E. Savona (Eds.), *Crime prevention in the 21st century: Insightful approaches for crime prevention initiatives* (pp. 377–388). Springer.

Farrell, G., Tilley, N., Tseloni, A., & Mailley, J. (2008). The crime drop and the security hypothesis. *British Society of Criminology Newsletter, 62*, 17–21.

Farrell, G., Tseloni, A., Mailley, J., & Tilley, N. (2011a). The crime drop and the security hypothesis. *Journal of Research in Crime and Delinquency, 48*(2), 147–175.

Farrell, G., Tseloni, A., & Tilley, N. (2011b). The effectiveness of vehicle security devices and their role in the crime drop. *Criminology and Criminal Justice, 11*(1), 21–35.

Farrell, G., Tilley, N., & Tseloni, A. (2014). Why the crime drop? *Crime and Justice, 43*, 421–490.

Farrell, G., Laycock, G., & Tilley, N. (2015). Debuts and legacies: The crime drop and the role of adolescent-limited and persistent offending. *Crime Science, 4*(16), 1–10.

Farrell, G., Tseloni, A., & Tilley, N. (2016). Signature dish: Triangulation from data signatures to examine the role of security in falling crime. *Methodological Innovations, 9*, 1–11.

Goold, B., Loader, I., & Thumala, A. (2010). Consuming security: Tools for a sociology of security consumption. *Theoretical Criminology, 14*(1), 3–30.

LeBeau, J., & Vincent, K. (1997). Mapping it out: Repeat address burglary alarms. In D. Weisburd & T. McEwan (Eds.), *Crime mapping and crime prevention, Crime prevention studies* (Vol. 8, pp. 289–310). Monsey: Criminal Justice Press.

Medawar, P. (1967). *The art of the soluble*. London: Methuen.

Nee, C., & Taylor, M. (2000). 'Examining burglars' target selection: Interview, experiment or ethnomethodology? *Psychology, Crime & Law, 6*, 45–59.

Neocleous, M. (2007). Security, commodity, fetishism. *Critique, 35*(3), 339–355.

Ouimet, M. (2002). Explaining the American and Canadian crime "drop" in the 1990s. *Canadian Journal of Criminology and Criminal Justice, 44*(1), 33–50.

Owen, N., & Cooper, C. (2013). *The start of a criminal career: Does the type of debut offence predict future offending?* London: Home Office.

Sampson, R. (2011). *False burglar alarms, Problem-oriented guides for police series, Guide No. 5* (2nd ed.). Washington, D.C.: US Department of Justice.

Skudder, H., Brunton-Smith, I., Tseloni, A., McInnes, A., Cole, J., Thompson, R., & Druckman, A. (2017). Can burglary prevention be low carbon and effective? Investigating the environmental performance of burglary prevention measures. *Security Journal, 31*, 111. https://doi.org/10.1057/s41284-017-0091-4.

Svensson, R. (2002). Strategic offences in the criminal career context. *British Journal of Criminology, 42*, 395–411.

Tilley, N. (2010). Shoplifting. In F. Brookman, H. Pierpoint, T. Bennett, & M. Maguire (Eds.), *Contemporary forms of crime: Patterns, explanations & responses* (pp. 48–68). Cullompton, Devon: Willan.

Tilley, N., & Tseloni, A. (2016). Choosing and using statistical sources in criminology – What can the crime survey for England and Wales tell us? *Legal Information Management, 16*(2), 78–90.

Tilley, N., Webb, J., & Gregson, M. (1991). Vulnerability to burglary in an inner-city area. *Issues in Criminological and Legal Psychology, 17*, 112–119.

Tilley, N., Thompson, R., Farrell, G., Grove, L., & Tseloni, A. (2015a). Do burglar alarms increase burglary risk? A counterintuitive finding and possible explanations'. *Crime Prevention and Community Safety, 17*(1), 1–19.

Tilley, N., Farrell, G., & Clarke, R. V. (2015b). Security quality and the crime drop. In M. Andresen & G. Farrell (Eds.), *The criminal act* (pp. 59–76). London: Routledge.

Tonry, M. (Ed.). (2014). *Why crime rates drop and why they don't* (Crime and Justice, Vol. 43, No. 1). Chicago: University of Chicago Press.

Tseloni, A., & Thompson, R. (2015). Securing the premises. *Significance, 12*(1), 32–35.

Tseloni, A., Mailley, J., Farrell, G., & Tilley, N. (2010). The cross-national crime and repeat victimization trend for main crime categories: Multilevel modeling of the International Crime Victims Survey. *European Journal of Criminology, 7*(5), 375–394.

Tseloni, A., Thompson, R., Grove, L., Tilley, N., & Farrell, G. (2014). The effectiveness of burglary security devices. *Security Journal, 30*(2), 646–664.

Tseloni, A., Farrell, G., Thompson, R., Evans, E., & Tilley, N. (2017). Domestic burglary drop and the security hypothesis. *Crime Science, 6*(3).

Van Dijk, J., Van Kesteren, J., & Smit, P. (2007). *Criminal victimisation in international perspective*. The Hague: Boom Juridische uitgevers.

Van Dijk, J., Tseloni, A., & Farrell, G. (2012). *The international crime drop: New directions in research*. Basingstoke: Palgrave Macmillan.

Wright, R. T., & Decker, S. H. (1994). *Burglars on the job: Streetlife and residential break-ins*. Boston: Northeastern University Press.

Zedner, L. (2003). Too much security? *International Journal of the Sociology of Law, 31*, 155–184.

Chapter 9
From Project to Practice: Utilising Research Evidence in the Prevention of Crime

Rebecca Thompson and Kate Algate

Abbreviations

AC	Advisory Committee
CSEW	Crime Survey for England and Wales
ESRC	Economic and Social Research Council
NCDP	Nottingham Crime and Drugs Partnership
NHWN	Neighbourhood and Home Watch Network (England and Wales)
RCUK	Research Councils United Kingdom
REF	Research Excellence Framework
SBD	Secured by Design

9.1 Introduction

This book makes a number of important contributions to knowledge regarding domestic burglary. We hope that at least some of this knowledge has found its way into practice. That said, much has been written about the challenges inherent in generating impact and exchanging knowledge with nonacademic partners, albeit when in relation to crime, this has predominantly been in a policing context (Murji 2010; Henry and Mackenzie 2012; Fyfe and Wilson 2012; Cockbain and Knutsson 2015; Knutsson and Tompson 2017; Goode and Lumsden 2018). It has been likened to an 'exercise in struggling uphill' (Ekblom 2002, p. 132; Tilley and Laycock 2000). It is argued knowledge exchange cannot be viewed as a one-way activity whereby research findings are simply 'packaged up' and presented to research users

R. Thompson (✉)
Quantitative and Spatial Criminology, School of Social Sciences, Nottingham Trent University, Nottingham, UK
e-mail: becky.thompson@ntu.ac.uk

K. Algate
Coventry Citizens Advice, Coventry, UK

© Springer Nature Switzerland AG 2018
A. Tseloni et al., *Reducing Burglary*,
https://doi.org/10.1007/978-3-319-99942-5_9

(Davies et al. 2008). Instead, it should be seen as a complex, two-way *process* (Innes 2010; Fyfe and Wilson 2012; Henry and Mackenzie 2012).

This chapter reflects upon the process of knowledge exchange undertaken as part of an Economic and Social Research Council (ESRC)-funded project which asked: 'which burglary security devices work, for whom and in what context?'[1] This research brought together a number of public and third sector organisations and academics for the purposes of a time-limited project. The chapter draws together the authors' collective personal reflections and discusses some of the ways in which the project influenced policy and practice. In order to capture the experiences of both the academics and practitioners involved in the research, the chapter is co-authored by a researcher from the burglary project team and a practitioner member of the project Advisory Committee.

9.2   Context

Over the last few decades, there has been a proliferation of research and commentary on knowledge exchange and collaboration more broadly (Mitton et al. 2007; Davies et al. 2008; Henry and Mackenzie 2012; Murji 2010; Fyfe and Wilson 2012; Cockbain and Knutsson 2015). Much of this work is concerned with how most effectively to communicate research evidence to practitioners and policymakers as well as the barriers and enablers to successful collaboration. In a policing context, the creation of the College of Policing in 2012 has placed growing emphasis upon the recognition and development of skills and knowledge within policing as well as improving the evidence base around 'what works' (College of Policing 2017; Crawford 2017). There are a number of contextual factors which may have facilitated this increased attention, of which two will briefly be discussed here.

Over the past 30 years, the quality of research undertaken in UK universities has been assessed to ensure the public investment of funds leads to impact. This assessment has been known under various guises, including the Research Selectivity Exercise, the Research Assessment Exercise or, more recently, the Research Excellence Framework (REF). According to Stern (2016), the REF has helped the sector to focus upon research quality and excellence whilst driving an awareness of the importance of research impact. The strategic partnership of the UK's seven Research Councils, Research Councils UK (RCUK), define research impact as 'the demonstrable contribution that excellent research makes to society and the economy' (ESRC 2017).

Prior to the assessment of research impact, different reward systems for academia and practitioners sometimes (and, to some extent, still) proved a barrier to working collaboratively (Mitton et al. 2007; Fraser 2004). Academics were largely rewarded on the basis of their peer-reviewed publications written for academic audiences (and, by the same token, often inaccessible to practitioners) (Buerger 2010). The REF has placed greater emphasis upon conducting impactful research and

---

[1] Hereafter referred to as 'the burglary project'.

engaging with research users which, in turn, has supported academics in being part of collaborative endeavours without limiting their academic careers (Stern 2016).

Conducting impactful research is increasingly important given the public sector climate. In October 2010, the British Government announced a reduction in the central funding to police forces in England and Wales by 20 percent over the period from March 2011 to March 2015 (HM Treasury 2010). The total number of people employed by police forces in England and Wales has consequently been falling since 2010 (Home Office 2017) which means forces are increasingly required to do more with fewer staff (HMIC 2014; House of Commons 2018). This period of austerity has placed increasing pressure upon public service organisations of all kinds to maintain previous levels of service and has left some systems overwhelmed (HMIC 2017; Bach and Cole 2017).

These contextual factors have also placed increasing pressure upon academics to produce high-quality research findings which add to the knowledge base whilst also sharing (or better, co-producing) that knowledge with policymakers, practitioners and the wider public (McAra 2017). As McAra (2017, p. 769) points out, academics must somehow '...sustain the requisite level of critical distance from emergent consumers of knowledge, without running the danger of being "absorbed" and "tamed"...' whilst, at the same time, influencing practice. So, although the increasing focus upon impact, in the authors' view, should be welcomed, it is not without its challenges in terms of academics maintaining what Crawford (2017, p. 208) terms their 'critical independence'.

9.3 The Project

The International Association of Chiefs of Police (n.d. cited in Rojek et al. 2012) classifies police-academic partnerships into three types (taking into account their commitment, formality and duration). These are cooperation, coordination and collaboration. Cooperation involves short-term, informal partnerships that may involve seeking advice or asking for data. Coordination includes more formal partnerships centred on a specific project where the partnership ends when the project ends. Finally, collaboration consists of a formal long-term partnership where academics and practitioners work on multiple projects over time. This chapter draws upon the authors' collective personal reflections of what was intended to be a specific, short-term project which involved formal *coordination* between academics and practitioners. The project therefore initially represented a 'coalition of temporary common interests' (Strang 2012, p. 211) as opposed to a partnership with common longer-term goals. This was particularly true for partner organisations whose dominant expertise related to burglary. However, the academic team have continued to *collaborate* on other projects with a number of partners with a wider organisational remit, for example, the Nottingham Crime and Drugs Partnership (NCDP), the Home Office, the Office for National Statistics and the Neighbourhood and Home Watch Network (England and Wales) (NHWN). This illustrates the evolving nature of these kinds of relationships and reinforces Tompson et al.'s (2017) point that partnerships may not neatly fall into one category.

Prior to submission of the funding bid, the academic project lead held discussions with the burglary lead for Nottinghamshire Police and NCDP. Both organisations helped formulate ideas and provided letters of support. The project was subsequently approved for funding by the ESRC over an 18-month period. The research was predominantly designed to assess the role of security in producing the dramatic falls in burglary since the mid-1990s. It also sought to determine the effectiveness of different security devices in various contexts. To this end, data were analysed from the Crime Survey for England and Wales (CSEW) and the UK Census. Some of the project findings are detailed in Chaps. 4, 5, 6 and 8 of this book.

To ensure the findings were both theoretically and practically relevant, the academic team convened a group of subject/practice experts to work with them – an Advisory Committee (AC). Individuals not already involved, by virtue of being part of the pre-bid submission discussions, were approached to be part of the AC via email after the project team had identified they had a specific remit either in preventing or in supporting victims of burglary. The AC mainly comprised practitioners from public and third sector organisations (for a full list, please see the NTU Burglary and Security website). Committee membership involved attending a number of face-to-face meetings, reading briefing notes and providing comments on emerging findings. Members were also asked to disseminate project findings amongst their own networks. The AC was an attempt to facilitate a process of knowledge exchange through face-to-face meetings and email contact. One of the primary justifications for having an AC stemmed from the fact that when organisations have a participatory role in shaping the research, they are more likely to view it as useful (Greene 2010).

The burglary project differs from most other research partnership endeavours in three ways. Firstly, the majority of the research involved analysis of publicly available data (the CSEW and the UK Census). This meant the project team were not reliant upon partners for data thus largely avoided data access/sharing issues. Secondly, the project did not involve directly reviewing or evaluating partner practices. Finally, the researchers were not reliant upon AC partners for funding. These three factors mean some of the more traditional barriers to partnership working that have been outlined in the existing literature were absent. However, there were a number of specific lessons to emerge from this project which will now be discussed.

9.4 Key Factors

This section will outline the factors we found to be particularly important in our experience of successfully exchanging knowledge between academics and practitioners – but how are we measuring *success*? As Frisch (2015, p. 52) states: '…the extant literature on researcher-practitioner partnerships fails to reach consensus in defining what constitutes a successful collaboration'. In this chapter, we have adopted the three criteria used by Frisch (2015, 2016) to assess success, namely, perceptions of the individuals involved, fulfilment of initial goals and the degree of translation into policy and practice. In order to achieve those objectives, we found two factors to be particularly important: first, the importance of good quality relationships and second, tailored communication. This is consistent with existing literature on academic-practitioner partnerships. These are by no means the only elements required to generate effective knowledge exchange but were those that really assisted the burglary project.

9.4.1 Relationships

One of the most frequently cited factors in the effective exchange of knowledge is the importance of good quality relationships (Mitton et al. 2007; Meagher et al. 2008; Wuestewald and Steinheider 2009; Marks et al. 2010; Foster and Bailey 2010; Reback et al. 2002; Strang 2012; Tartari et al. 2012; Hara et al. 2003; Steinheider et al. 2012; Oliver et al. 2014). Crawford (2017, p. 206) goes as far as to state that relationships are the 'backbone' of co-production. The burglary project AC brought together individuals from organisations with different needs, interests, objectives, cultures and decision-making processes (Rosenbaum and Roehl 2010; Crawford 2017). Greene (2015, p. 117) likens this type of research with learning to dance the Tango – '......police researchers seeking to deliver academically valid, policy-relevant and effective work must, like those learning the Tango, navigate the complexities of another world, learn its symbols and language and maintain the "dance" with their partners'. With the burglary project, we were required to learn the symbols and language of multiple worlds.

For this reason, it was especially important to meet face-to-face in order to develop relationships and build rapport (Grieco et al. 2014; Innvaer et al. 2002; Cockbain 2015). This relationship building began before the submission of the bid to the ESRC. By involving a number of partners in the development of the bid, the academic team hoped to increase the usefulness and practical applicability of the project. Once funded, the AC was scheduled to formally meet four times over the 18-month period. Ideally, there would have been more face-to-face interaction. However, this was not practical given those involved were not within close proximity geographically. This placed greater importance on communication via email, but this was not without its issues (see Sect. 9.4.2).

During all interactions (whether in person or via email), trust, mutual respect and honesty were crucial. This is very well supported by previous research (Nutley et al. 2007; Innvaer et al. 2002; Foster and Bailey 2010; Burkhardt et al. 2017; Reback et al. 2002; Lane et al. 2004; Laycock 2015; Crawford 2017; Tompson et al. 2017). As Greene (2010, p. 124) states: '...trust provides a level of comfort in the sharing of information, some of which may be sensitive, but nonetheless necessary to understand the nature of the problem confronted'. The burglary project was the first time some partners had worked together. One drawback of this kind of short-term project is that it can take time to develop trust.

With regard to trust, years of what Rosenbaum (2010, p. 144) refers to as 'hit-and-run' research may have left some practitioners cautious about collaborating with academics. In this case, AC practitioners trusted the project team not to disclose any sensitive meeting contributions/reflections without being fully anonymised or without their prior consent. The vast majority of AC contributions were either suitable for public consumption or already in the public domain. However, the project team took a cautious approach whenever using information from AC members so as not to make them feel their trust was misplaced. From an academic perspective, the project team were sharing new, unpublished findings, and trusted practitioners would not publish them ahead of time, without permission or without appropriate citation. It was therefore vital to create a space within which experiences could be freely shared (and, in some cases, respectfully challenged) without fear of wider publication or reprisal. Open and honest communication between partners was fundamental to the success of the project, thus the 'Chatham House Rule'[2] was enforced during meetings. Minutes were taken during all AC meetings, and these were shared with the group for approval.

The importance of trust is highlighted by the project findings regarding burglar alarms. In short, burglar alarms were found to be associated with an increased (rather than decreased) risk of burglary with entry (for more information, see Chaps. 4 and 8 of this book and Tilley et al. 2015). These somewhat counter-intuitive results were likely to gain interest from the wider media. The AC acted as a sounding board to discuss the analysis in a relatively safe environment and provided an invaluable opportunity to obtain feedback as to how the team might proceed to disseminate findings. Practitioners were also able to draw on their own experiences to provide insights as to why such an effect may have been found.

Perhaps most importantly, the AC was founded upon '…mutual respect for the knowledge that each partner brings to the collaboration' (Marks and Sklansky 2008, p. 92). There was no hierarchy in relation to knowledge – knowledge obtained via academic research did not trump experiential knowledge – both were viewed as equally valid. The purpose of the AC was to draw on the respective strengths of each individual in order to generate meaningful dialogue and impact (Crawford and L'Hoiry 2017; Innes and Everett 2008). Thus, the AC was seen as a forum for 'mutual education' (Burawoy 2005, p. 8). As an example, the AC provided a platform to discuss and respond to a Department for Communities and Local Government (DCLG) building regulations review, specifically to proposed changes to the Secured by Design (SBD) standard. These discussions culminated in a letter to the DCLG published in *The Times* (23rd October 2013). This would not have been possible without both academic and practitioner involvement.

---

[2] 'When a meeting, or part thereof, is held under the Chatham House Rule, participants are free to use the information received, but neither the identity nor the affiliation of the speaker(s), nor that of any other participant, may be revealed' (Chatham House 2018).

All this said, we recognise the relationship-building process was perhaps easier for the burglary project due to the distinctive nature of the partnership (see Sect. 9.3). Not being reliant upon each other for data or funding undoubtedly assisted the academic members of the project team in maintaining their independence and impartiality (Crawford 2017; McAra 2017). However, the short-term nature of the project did mean it was, in some cases, more difficult to build meaningful, long-term relationships.

9.4.2   Communication

9.4.2.1   Communication with the Advisory Committee

Oliver et al.'s (2014) systematic review of knowledge exchange research identified a number of barriers and facilitators to evidence use by policymakers. Being unable to find (or access) good quality, timely and relevant research outputs was a frequently cited barrier to evidence uptake (ibid). Relatedly, previous research suggests regular face-to-face interactions between academics and potential users increase the likelihood of research use and can help to establish rapport (Grieco et al. 2014; Weiss 1995; Nutley et al. 2007; Cockbain 2015). In this case, interaction took the form of face-to-face meetings consisting of the project team and AC stakeholders and email contact. The face-to-face meetings usually involved the research team presenting emerging findings followed by a group discussion. These meetings were invaluable for the project team – not only in the discussions that were generated but also in forcing the team to reflect upon the relevance and meaning of the findings to the world outside of academia. Without the AC meetings, the team would have been at risk of becoming isolated and insular. The AC therefore provided a fresh, 'outside' perspective. The presentations and briefing notes required the team to synthesise project findings into a digestible format. Producing one version of the briefing note for the range of different AC organisations was not ideal (nor straightforward), but the team tried to appeal to a wide audience by capturing the *general* project findings and potential practical implications in these notes. Producing multiple briefings tailored to each AC organisation was not feasible within the project timescales. However, the project team tried to be responsive to AC need by producing specific documents or giving presentations on request.

The frequency of communication with the AC was an important consideration for the team. The importance of regular communication between academics and practitioners is an oft-cited factor in effective knowledge exchange (Fleming 2010). As mentioned, the frequency of AC face-to-face contact was not optimal. This direct contact was therefore supplemented with emails and an online blog. However, there is a fine balance between keeping people updated and overwhelming them. We are in agreement with the existing literature regarding the importance of face-to-face contact. However, in this case, meeting regularly in person was simply not feasible due to geographical and diary constraints. With this in mind, we argue that a more effective way to maintain regular contact with multiple partners might be to supplement face-to-face communication with other virtual means, such as blogs, vlogs, live discussions via Twitter (e.g. see #wecops) and webinars.

Relatedly, academic research takes time. Practitioners may expect more regular updates than can be realistically delivered (Lane et al. 2004; Fleming 2010; Crawford 2017). There can sometimes therefore be tension between the need to provide research findings when they are 'ready' to be more widely disseminated and the regularity with which partners would like to be informed. Added to this, the policy environment is ever-changing; thus, there is a danger that research can lose its relevance and appeal quickly (Weiss and Weiss 1981; Ritter 2009). As mentioned, the project team tried to address this by producing briefing notes and presentations on request.

9.4.2.2   Communication Beyond the Advisory Committee

The burglary project had a number of different audiences beyond the AC – other academics, police forces, victim support organisations, insurance companies, government departments, security providers and house builders to name a few. Practitioners often state they struggle to find the time to search for and read what can sometimes be lengthy 'academic' documents (Oliver et al. 2014; Rosenbaum 2010; Ritter 2009). Added to this, they may be physically unable to obtain particular documents due to paywalls and expensive subscription fees and/or because they do not have access to university library facilities (Ritter 2009). To maximise impact and address some of these accessibility issues, the project team opted to disseminate research findings using a range of media.

In order to disseminate the findings to these groups, the team produced a range of 'user-friendly', freely available outputs. This was important due to the apparent tension between information most suited to policymakers (summative and accessible) and information produced for academics (often nuanced, lengthy and complex) (Ritter 2009). This is highlighted by a study of Australian government drug policymakers which found that academic literature was only used in 28 percent of cases where decisions were required (ibid).

Towards the end of the project, the team hosted a conference which brought together stakeholders from a range of organisations. During the event, participants were asked for their views on the most effective way to disseminate research findings. Here is a summary of the most common answers received:

- Provide accessible findings in plain English (simple messages suitable for a variety of audiences, e.g. easy-to-remember catchphrases).
- Create brief overviews which provide links to access more details if needed.
- Facilitate training days.
- Create a 'Virtual Learning Environment' and/or central repository through which to store research findings.
- Establish a 'research contact' in each organisation.

The suggestions received are very much in agreement with previous research. A consistent message across research of this type is that practitioners and policymakers express a strong preference for short bulletins or single-page summaries with clearly worded recommendations (Ritter 2009; Mitton et al. 2007; Reimer et al. 2005; Tilley and Laycock 2000; Innvaer et al. 2002; Bowers et al. 2017). Personal contacts are also a frequently cited source of information (Nutley et al. 2007; Ritter 2009). Importantly, for both written and spoken outputs, the likelihood of use is increased when products are tailored to the intended audience (Nutley et al. 2007). In particular, Mitton et al. (2007, p. 737) note the importance of actionable messages, i.e. 'information on what needs to be done and the implications'.

A number of the resources listed above already exist in the UK. In a policing context, there is the National Police Library, College of Policing website, the What Works Centre for Crime Reduction and the Global Policing Database to name a few. In some cases, there may be a need better to publicise available resources. In others, there may be a need for academics to produce more accessible, user-friendly outputs. For the authors of this chapter, what is important in accessibility is making academic research both easy to find and easily understood. This call for accessible research outputs is not new, but, in our view, it does no harm to reinforce this point (Tilley and Laycock 2000).

The project team published findings in a number of practitioner-focused outlets. For example, a blog was written for *Policing Insight* (an online policing magazine) (Thompson and Tseloni 2016) and an article published in *Significance* magazine (Tseloni and Thompson 2015). In addition, findings were presented to national NHWN stakeholders (Tseloni 2014a, 2015), to two House of Commons Select Committees (Tseloni 2014b, 2017), the Home Office (Thompson 2016), at the What Works in Crime Reduction Conference (Thompson 2017) and at a variety of other practitioner forums.

The team also published via traditional academic dissemination routes, for example, peer-reviewed academic papers and academic conference presentations. Helpfully, since April 2013, all ESRC grants must comply with the RCUK Policy on Open Access, meaning peer-reviewed research articles generated from ESRC-funded research should be freely available online. Therefore, all journal articles from the burglary project are free to read online. In terms of knowledge exchange, previous research has shown that when research is freely available, it is more likely to be used in policy decision-making (Ritter 2009).

Asking AC members to disseminate findings within their own networks meant they could use their credibility within their own organisations to share results (Meagher et al. 2008). To this end, project findings were incorporated into NHWN guidance. This was disseminated via leaflets at national NHWN events, in the NHWN Members' Guide (of which 500,000 copies were distributed across England and Wales) and across the NHWN via their national communication system, Neighbourhood Alert. This information was presented to the Home Office and Minister of State for Crime Prevention in February 2015. In addition, the charity made reference to the WIDE security combination findings (see Chap. 4) in a number of television and radio interviews. The research findings ultimately influenced the charity's interactions with other external partners and continue to help in the delivery of the NHWN 2015–2020 Strategic Plan. In particular, it assisted the prevention of crime by increasing '…the ability for individuals and communities to be able to identify threats as well as protect themselves and others' as well as helping '…communities be safer and more resilient' (NHWN 2015, pp. 7–8). Without the involvement of the charity in the AC, this level of dissemination and impact may not have occurred.

As well as practitioners, the research findings were likely to be of interest to the general public. To this end, the academic project lead gave a number of radio, newspaper and television interviews. Within these forums, the team were required to synthesise complex statistical information into short summaries that would be understood by a lay person. In addition, some of the findings to emerge from the project could be seen as being of a sensitive nature. For example, the project involved ascertaining the most effective security devices (see Chap. 4) as well as the types of individuals, households and areas at greatest risk of experiencing a burglary (see Chap. 5). Due consideration therefore had to be given to the potential adverse effects of publishing these findings, such as increasing fear of crime. An additional benefit of the AC was that many of the practitioners involved had vast experience in communicating messages of this nature to the general public. The project team therefore drew on their experiences and knowledge to tailor dissemination. To summarise, the team found great value in publishing in both academic and practitioner outlets.

9.5   Challenges in Exchanging Knowledge and Facilitating Impact

There remain two unresolved challenges in our attempts to exchange knowledge and facilitate impact from the burglary project. These relate to how the research team articulated the potential practical benefits to AC members in the early stages of the project and, secondly, how the team accurately traced and documented impact.

9.5.1   Articulating the Potential Practical Benefits of Involvement

As mentioned, the nature of this partnership endeavour was somewhat different to others in that the academic team were not reliant upon partners for data or funding, neither were they directly commissioned to carry out the research on behalf of an AC organisation. Prior to the receipt of funding, the project team worked closely with a small number of partners to develop and co-design the research. Upon commencement of the project, the team identified a range of additional organisations which they felt would be of benefit to, and benefit from, being part of the AC – both in terms of the knowledge the team would share with them and the insights they would be able to offer the team. Individuals were asked to join the AC on a voluntary basis.

The vast majority agreed to be members of the AC, but there were some who were fully committed and others who did not respond to emails. In cases where individuals were interested in the project but could not commit to attending face-to-face meetings, the team tried to be flexible in sharing briefing notes and requesting feedback via email instead. There are a number of reasons why individuals may not have wanted to participate in the research. One frequently cited barrier to participating in knowledge exchange activities is lack of time and resources (Nutley et al. 2007; see Sect. 9.4.2). This was the most common reason relayed to the team for not participating in the burglary project AC. Existing literature suggests other barriers to engaging in research include individuals viewing the process as a threat to professional expertise (Bullock and Tilley 2009), mistrusting academia (Mitton et al. 2007; Wilkinson 2010) and hostility to research more generally (Fyfe and Wilson 2012).

This leads us onto our first challenge – better articulating the potential benefits of involvement to practitioners. As Stephens (2010, p. 151) states: 'everyone in the partnership has to understand why they are engaged, what they want out of the project and the risks involved'. Although the original research proposal required the research team very clearly to outline the potential impact of the project, it can be very difficult accurately to anticipate this before a project has started (Crawford 2017). In other words, it is hard to specify to practitioners exactly 'what's in it for them' in the early stages of research. As Kleemans (2015, p. 60) observes, '…there has been a constant struggle for realistic claims about what empirical research could produce for politicians and policymakers'. A number of partners were involved in the writing of the bid and the research planning phase which helped to secure their support from the outset (Eagar et al. 2003; Nutley et al. 2007). However, in most cases, practitioners still, quite rightly, needed to justify the time spent participating in project meetings, reading briefing notes and commenting upon findings. Braga and Hinkle (2010, p. 116) argue it is easier to justify the time spent when practitioners have '…something good in [their] hands'. This may be in the form of accessible briefing documents or past research impact examples.

We know from previous research that policymakers are more likely to be involved when the following three conditions are met: first, there is a small investment of time; second, they feel they would gain from being involved; and finally, their expertise is closely aligned with project requirements (Ross et al. 2003). More consideration therefore needs to be given to these three conditions in order to help practitioners justify the time they spend on the project to those to whom they report. At the beginning of a project, there is a danger that research of this nature can feel quite abstract to practitioners who are used to working in fast-paced environments where decisions are often made quickly. Given the varying paces at which different organisations work (Canter 2004; Foster and Bailey 2010; Reback et al. 2002), it was important to manage expectations in terms of the exact nature of the project and when findings could be expected. As has been noted elsewhere, academia has a tendency to move more slowly than many policy environments. These differing speeds can prove a barrier to effective knowledge exchange if not managed closely (Foster and Bailey 2010). As an example, AC meetings were used to share and discuss preliminary findings – this often resulted in requests for further information outside of the meetings. In most cases, it was possible to provide this information. However, on occasions it wasn't due to the research being at a very early, exploratory phase. In an attempt to minimise misunderstanding and these types of issues, the aims and purpose of the burglary project AC as well as the project timescales were agreed during the first meeting.

9.5.2   How to Trace and Document Impact?

As mentioned, academics are increasingly required to evidence the use and wider impact of their research. This poses its own challenges, particularly in the operationalisation of the terms 'use' and 'impact'. In relation to the burglary project, the team hoped that by adopting a range of dissemination methods (see Sect. 9.4.2) and by making the research accessible, its results and implications would at least be discussed and debated. Notwithstanding the wider issues with the term impact, one challenge relates to how to document those corridor debates and email exchanges. In an academic REF context, being able to *evidence* the impact of research is almost as important as generating it since impact is defined as the '…*demonstrable* contribution…' that research makes to wider society and the economy (ESRC 2017). For example, there may be burglary prevention publicity campaigns which have been directly informed by findings from the project. However, unless the research is cited, it is very difficult to evidence this (even if it appears obvious). This is one challenge the project team are yet to overcome.

As Nutley et al. (2007, p. 295) observe, impact is '…difficult to operationalise, political in essence, and hard to assess in a robust and widely accepted manner'. Documenting the exact uses of research findings and understanding how knowledge is used are not easy (Mitton et al. 2007). 'Knowledge use is a complex change process in which "getting the research out there" is only the first step' (Nutley et al. 2003, p. 132). Research findings may add to/challenge existing knowledge, change attitudes and/or alter behaviour, but this is by no means simple to document or measure. Research impact is often non-linear, serendipitous and indirect (Molas-Gallart et al. 2000; Crawford 2017). It may also take a number of years for impact to be realised. There is a wide range of factors which influence decision-making in policy and practice contexts, of which research evidence is only one (Tilley and Laycock 2000; Lum et al. 2012). That said, being able to document the policy and practice changes that occur as a result of research is incredibly useful in terms of justifying research funding, designing future projects and developing a better understanding of the most effective pathways to impact.

9.6 Discussion

Armed with the most comprehensive long-term measure of crime trends available (the CSEW) and detailed population characteristics (the UK Census), the burglary project team sought to answer the question: 'which burglary security devices work, for whom and in what context?' They believed answering this question was useful from both an academic *and* practice standpoint. The research therefore not only set out to fill a gap in the academic literature but also to produce findings which had practical relevance. The research was designed to be undertaken over an 18-month period and involved a small team of academics working alongside an advisory group of practitioners with specific expertise in burglary prevention. As discussed, academics are increasingly required to demonstrate how their research has been used in practice. This, combined with an unprecedented period of austerity in policing (and the public sector more broadly), means huge benefits can be drawn from working together. This chapter drew upon the authors' collective personal reflections on this process of knowledge exchange.

A number of scholars are critical of the supposed one-sided nature of traditional 'outreach' or 'hit and run' (Rosenbaum 2010) research of this kind. They argue for a more holistic, enduring arrangement (Steinheider et al. 2012) whereby there is sustained engagement between academics and practitioners rather than a series of one-off events (Henry and Mackenzie 2012; Fyfe and Wilson 2012; Engel and Henderson 2013). They suggest that to be successful, partnerships should focus on knowledge use, transfer and exchange rather than individual projects or activities (ibid). The current authors agree that generating a sustained 'culture of collaboration' is important and that often the most effective way to do this is through long-term, formal engagement between multiple partners. With our short-term arrangement, issues arose in relation to the infrequency of face-to-face contact. In some instances, individuals were not able to attend every scheduled meeting and thus missed out on key information. The importance of attendance was accentuated when there were only a small number of meetings. It can also take time to develop trusting relationships.

Having said this, there were a number of advantages to our short-term partnership. In this case, we convened a small number of subject/practice experts on a particular topic. We were fortunate not to face a number of the barriers encountered in longer-term projects or those where one partner is reliant upon the other for data or funding. For example, we faced fewer issues with regard to staff turnover (Burkhardt et al. 2017; Cordner and White 2010) and no issues in relation to data sharing. We were also asking partners to make a smaller, very specific commitment as opposed to a more general commitment to a partnership arrangement over a number of years. Having such a focused project also meant that our AC was comprised of committed individuals who had extensive knowledge of burglary prevention. In this sense, the burglary project met all three of the conditions Ross et al. (2003) outline as increasing the likelihood of policymaker involvement; we were asking for a small investment of time from partners with relevant expertise who would (hopefully) benefit from their involvement.

Fortuitously, having developed mutually respectful, trusting relationships, the project team have continued to work with a number of the burglary project AC members on other pieces of research. For example, members of the burglary project team and AC have since received further funding from the ESRC to explore violence (ref: ES/L014971/1) and anti-social behaviour (ref: ES/P001556/1) as well as a Knowledge Transfer Partnership grant to conduct research in relation to shop theft (ref: KTP009423). Working with the Office for National Statistics, the academic team suggested improvements to the wording of certain questions in the CSEW. These related to burglar alarms, Neighbourhood Watch and double glazing. The suggestions were made as a direct result of findings from various projects and have since been adopted. The burglary project also involved working closely with the NCDP. As a result, Professor Andromachi Tseloni and Dr. James Hunter have since worked alongside NCDP, Nottingham City Homes and Nottinghamshire Police to further test the WIDE findings in practice (see Chaps. 4 and 6).

9.7   Conclusion

The project team is indebted to the AC in acting as invaluable critical friends who provided useful challenges, steered the project, offered suggestions, assisted in the development of theory and helped to disseminate findings to the most appropriate audiences. As Stanko (2007) suggests, researchers often do not know how best to disseminate findings to potential research users. Developing a trusting and mutually respectful arrangement from the outset allowed the team more effectively to disseminate findings from the project and more easily to identify opportunities to influence practice (in other words, to generate impact). This is evidenced by the burglary pilot project which would not have happened without the support of the AC (see Chap. 6). On the whole, we found having an AC to be an effective way to exchange knowledge.

We accept Murji's (2010) point that 'off-the-shelf' guides have their limitations in that academic-practitioner partnerships are often unique endeavours. However, an enduring aspect of our own reflections (in agreement with much of the existing academic literature) is the importance of good quality relationships, in particular trust and mutual respect. There is great value in listening to (and respecting) the opinions of others. As Fleming (2010, p. 139) states, it is important to 'stand in the other person's shoes'. The burglary project team had enormous respect for the practitioners working in the field and vice versa.

From the project team's perspective, the intention to generate impact was not a box-ticking exercise but stemmed from a genuine commitment to make a difference. This commitment was shared by the AC partners and formed a perfect foundation for the relationship. It is by no means an easy task to produce scientifically credible research which is also timely and practically useful (Greene 2015; McAra

2017). Exchanging knowledge (especially with multiple partners with different organisational interests, cultures and priorities) is a complex task and one which, we would readily admit, we did not always get right. With hindsight, we should have had more frequent, more effective communication outside of formal meetings of the AC – making better use of virtual methods. In addition, although accepting the often non-linear, serendipitous nature of research impact, we should try to develop a more effective means of tracing and documenting changes to policy and/or practice that arise as a direct result of the research. However, we hope it is clear from the examples we have provided that the findings generated from the project have helped to shape policy and practice. We also hope the learning we have shared within this chapter will be useful for others.

References

Bach, W., & Cole, S. (2017). Police on the frontline bear the brunt of cuts to their service and others. https://www.theguardian.com/uk-news/2017/jun/28/police-on-the-frontline-bear-the-brunt-of-cuts-to-their-service-and-others. Accessed 31 Jan 2018.

Bowers, K., Tompson, L., Sidebottom, A., Bullock, K., & Johnson, S. D. (2017). Reviewing evidence for evidence-based policing. In J. Knutsson & L. Tompson (Eds.), *Advances in evidence-based policing* (pp. 98–116). Oxon: Routledge.

Braga, A. A., & Hinkle, M. (2010). The participation of academics in the criminal justice working group process. In J. M. Klofas, N. Kroovand Hipple, & E. F. McGarrell (Eds.), *The new criminal justice* (pp. 114–120). New York: Routledge.

Buerger, M. E. (2010). Policing and research: Two cultures separated by an almost-common language. *Police Practice and Research, 11*(2), 135–143.

Bullock, K., & Tilley, N. (2009). Evidence-based policing and crime reduction. *Policing: A Journal of Policy and Practice, 3*(4), 381–387.

Burawoy, M. (2005). For public sociology. *American Sociological Review, 70,* 4–28.

Burkhardt, B. C., Akins, S., Sassaman, J., Jackson, S., Elwer, K., Lanfear, C., Amorim, M., & Stevens, K. (2017). University Researcher and Law Enforcement Collaboration: Lessons from a study of justice-involved persons with suspected mental illness. *International Journal of Offender Therapy and Comparative Criminology, 61*(5), 508–525.

Canter, D. V. (2004). A tale of two cultures: A comparison of the cultures of the police and academia. In R. Adlam & P. Villiers (Eds.), *Policing a safe, just and tolerant society: An international model* (pp. 109–121). Winchester: Waterside Press.

Chatham House (2018). Chatham House Rule. https://www.chathamhouse.org/about/chatham-house-rule#. Accessed 9 Feb 2018.

Cockbain, E. (2015). Getting a foot in the closed door: Practical advice for starting out in research into crime and policing issues. In E. Cockbain & J. Knutsson (Eds.), *Applied police research: Challenges and opportunities* (pp. 21–33). Oxon: Routledge.

Cockbain, E., & Knutsson, J. (Eds.). (2015). *Applied police research: Challenges and opportunities.* Oxon: Routledge.

College of Policing. (2017). About us. http://www.college.police.uk/About/Pages/default.aspx. Accessed 30 May 2018.

Cordner, G., & White, S. (2010). Special issue: The evolving relationship between police research and police practice. *Police Practice and Research: An International Journal, 11*(2), 90–94.

Crawford, A. (2017). Research co-production and knowledge mobilisation in policing. In J. Knutsson & L. Tompson (Eds.), *Advances in evidence-based policing* (pp. 195–213). Oxon: Routledge.

Crawford, A., & L'Hoiry, X. (2017). Boundary crossing: Networked policing and emergent 'communities of practice' in safeguarding children. *Policing and Society, 27*(6), 636–654.

Davies, H., Nutley, S., & Walter, I. (2008). Why 'knowledge transfer' is misconceived for applied social research. *Journal of Health Services Research & Policy, 3,* 188–190.

Eagar, K., Cromwell, D., Owen, A., Senior, K., Gordon, R., & Green, J. (2003). Health services research and development in practice: An Australian experience. *Journal of Health Services Research & Policy, 8*(2), 7–13.

Economic and Social Research Council (ESRC). (2017). What is impact? http://www.esrc.ac.uk/research/impact-toolkit/what-is-impact/. Accessed 15 June 2017.

Ekblom, P. (2002). From the source to the mainstream is uphill: The challenge of transferring knowledge of crime prevention through replication, innovation and anticipation. *Crime Prevention Studies, 13,* 131–203.

Engel, R. S., & Henderson, S. (2013). Beyond rhetoric: Establishing police-academic partnerships that work. In J. M. Brown (Ed.), *The future of policing* (pp. 217–236). Oxon: Routledge.

Fleming, J. (2010). Learning to work together: Police and academics. *Policing: A Journal of Policy and Practice, 4*(2), 139–145.

Foster, J., & Bailey, S. (2010). Joining forces: Maximising ways of making a difference in policing. *Policing: A Journal of Policy and Practice, 4*(2), 95–103.

Fraser, I. (2004). Organizational research with impact: Working backwards. *Worldviews on Evidence-Based Nursing, 1*(1), S52–S59.

Frisch, N. E. (2015). Exploring the nature and success of an embedded Criminologist partnership. Masters thesis submitted to the Faculty of the Graduate School of the University of Maryland. https://drum.lib.umd.edu/handle/1903/17203 Accessed 30 May 2018.

Frisch, N. E. (2016). Examining the success of an embedded criminologist partnership. *Translational Criminology, 2016*(Spring), 24–26.

Fyfe, N. R., & Wilson, P. (2012). Knowledge exchange and police practice: Broadening and deepening the debate around researcher-practitioner collaborations. *Police Practice and Research, 13*(4), 306–314.

Goode, J., & Lumsden, K. (2018). The McDonaldisation of police-academic partnerships: Organisational and cultural barriers encountered in moving from research *on* police to research *with* police. *Policing and Society: An International Journal of Research and Policy, 28*(1), 75–89.

Greene, J. R. (2010). Collaborations between police and research/academic organizations: Some prescriptions from the field. In J. M. Klofas, N. Kroovand Hipple, & E. F. McGarrell (Eds.), *The new criminal justice* (pp. 121–127). New York: Routledge.

Greene, J. R. (2015). Police research as mastering the tango: The dance and its meaning. In E. Cockbain & J. Knutsson (Eds.), *Applied police research: Challenges and opportunities* (pp. 117–128). Oxon: Routledge.

Grieco, J., Vovak, H., & Lum, C. (2014). Examining research-practice partnerships in policing evaluations. *Policing, 8*(4), 368–378.

Hara, N., Solomon, P., Kim, S.-L., & Sonnenwald, D. H. (2003). An emerging view of scientific collaboration: Scientists' perspectives on collaboration and factors that impact collaboration. *Journal of the American Society for Information Science and Technology, 54*(10), 952–965.

Henry, A., & Mackenzie, S. (2012). Brokering communities of practice: A model of knowledge exchange and academic-practitioner collaboration developed in the context of community policing. *Police Practice and Research, 13*(4), 315–328.

HM Treasury. (2010). *Spending review 2010.* London: Her Majesty's Stationery Office.

HMIC. (2014). *Policing in austerity: Meeting the challenge.* London: HMIC.

HMIC. (2017). *State of policing – The annual assessment of policing in England and Wales 2016: Her Majesty's Chief Inspector of Constabulary.* London: HMIC.

Home Office. (2017). *Police workforce, England and Wales, 31 March 2017.* London: Home Office. https://www.gov.uk/government/uploads/system/uploads/attachment_data/file/630471/hosb1017-police-workforce.pdf. Accessed 30 May 2018.

House of Commons. (2018). *Policing for the future: Tenth report of session 2017–19. HC515.* London: House of Commons.

Innes, M. (2010). A 'Mirror' and a 'Motor': Researching and reforming policing in an age of austerity. *Policing: A Journal of Policy and Practice, 4*(2), 127–134.

Innes, C. A., & Everett, R. S. (2008). Factors and conditions influencing the use of research by the Criminal Justice System. *Western Criminology Review, 9*(1), 49–58.

Innvaer, S., Vist, G., Trommald, M., & Oxman, A. D. (2002). Health policy-makers' perceptions of their use of evidence: A systematic review. *Journal of Health Services Research & Policy, 17*(4), 239–244.

International Association of Chiefs of Police. (n.d.). *Establishing and sustaining law enforcement-researcher partnerships: Guide for researchers.* Washington, D.C.: IACP.

Kleemans, E. R. (2015). Organized crime research: Challenging assumptions and informing policy. In E. Cockbain & J. Knutsson (Eds.), *Applied police research: Challenges and opportunities* (pp. 57–67). Oxon: Routledge.

Knutsson, J., & Tompson, L. (Eds.). (2017). *Advances in evidence-based policing.* Oxon: Routledge.

Lane, J., Turner, S., & Flores, C. (2004). Researcher-practitioner collaboration in community corrections: Overcoming hurdles for successful partnerships. *Criminal Justice Review, 29*(1), 97–114.

Laycock, G. (2015). Trust me, I'm a researcher. In E. Cockbain & J. Knutsson (Eds.), *Applied police research: Challenges and opportunities* (pp. 45–56). Oxon: Routledge.

Lum, C., Telep, C. W., Koper, C. S., & Grieco, J. (2012). Receptivity to research in policing. *Justice Research and Policy, 14*(1), 61–94.

Marks, M., & Sklansky, D. (2008). Voices from below: Unions and participatory arrangements in the police workplace. *Police Practice and Research, 9*(2), 85–94.

Marks, M., Wood, J., Ally, F., Walsh, T., & Witbooi, A. (2010). Worlds apart? On the possibilities of police/academic collaborations. *Policing: A Journal of Policy and Practice, 4*(2), 112–118.

McAra, L. (2017). Can Criminologists change the world? Critical reflections on the politics, performance and effects of criminal justice. *British Journal of Criminology, 57*(4), 767–788.

Meagher, L., Lyall, C., & Nutley, S. (2008). Flows of knowledge, expertise and influence: A method for assessing policy and practice impacts from social science research. *Research Evaluation, 17*(3), 163–173.

Mitton, C., Adair, C. E., McKenzie, E., Patten, S. B., & Perry, B. W. (2007). Knowledge transfer and exchange: Review and synthesis of the literature. *The Milbank Quarterly, 85*(4), 729–768.

Molas-Gallart, J., Tang, P., & Morrow, S. (2000). Assessing the non-academic impact of grant-funded socio-economic research: Results from a pilot study. *Research Evaluation, 9*(3), 171–182.

Murji, K. (2010). Introduction: Academic-police collaborations – Beyond 'two worlds'. *Policing: A Journal of Policy and Practice, 4*(2), 92–94.

Neighbourhood and Home Watch Network (England & Wales) (NHWN). (2015). *Neighbourhood and Home Watch Network 2015-2020 Strategic Plan.* https://www.ourwatch.org.uk/knowledge/nhwn-strategic-plan-2015-20/. Accessed 9 Feb 2018.

Nutley, S. M., Walter, I., & Davies, H. T. O. (2003). From knowing to doing: a framework for understanding the evidence-into-practice agenda. *Evaluation, 9*(2), 125–148.

Nutley, S. M., Walter, I., & Davies, H. T. O. (2007). *Using evidence: How research can inform public services.* Bristol: The Policy Press.

Oliver, K., Innvar, S., Lorenc, T., Woodman, J., & Thomas, J. (2014). A systematic review of barriers to and facilitators of the use of evidence by policymakers. *BMC Health Services Research, 14*(2).

Reback, C. J., Cohen, A. J., Freese, T. E., & Shoptaw, S. (2002). Making collaboration work: Key components of practice/research partnerships. *Journal of Drug Issues,* 837–848.

Reimer, B., Sawka, E., & James, D. (2005). Improving research transfer in the addictions field: A perspective from Canada. *Substance Use and Misuse, 40*(11), 1707–1720.

Ritter, A. (2009). How do drug policy makers access research evidence? *International Journal of Drug Policy, 20*, 70–75.

Rojek, J., Alpert, G., & Smith, H. (2012). The utilization of research by the police. *Police Practice and Research, 13*(4), 329–341.

Rosenbaum, D. P. (2010). Police research: Merging the policy and action research traditions. *Police Practice and Research, 11*(2), 144–149.

Rosenbaum, D. P., & Roehl, J. (2010). Building successful anti-violence partnerships: Lessons from the Strategic Approaches to Community Safety Initiative (SACSI) model. In J. M. Klofas, N. Kroovand Hipple, & E. F. McGarrell (Eds.), *The new criminal justice* (pp. 39–50). New York: Routledge.

Ross, S., Lavis, J., Rodriguez, C., Woodside, J., & Denis, J.-L. (2003). Partnership experiences: Involving decision makers in the research process. *Journal of Health Services Research & Policy, 8*(2), 26–34.

Stanko, B. A. (2007). From academia to policy making: Changing police responses to violence against women. *Theoretical Criminology, 11*(2), 209–219.

Steinheider, B., Wuestewald, T., Boyatzis, R. E., & Kroutter, P. (2012). In search of a methodology of collaboration: Understanding researcher-practitioner philosophical differences in policing. *Police Practice and Research, 13*(4), 357–374.

Stephens, D. W. (2010). Enhancing the impact of research on police practice. *Police Practice and Research, 11*(2), 150–154.

Stern, N. (2016). *Building on success and learning from experience: An independent review of the Research Excellence Framework.* https://www.gov.uk/government/uploads/system/uploads/attachment_data/file/541338/ind-16-9-ref-stern-review.pdf. Accessed 31 Jan 2018.

Strang, H. (2012). Coalitions for a common purpose: Managing relationships in experiments. *Journal of Experimental Criminology, 8*(3), 211–225.

Tartari, V., Salter, A., & D'Este, P. (2012). Crossing the Rubicon: Exploring the factors that shape academics' perceptions of the barriers to working with industry. *Cambridge Journal of Economics, 36*, 655–677.

The Times. (2013, October 23). Letters to the Editor: New home security.

Thompson, R. (2016). *Improving the crime reduction evidence base: Acquisitive crime.* Presentation to the Home Office, 27 Apr 2016.

Thompson, R. (2017, January 24). *The WIDE benefits of burglary security.* Presentation to the what works in crime reduction conference, British Library.

Thompson, R., & Tseloni, A. (2016). What works: Which security devices best protect homes against burglary? *Policing Insight.* https://policinginsight.com/analysis/protecting-homes-burglary-effective-security-devices/. Accessed 31 Jan 2018.

Tilley, N., & Laycock, G. (2000). Joining up research, policy and practice about crime. *Policy Studies, 21*(3), 213–227.

Tilley, N., Thompson, R., Farrell, G., Grove, L., & Tseloni, A. (2015). Do burglar alarms increase burglary risk? A counter-intuitive finding and possible explanations. *Crime Prevention and Community Safety, 17*(1), 1–19.

Tompson, L., Belur, J., Morris, J., & Tuffin, R. (2017). How to make police-researcher partnerships mutually effective. In J. Knutsson & L. Tompson (Eds.), *Advances in evidence-based policing* (pp. 175–194). Oxon: Routledge.

Tseloni, A. (2014a, November 19). *Which burglary security devices work for whom and in what context? Burglary – What Works and Why?* Presentation to the National Neighbourhood and Home Watch Network Stakeholders Event, London.

Tseloni, A. (2014b, January 28). Evidence given to the Justice Committee on 'Crime reduction policies: A co-ordinated approach?' http://www.publications.parliament.uk/pa/cm201415/cmselect/cmjust/307/140128.htm. Accessed 30 May 2018.

Tseloni, A. (2015, December 2–3). Participation in 'The way ahead for Neighbourhood Watch'. National Neighbourhood and Home Watch Network Stakeholders Event, Birmingham.

Tseloni, A. (2017, March 28). Evidence given to the Home Affairs Committee: 'Policing for the future: changing demands and new challenges'. Portcullis House. http://data.parliament.uk/writtenevidence/committeeevidence.svc/evidencedocument/home-affairs-committee/policing-for-the-future-changing-demands-and-new-challenges/oral/49475.html. Accessed 30 May 2018.

Tseloni, A., & Thompson, R. (2015). Securing the premises. *Significance, 12*(1), 32–35.

Weiss, C. H. (1995). The haphazard connection: Social science and public policy. *International Journal of Educational Research, 23*(2), 137–150.

Weiss, J. A., & Weiss, C. H. (1981). Social scientists and decision makers look at the usefulness of mental health research. *American Psychologist, 36*(8), 837–847.

Wilkinson, S. (2010). Research and policing – Looking to the future. *Policing: A Journal of Policy and Practice, 4*(2), 146–148.

Wuestewald, T., & Steinheider, B. (2009). Practitioner-researcher collaboration in policing: A case of close encounters. *Policing, 4*(2), 104–111.

Chapter 10
Conclusions: Reducing Burglary – Summing Up

Andromachi Tseloni, Rebecca Thompson, and Nick Tilley

Domestic burglary has fallen substantially over the last 20 years in many countries but remains a high-volume crime affecting many households. As well as financial loss and damage to property, the psychological impact of a burglary can be considerable (Dinisman and Moroz 2017). For this reason, burglary consistently ranks as a top public concern in relation to crime and disorder and is likely to remain an important area of crime prevention.

This book has reported a range of original research that speaks to physical security measures that are installed with the aim of reducing risks of domestic burglary. The book sheds new light on the impact that physical security has on burglars' decision-making processes as well as burglary patterns and trends which directly inform burglary prevention. This last chapter collates the main points made in this book into three sections:

- Burglary trends and patterns (Sect. 10.1).
- Which security devices work and how (Sect. 10.2)?
- Burglary prevention lessons (Sect. 10.3).

The chapter summarises the main lessons that emerge from the research we have undertaken, alongside other cognate work that also speaks to the patterns of impact that security measures have had, and can be expected to continue to have in the future, on domestic burglary. In each case we flag the major points in this book where the relevant arguments and research findings are described in detail. In a few

A. Tseloni (✉) · R. Thompson
Quantitative and Spatial Criminology, School of Social Sciences, Nottingham Trent University, Nottingham, UK
e-mail: andromachi.tseloni@ntu.ac.uk

N. Tilley
Jill Dando Institute, Department of Security and Crime Science, University College London, London, UK

© Springer Nature Switzerland AG 2018
A. Tseloni et al., *Reducing Burglary*,
https://doi.org/10.1007/978-3-319-99942-5_10

cases, where the research reported here does not address key issues in any detail, we cite other research the interested reader might like to consult.

Most of the points overviewed below relate to research findings on which we can have some confidence. We also note the major data sources that can be used in analysing overall burglary patterns, highlight areas where there is urgent need for specific areas of future research and spell out some important policy and practice implications if the welcome reductions in burglary widely observed over the past quarter century in many countries are to be maintained and extended.

Readers need to bear in mind that the data analysed in most of the original research reported here relate to England and Wales in particular, albeit that one chapter focuses specifically on France. We would certainly hope that the findings we report would apply also in other jurisdictions, although of course we cannot be certain.

10.1   Burglary Trends and Patterns

Against expectations, dramatic falls in many crimes, including burglary, have been witnessed across many countries since the mid-1990s, generally referred to as the 'crime drop' (Tseloni et al. 2010). Burglary trends and patterns such as this have been best understood with the use of victimisation surveys that overcome many of the weaknesses in recorded crime data. They often include supplementary questions that can help in the identification and analysis of patterns and trends (Chap. 1). The (international and across crime types) reach, timing and trajectory of the crime falls (Tseloni et al. 2010) imply that '…changes in the quantity and quality of security have played a major part in driving crime falls in most industrial societies' (Farrell et al. 2011, p. 151). This book provides further evidence in support of this hypothesis in relation to burglary.

Two national crime surveys, the Crime Survey for England and Wales (CSEW) and the French Cadre de Vie et Sécurité (CVS), have been used in this book (Chaps. 1, 4, 5, 7 and 8). Through this data, we find that burglaries are not uniformly distributed: some households, neighbourhoods, regions and countries are more affected than others. Both the fall in burglary and the uptake of security were uneven across population groups and area types. Burglary became more concentrated against households which are less likely to have the most effective security combination (Window locks, Internal lights on a timer, Double door locks or deadlocks and External lights on a sensor – WIDE). The gap between households who do and do not have WIDE has widened over time meaning certain groups have not felt the positive impact of the national drop in burglary (Chap. 5).

Part of the drop in levels of burglary was a consequence of physical security improvements that have not been systematically documented over the period and therefore cannot be directly measured alongside burglary falls (Chap. 8). These include, for example, security improvements and increased surveillance in public spaces of residential neighbourhoods (Chap. 2); modern building standards for new housing developments which incorporate high-quality windows, doors and frames originally for conserving heating energy, also to comply with SBD standards (Chap. 3); and similar improvements to the existing housing stock undertaken by home owners and landlords. The evidence on the security hypothesis for the burglary fall in this book refers to one jurisdiction, England and Wales. Similar proliferation of physical security and CPTED policies occurred across many industrialised countries.[1]

10.2 Which Security Devices Work and How?

Burglars' accounts on the deterrent role of physical security and surveillance (Chap. 3) are in full agreement with the kinds of interventions that made burglary prevention projects successful (Chap. 2). Burglars target properties with low natural surveillance, easy access and escape routes and poor physical security in locations which seemingly lack community spirit (Chap. 3). This book presents new research evidence in relation to physical security and in particular how this plays out in different community conditions (Chaps. 4, 5, 6, 7 and 8). Physical security is the most straightforward housing feature to be investigated not least because of data availability – the CSEW in England and Wales and the CVS in France. The type and prevalence of devices partly differ across countries, and in France physical security features also differ between houses and apartments (Chap. 7).[2]

Burglars can assess the quality and robustness of doors, windows, their locks and other physical security features, including type and brand of burglar alarms. In addition, the evidence presented in this book suggests they are not deterred by most burglar alarms and perceive excessive visual security, such as gated developments and window grills, as an indication of high-value possessions (Chap. 3). They may therefore find properties with these specific devices attractive. The most effective device combination (in terms of protection, safety *and* cost) in England and Wales was window locks, internal lights on a timer, double door locks or deadlocks and external lights on a sensor. This is captured in the acronym, 'WIDE' (Chap. 4). Window and double door locks formed the basis of all effective security combinations highlighting the importance of restricting access through the use of good quality windows and doors as well as simulating occupancy and increasing surveillance potential through security lighting.

In France security doors (alone) offer the second highest (after alarms) protection against burglary with entry for houses (Chap. 7). Digital locks (alone) offer the highest protection for apartments. The most effective combination for both housing types includes alarms, digital locks and security doors (Chap. 7).

---

[1] Please see evidence for the Netherlands by Vollaard and Van Ours (2011) and De Waard (2015) and for Chile by Ojeda (2015).

[2] The most prominent were digital locks and caretakers which are more common in French apartments (than houses and hardly exist in the UK).

The evidence from burglars' accounts (Chap. 3) and from previous research (e.g. Cromwell and Olson 2004) suggest that burglary is a process of distinct hurdles and decisions. The entire sequence of these decisions was introduced and tested in Chap. 7 of this book as follows:

1. Selection of neighbourhood (Chaps. 5 and 7)
2. Selection of a property (Chaps. 3, 4, 5 and 7)
3. Burglary with entry (Chaps. 4 and 7)
4. Property stolen (Chap. 7)

It is proposed burglars assess the situation at each stage of the above and accordingly move to the next stage or abandon the process. There are 'transition points' between each stage – to move from one to the next, a burglar must not be (a) deterred (i.e. discouraged from selecting the property), (b) thwarted (i.e. physically prevented from entering) or (c) interrupted (i.e. leave the house without having taken anything) (Chaps. 4 and 7). Different security devices have distinctive 'deter' or 'thwart' mechanisms highlighting the importance of considering different 'security packages' and their relative effectiveness in order to provide more accurate crime prevention advice (Chap. 4).

The most intriguing findings of this book were in relation to burglar alarms which according to burglars, with one exception, do not deter them (Chap. 3). Burglar alarms alone increase the risk of both burglary with entry and attempted burglary and, in combination with other devices, reduce the overall level of protection against burglary with entry in England and Wales (Chaps. 4 and 8). The increased risk of attempted burglary associated with alarms is also supported by evidence from France: an offender may try and fail to enter a property due to being disrupted by the sound of a burglar alarm or someone responding to the alarm (Chaps. 4 and 7). However, the evidence from France with regard to burglary with entry partly contradicts what was found for England and Wales. In France alarms (alone and in combination) are effective in preventing burglaries against houses but alone do not protect apartments (Chap. 7).[3] There might be a proliferation of burglar alarms in England and Wales partly fuelled by their low cost and, consequently, low-quality products which may often sound due to faulty technical problems rather than to alert about break-ins. They can also be perceived as a nuisance and thus be ignored by neighbours and passers-by (see Tilley et al. (2015); Chaps. 4 and 8).

Weak community relations might play a role in alarms' ineffectiveness as suggested from the evidence in relation to French apartments: unlike houses, apartment blocks do not encourage meaningful social interactions, and neighbours may be indifferent or reluctant to respond when alarms go off. Households in urban areas have consistently higher burglary risk and greater levels of effective physical security than others (Chap. 5). Conversely, households in rural areas are generally less likely to have effective security but have sustained low exposure to burglary (Chaps. 5 and 7). Environmental factors, such as living in an urban area, with high population

---

[3] As 85 percent of households in England and Wales live in houses, the contradictory finding in relation to this type of housing between the two countries is not a statistical artefact.

density and high crime levels (which may affect burglars' familiarity and accessibility) have a stronger effect than physical security when targeting properties (Chap. 7).

A further factor that may limit the effectiveness of physical security is target attractiveness (Chaps. 3, 5 and 7). Houses and apartments over 100m^2 in France are more targeted than smaller ones and so are wealthy houses (independently of size) (Chap. 7). Similarly (over the crime drop), in England and Wales, affluent households had the highest security increases without necessarily the highest burglary drops – which were actually enjoyed by middle-income households, earning £20,000–£29,999 per annum (Chap. 5). The above evidence tells us how and under which conditions physical security works to prevent burglary. The next question is how this evidence can be used for burglary prevention by householders, landlords and the public, voluntary and private sectors in their policies, guidelines and regulations.

10.3   Burglary Prevention Lessons

Domestic burglary is a high-volume crime, which can cause substantial distress to its victims. As a result of its high volume and impact, preventing domestic burglary has been a sustained focus of policy attention. The research reported in this book takes us beyond current theoretical knowledge as well as being transferable to burglary prevention in practice. It provides insights about measures that householders and landlords can take to protect their homes and properties.

With respect to community protection, the research findings reported in this book can be translated into practical advice about specific interventions the police, Police and Crime Commissioners, Crime and Safety Partnerships, victim support organisations, Neighbourhood Watch, the Home Office and other responsible agencies can implement to reduce burglary rates in their jurisdiction. Burglary levels can be reduced efficiently and effectively by prompt improvement to the security of dwellings where burglaries have taken place and the dwellings close to them. The use of WIDE security measures focused on burgled premises, and those nearby, has produced promising burglary reduction outcomes without displacement of burglary risks to nearby neighbourhoods in a demonstration project in Nottingham (Chap. 6).

The findings also have practical implications for the private sector (insurance companies, the security industry, the building and planning sector) and government bodies that oversee and/or regulate their activities. Burglar alarms do not necessarily deter burglars (Chaps. 3, 4, 7 and 8) – the industry can clearly either rethink their approach to alarms (and their design) or become outdated. Home insurers' requirements that homes should be equipped with doors and windows that lock with a key and a fully operating alarm for cover eligibility are partly contested by the book's findings. Without further insights on specifications and contexts within which alarms fulfil their role, insurers' policies are responsible for potentially misleading the public into a false sense of security.

The enduring high burglary risks to specific households which are unlikely to acquire effective physical security (Chap. 5) have implications for the way crime prevention agencies respond to victims. It also has implications for housing policy and the use of grants for security upgrades to those most in need. Protecting the most vulnerable households, by offering effective physical security upgrades in the first place, brings down overall burglary rates. As burglary has fallen substantially, its prevention is now easier than it was two decades ago, precisely because it has become highly concentrated on a small number of household types (Chaps. 1, 5 and 8). Physical security combinations that effectively deter burglars directly speak to social housing standards for local authorities and housing associations as well as licensing policies for rented accommodation, HMOs and student landlords (Chaps. 4 and 7).

Physical security alone is not always enough to deter burglars, as demonstrated in the case of households living in urban areas (Chaps. 5 and 7). 'Design Against Crime' emerged as a practical and effective programme for crime prevention based on Crime Prevention through Environmental Design (CPTED). It has comprised a major framework for designing and delivering crime prevention into new developments or making changes to existing ones to reduce the risk of burglary, especially between 1998 and 2011. After 2011, however, SBD planning and building requirements have become localised despite evidence that new or refurbished developments with SBD standards have lower household crime rates, including repeats (Chap. 3). SBD needs national implementation if new developments are not to risk high rates of burglary.

Central (and local) government could regulate or provide incentives encouraging the nationwide adoption of SBD standards, combining physical security with ample informal surveillance opportunities, for planning and building new or renewing existing housing (Chaps. 3, 4 and 7). Surveillance opportunities need not be solely based on the physical layout, architecture and landscaping of houses and their surroundings, but also enlist community support elements (Chap. 2). Burglars can adapt and so should prevention. In order to succeed, interventions require residents' buy-in and effective collaboration between practitioners and academic researchers on equal footing (Chaps. 2 and 9).

10.4   Future Opportunities

The evidence presented in this book advanced our understanding of which, how and when security works to deter burglars. However, there is still a lot we do not know. This last subsection attempts to identify gaps in knowledge and potential avenues for future research and to outline the information/data this work would require. Indeed, a prerequisite for the success (and initial step) of any form of intervention is gathering information about the problem in hand and the areas and people most affected (Chap. 2). Keeping good records of interventions and outcomes facilitates constructive evaluations of what worked and what did not and the conditions needed for the measures to work. Findings can then inform decisions about what measures to replicate and where to try them in the future (Chaps. 2 and 6).

The findings reported in this book contribute new cross-national understanding of the preventive strength of specific security devices and their combinations. A major limitation however is that the number and type of security devices examined are constrained by the available data. The CSEW could usefully ask *both the entire sample and, at the time of the incident, burglary victims* questions about the presence of a wider range of security devices (including dogs, Chap. 7) to allow their effectiveness to be tested. Additional questions (some of which have already been adopted in the CSEW as a result of this research) include whether the security devices (e.g. burglar alarms) were activated at the time of the burglary for victims and for the entire sample how often/when they are activated. Such knowledge could subsequently inform security investments and help produce further falls in burglary. We suggest that other national (e.g. the National Crime Victimisation Survey in the USA) and international (notably the International Crime Victims Survey) crime surveys follow the structure of the CSEW questionnaires on (a) crime security and prevention and (b) detailed information about the reported crime and modus operandi to inform similar analyses elsewhere with potential policy impacts.

The role informal surveillance and physical security plays at the different stages of burglars' decision-making during the commission of this crime is the natural extension of the research discussed in this book. This avenue of enquiry again necessitates large sample sizes in order to examine single devices and combinations. It also requires contextual information about the neighbourhoods of respondents which can be gauged from the Census and other surveys offering possibilities for data linkage and hierarchical and/or hurdle modelling methodology[4] (McLachlan and Peel 2000; Mullahy 1986; Osborn et al. 1996).

Future applied research that promises relevance to the prevention of burglary in practice will require close collaboration from those in policy and practice alongside those in academe. The challenges in achieving this are substantial (Chap. 9). For example, the findings on alarms warrant further research to better understand their potential effects which offer one opportunity for industry-academic collaboration. Another avenue for advancing knowledge in the burglary prevention field is close collaboration across the public, voluntary and academic research sectors. For example, in order to build a sound knowledge base, delivering and evaluating the impact of crime reduction initiatives require (time and/or financial) commitment, regular, tailored and accessible communication and the development of trusting, mutually beneficial collaborative arrangements between national and local government, practitioners, data providers and academic researchers (Chaps. 2, 6 and 9).

---

[4] Apart from a conference presentation mentioned in Chap. 5 (Tseloni 2011), to date such analyses have tested the effects of routine activities and social disorganisation on burglary victimisation but have not specifically examined the independent effects of particular security devices and their combinations (Tseloni 2006).

References

Cromwell, P., & Olson, J. N. (2004). *Breaking and entering: Burglars on burglary*. Belmont: Wadsworth.

De Waard, J. (2015, June 8–10). Explaining the crime drop in The Netherlands: The importance of comparisons with other industrialised countries. Session on Towards a more effective and efficient judicial chain: Making use of international data. Stockholm Criminology Symposium 2015.

Dinisman, T., & Moroz, A. (2017). *Understanding victims of crime*. London: Victim Support.

Farrell, G., Tseloni, A., Mailley, J., & Tilley, N. (2011). The crime drop and the security hypothesis. *Journal of Research in Crime and Delinquency, 48*(2), 147–175.

McLachlan, G. J., & Peel, D. (2000). *Finite mixture models*. New York: Wiley.

Mullahy, J. (1986). Specification and testing of some modified count data models. *Journal of Econometrics, 33*, 341–365.

Ojeda, H. S. (2015, June 8–10). Testing security hypothesis to explain burglary downward trends in Chile. Session on The crime drop. Testing hypotheses. Stockholm Criminology Symposium 2015.

Osborn, D. R., Ellingworth, D., Hope, T., & Trickett, A. (1996). Are repeatedly victimised households different? *Journal of Quantitative Criminology, 12*, 223–245.

Tilley, N., Thompson, R., Farrell, G., Grove, L., & Tseloni, A. (2015). Do burglar alarms increase burglary risk? A counter-intuitive finding and possible explanations. *Crime Prevention and Community Safety, 17*, 1–19.

Tseloni, A. (2006). Multilevel modelling of the number of property crimes: Household and area effects. *Journal of the Royal Statistical Society Series A-Statistics in Society, 169*(Part 2), 205–233.

Tseloni, A. (2011, December 13). Household burglary victimisation and protection measures: Who can afford security against burglary and in what context does it matter? Crime Surveys Users Meeting, Royal Statistical Society, London. Available online: http://www.ccsr.ac.uk/esds/events/2011-12-13/index.html.

Tseloni, A., Mailley, J., Farrell, G., & Tilley, N. (2010). Exploring the international decline in crime rates. *European Journal of Criminology, 7*(5), 375–394.

Vollaard, B., & van Ours, J. C. (2011). Does regulation of built-in security reduce crime? Evidence from a natural experiment. *The Economic Journal, 121*, 485–504.

Index

A
Academic-practitioner partnerships, 249
Academics
  academic-practitioner partnerships, 258
  audiences, 246
  careers, 247
  collaborating with, 247, 250
  critical independence, 247
  dissemination routes, 253
  documents, 252
  exchanging knowledge, 248
  high-quality research, 247
  literature, 252
  police-academic partnerships, 247
  and practitioners (*see* Practitioners)
  REF context, 256
  time-limited project, 246
Advisory Committee (AC)
  burglary project, 249, 255
  communication beyond, 252–254
  communication with, 251–252
  meetings, 250, 256, 259
  members, 250, 254, 258
  mutual education, 250
  organisation, 254
  partners for funding, 248
  practitioners, 248, 250
  process of knowledge exchange, 248
  project team, 258
  on voluntary basis, 255
Alley gates, 33–34
Annual household income, 135–137
Anti-burglary security devices, 112,
    113, 144
Area of residence, households, 139–141

B
Bivariate logit regression models, 151,
    157–160
British Crime Survey (BCS), 3, 97, 111, 229
Burglar alarms, 239–241, 250, 258, 268, 269
Burglary
  analysis, 217
  binary event, 196
  causes and characteristics, 196
  causing factors, 195
  counts, 2–4
  CSEW, 201
  data sources, 2–4, 266
  distribution
    crime survey data, 4
    household, 5
    ICVS, 5
    incidence, 4
    prevalence, 4, 5
    UNODC, 5
    variations, 8
  forced entry, 200
  future opportunities, 270–271
  high-volume crime, 1
  impacts, 15
  incidence rates, 9, 10, 14
  local crime and delinquency, 217
  offenders, 16
  opportunist nature, 196
  police services, 197
  prevalence, 7
  prevention, 269–270
  process, 199, 200
  psychological impact, 265
  responding, 15, 16

Burglary (*cont.*)
  RV, 12–14
  SBD (*see* Secured by Design (SBD))
  security devices, 267–269
      (*see also* Security devices)
  security measures, 265
  spatial and environmental analysis, 196
  target selection, 197
  targeting, 86, 200
  theft, 200
  trends and patterns, 266–267
  variations, 6, 10–11
  victimisation rates, 201
  victims, 1, 200, 201
Burglary pilot (target hardening) process, 188
Burglary prevention
  approaches and possible activities, 23, 24
  contexts, 39, 40
  CPU, 24
  crime survey, for England and Wales,
      21, 22
  detection rates, 23
  displacement, 36
  in England and Wales, 22
  evaluation, 39
  multi-agency approach, 24
  policing, 39
  primary prevention, 24
  programmes
      alley gating, 33
      CRP, 32–33
      design against crime, 33
      Huddersfield project, 31, 32
      Kirkholt Burglary Prevention Project,
          27, 28
      Neighbourhood Watch, 25
      Olympic model, 31
      PEP model, 34, 35
      property marking, 26
      publicity schemes, 26, 27
      Safer Cities, 28–30
      street lighting, 35
  strategies, 40–41
  'what works', 37–39
Burglary risks
  BCS data, 111
  benefits, 114–116
  CSEW data, 111
  data and sample sizes, 147, 149–151
  of dwellings with no security, 236
  and effective security, 156, 157
  household and area characteristics,
      148–149
  household types, 111

limitations, 116, 117
  population heterogeneity, 111
  socio-economic characteristics, 111
  standard error of covariance, 152
  statistical model and modelling strategy,
      151, 152, 156, 205
  variables, 146, 147
  victimisation, 111, 112
  vulnerability, 112
  and WIDE security, 157–160
Burglary Task and Finish group (BTF), 171
Burglary-dwelling, 2

C
Cadre de Vie et Sécurité (CVS), 2, 87, 198, 199
Centre d'études sociologiques sur le droit et les
      institutions pénales (CESDIP), 198
College of Policing website, 253
Communication
  beyond the AC, 252–254
  with AC, 251–252
Community safety partnership, Nottingham, 170
Computer-assisted personal interviewing
      (CAPI), 198
Computer-assisted self-interviewing (CASI), 198
Control variables, 204
Cost-benefit perspective, 184
Council houses, 133, 238, 239
Crime and Disorder Act, 24, 170
Crime drop, 41, 266
  cross-national test, 226
  e-crimes and phone theft test, 226
  prior increase test, 226
  security hypothesis
      (*see* Security hypothesis)
  uneven (*see* Uneven crime drop)
  variable trajectories test, 226
  violence in the United States, 226
Crime levels, 225
Crime measurements, 14
Crime prevention, 144, 145, 166
  ACPO, 47
  CPTED, 47
  household factors, 45–46
  place-based approaches, 46
  within planning system
      Crime and Disorder Act, 48
      Design and Access Statements, 49
      Localism Act, 51
      National Planning Policy Framework, 51
      policy documents, 50
      Quality Standards, 51
      safe developments, 49

Safer Places, 49
SBD scheme, 48
Scheme Development Standards, 50
Social Housing Grants, 50
urban planning, 48
situational crime prevention intervention, 46
Crime Prevention and Victimisation modules, 85
Crime Prevention module, 118
Crime Prevention through Environmental
Design (CPTED), 33, 47, 49, 51, 270
Crime Prevention Unit (CPU), 24–26
Crime Reduction Programme (CRP), 32–33, 84
Crime repetition, 167
Crime Survey for England and Wales (CSEW),
2, 21, 25, 78, 111, 112, 114,
118–128, 133, 135, 136, 138, 139,
170, 196, 201, 229, 248
deter/thwart calculations, 100–102
face-to-face victimisation survey, 97
methodology, 102
questionnaire structure, 97, 98
respondents, 98
sample selection, 97
security information, 98
victims, attempted burglary and burglary
with entry, 100
Criminological theories of lifestyle, 109
Cross-national test, 226

D
DAPPER test, 241, 242
Data signatures
CSEW/BCS data, 229, 230
domestic burglary, 229
security-led burglary drop, in England and
Wales, 230
forced entry and unforced entry, 236, 237
over time SPF, 233–235
SPFs, 232, 233
trends, 233, 234
trends in security devices, 230
Decay, 13
Defensible space, 53, 61, 66, 67, 71, 72
Department for Communities and Local
Government (DCLG), 250
Digital locks, 206, 208, 219
Displacement, 36, 175
Domestic burglary, 77, 170, 245
in England and Wales, 237, 242
reductions, 224, 225
role of security, 225
security hypothesis, 229
trends, 108, 144

E
Ecological factors, 196
Economic and Social Research Council
(ESRC), 246, 248, 249, 253, 258
e-crimes and phone theft test, 226
Effectiveness, 196, 197, 211, 218
Elegant security, 225, 241
'Encounter versus enclosure' debate, 53
Enhanced security, 113
Environmental factors, 204, 205, 217
Equity, 115
ESRC-SDAI-funded academic research, 172
ESRC-SDAI project, 172
Estate Action, 29, 34
Ethnic groups
security availability and burglary risks,
127–129
Evidence-based prevention, 267, 269
Explanatory factors, 205

F
Face-to-face interaction, 249, 251
Familiarity, Accessibility, Visibility, Occupancy
and Rewards (FAVOR), 82
Financial costs, 15
French National Institute of Statistics and
Economic Studies, 198
French National Observatory of Crime and
Criminal Justice (ONDRP), 198
French Victimisation Survey, 87

G
Geographical displacement, 169
Group composition, 114

H
High income households, 135
High level of security, 113
Hit-and-run research, 250
Home Security Assessment, 175, 178, 179, 188
Horizontal equity, 142
Household car ownership, 138–139
Household composition
security and burglary risk
children under 16 years old *vs.*
households without children, 130
earlier time periods, 131
family units, 130
HMOs, 130
lone parent households, 130, 132
single adult households, 131

Household composition (*cont.*)
    single adult with children under 16
       years old households *vs.* other
       households, 130
    single and more than two adult
       households *vs.* two adult
       households, 130
Household crime, 6
Household reference person (HRP), 119,
       122, 125, 128, 129, 135, 142–144,
       146, 161
Household security, 99
Household tenure, 133–135
Household victimisation, 198, 199
Householders, 80
Houses in Multiple Occupation (HMOs),
       130, 135
Housing associations, 80
Housing unit, 196, 197, 199–206, 208, 209,
       211–214, 215, 217–219
Huddersfield project, 31, 32, 39

I
Impact
    and dialogue, 250
    and exchanging knowledge, 245
    facilitating, 254–256
    maximise, 252
    research, 246
    trace and document, 256
Inner city/urban households, 139, 141
International Association of Chiefs of Police, 247
International crime drop, 225
International Crime Victims Survey
       (ICVS), 2, 225

K
Kirkholt Burglary Prevention Project, 27, 28, 84
Knowledge exchange
    AC (*see* Advisory Committee (AC))
    academic-practitioner partnerships, 249
    academics (*see* Academics)
    challenges
       practical benefits of involvement,
        254–256
       trace and document impact, 256
    collaboration, 247
    communication
       beyond the AC, 252–254
       with AC, 251–252
    cooperation, 247
    coordination, 247

CSEW, 248
    culture of collaboration, 257
    domestic burglary, 245
    ESRC, 246, 248, 249, 253, 258
    funding bid, 248
    International Association of Chiefs of
       Police, 247
    Knowledge Transfer Partnership, 258
    with non-academic partners, 245
    policymakers, 247, 257
    population characteristics, 257
    practitioners (*see* Practitioners)
    public sector climate, 247
    publicly available data, 248
    REF, 246
    relationships, 249–251
    Research Assessment Exercise, 246
    Research Selectivity Exercise, 246
    role of security, 248
    scholars, 257
    short-term partnership, 257
    success, 248
    time-limited project, 246
Knowledge Transfer Partnership, 258

L
Lifestyle theory, 195
Local authorities, 133
Localism Act, 51
Logistic (logit) regression, 151–152, 205
Lower Super Output Areas (LSOAs), 174,
       175, 177, 178, 183
Low-income households, 135, 207, 208, 212

M
Maryland Scale of Scientific Methods for
       Evaluating Crime Prevention,
       168, 186
Middle and upper income groups, 136, 207,
       208, 212
Movement control, 61
    benefits, 64
    connectivity, levels of, 64
    footpaths benefiting, 64, 65
    walkability, 64
Multivariate response models, 157
Mutual education, 250

N
National Crime Statistics, 8
National crime survey data, 167

National Crime Victimisation Survey (NCVS), 2
National Observatory of Crime and Criminal Justice, 17
National Planning Policy Framework, 51
National Police Library, 253
Near repeat victimisation (NRV), 173, 174
Negative correlation, 120, 121
Neighbourhood and Home Watch Network (England & Wales) (NHWN), 253, 254
Neighbourhood permeability, 54
Neighbourhood Watch, 25
Non-academic partners, 245
Non-rural households, 140
Nottingham
    burglary profile, 170
    City of, 169
    NCDP, 170
Nottingham City Homes (NCH), 171
Nottingham Crime and Drugs Partnership (NCDP), 166, 170, 247, 258
    for recognition, 187
    history, 187
    statement, 186
Nottingham Drug & Alcohol Action Team (DAAT), 187
Nottingham pilot burglary target hardening initiative
    aerial map, 177, 178
    control area, 178
    Home Security Assessment, 175, 178, 179
    inception and operational framework, 171
    LSOA, 177, 178
    OPCC funding, 179
    participating area selection, 173–175
    PCSOs, 179
    PCU, 175, 177–179
    property landlords/owners of rented homes, 179
    research evidence, 171–173
    Security Survey forms, 177, 190
    solar lights, 177, 179
    test and control areas, 175, 176
    test areas, 177

O
Offence displacement, 169
Offender interviews
    burglar alarms, 68–69, 82, 83
    FAVOR, 82
    locks, 68, 83
    occupancy cues/proxies, 83, 84
    target selection process, 58–61, 82

Offenders, 56, 107
Office for National Statistics (ONS), 5, 78, 79, 97
Office of the Police and Crime Commissioner (OPCC), 166
Olympic model, 31
Over time SPF, 233–235
Owner-occupiers, 134

P
Partnership burglary pilot, 189
Penal Code, 197
Personal victimisation, 198
Physical security, 46, 50, 67, 68, 71, 72, 135, 267, 269–271
Pilot project, 166, 170, 172, 177
Pilot target hardening initiative, 166
Police
    ADT alarms, 69
    budgets, 51
    defensible space, 53
    footpaths benefiting, 64, 65
    Police Crime Prevention Initiatives, 47
    responsibility, 48
    SBD (see Secured by Design (SBD))
Police Community Support Officers (PCSOs), 179
Police recorded crime data, 2, 167, 195
Police's Pre-Crime Unit (PCU), 175, 177–179
Police-academic partnerships, 247
Policing context, 253
Policing Insight, 253
Policy instruments, 169
Policy interventions, 168, 169
Policymakers, 246, 247, 251–253, 255, 257
Population heterogeneity, 31, 110, 111
Practice-embedded research, 269, 271
Practitioners
    AC, 250
    academic-practitioner partnerships, 249, 258
    and academics (see Academics)
    advisory group, 257
    assessment of research, 246
    communication, 251
    exchanging knowledge, 248
    involvement, 250
    member of project Advisory Committee, 246
    outlets, 253, 254
    and policymakers, 246, 247, 253
    public and third sector organisations, 248
    trusted, 250
Prior crime increase test, 226
Priority Estates Projects (PEP), 34, 35

Private housing, 9
Private landlords, 80, 133
Private road, 67
Private sector organisations, 17
Property crime, 115, 195
Property landlords/owners of rented homes, 179
Property marking, 25–27, 39, 177, 180–182
Protection, 196, 205, 211, 215
Public sector climate, 247
Public/social/council housing, 9, 130, 133–134
Publicity schemes, 26, 27

Q
Qualitative variable, 119
Quantitative effect, 205

R
Reducing Burglary Initiative (RBI), 32, 33,
        84, 165
Reference household (RH), 119
Regression analysis, 119, 205
Relationship-building process, 251
Relationships, knowledge exchange, 248–251
Repeat and Near Repeat Burglary Pilot Project
        Protocol, 172
Repeat and near repeat victimisation, theory
        of, 167
Repeat burglaries, 173
Repeat victimisation (RV), 5, 12–14, 28, 31,
        32, 37, 38, 40, 214–217, 219
Research Assessment Exercise, 246
Research Excellence Framework (REF), 246
Research Selectivity Exercise, 246
Residential burglary, 2
Residential properties, 115
Responsible Authorities, 186
Role of security
        burglar alarms, 239–241
        capable guardian, 224
        council houses, 238, 239
        crime categories, 225
        crime drop (see Crime drop)
        crime levels, 225
        crime trends, 224
        data signatures (see Data signatures)
        design, 241, 242
        domestic burglary, 224
        double glazing, 238
        elegant, 225
        ICVS, 225
        improvements, 237, 238
        methodology, 225

        potential offender, 224
        quality, 237, 239
        reporting and recording practices, 225
        routine activities, 224
        security-led burglary drop
                (see Security-led burglary drop, in
                England and Wales)
        social conditions in Britain, 223
        target suitability, 224
Routine activities, 109, 110, 119, 224
Routine activity theory, 195–197

S
Safer Cities, 28–30
Safer Cities Programme, 84, 165
SBD New Homes, 70
Secured by Design (SBD), 250
        CPTED, 47
        crime prevention (see Crime prevention)
        defensible space, 53
        development, management and
                implementation, 47–48
        effectiveness, 51–52
        housing types and styles, social and
                privately owned, 55–57
        limitations, research, 56–58
        limiting through movement, 53
        methodology, 55
        neighbourhood permeability, 54
        opportunity, for guardianship, 54
        place-based crime reduction, 46
        principles, 62
        'private' sign, 72
        surveillance, 54, 55
        target attractiveness
                defensible space, 66–67
                management and maintenance, 70–71
                movement control
                        (see Movement control)
                physical security, 67–69
                suitable target, 58–59
                surveillance, 61–63
                unsuitable target, 59–61
Security
        devices, 268
        improvements, 266
        physical, 265, 267, 270
        security measures, 265
        upgrades, 270
        visual, 267
        WIDE security, 269
Security availability
        and burglary risk, 113–114

anti-burglary security devices, 112, 113
benefits, 114–116
limitations, 116, 117
in residential properties, 112
Security devices, 16
  alarms, 87
  analysis
    effectiveness, 211
    forced entry, 211
    houses and apartments, 211
    housing unit, 211–213
    theft, 211, 213, 214
  assessment, 196
  attempted burglary, 86, 92, 94
  availability, 78–80
  break-in, 197
  cameras and digital locks, 87
  categories, 196
  CSEW, 79, 97
  data sources, 198, 199
  demographic characteristics, 95
  descriptive statistics, 96
  deterred and thwarted, 87, 90–94
  dissuasive effect, 218
  doors, 87
  double locks/deadlocks, 93
  effectiveness, 78, 196, 197, 218
    large-scale initiatives, 84–85
    offender interviews, 82–84
    victimisation survey data, 80–81
  environmental cues, 83
  environmental factors, 204, 205, 217
  expected mechanism, 88
  FAVOR-able cues, 88, 90
  forced entry, 208, 209
  French victimisation data, 196
  householder vigilance, 94
  housing unit, 206
  human activation, 94
  individual and incident factors, 95
  insurance companies, 78, 269
  lifestyle, 204, 205
  modelling, 151, 205
  non-burgled households, 85
  Penal Code, 197
  perpetrator, 197
  physical, 78
  police forces, 94
  population groups and areas, 95
  in property and public awareness, 130
  property crime, 195
  protection, 196
  protective role, 218
  reference framework, 195

in residential properties, 115
routine activity theory, 197
RV, 197, 214–217
security features within housing units,
    202, 203
SIAT, 85
SPF value, 86
stages of burglary, 86, 199–201
statistical modelling, 93, 150
targeted and untargeted house, 87
targeting, 86, 206–208
theft, 197, 209, 210
transition points, 87
types, 201
unanticipated guardianship, 86
Security door, 208
Security hypothesis
  burglary drop (see Data signatures)
  commentary, 228
  crime drop, 142, 226, 227
  criminal careers, 227
  criticisms, 228
  and improvements, 227
  indirect crime prevention effects, 227
  negative attributes, 229
  prevention, 228
  proponents, 228
  semi-public places, 228
  testing, 229
  vehicle thefts, 228, 229
Security Impact Assessment Tool (SIAT), 85
Security packages, 95
Security Protection Factors (SPFs), 86, 89,
    232, 233
Security Survey, 177, 190
Security upgrades, 17
Security-led burglary drop, in England and
    Wales
  average protective effects vs. burglary, 232
  forced entry and unforced entry, 236, 237
  installation/no security, 230, 231
  no security relative to national risk, 236
  over time SPF, 233–235
  security relative, to national risk, 236
  and single security, 233, 234
  SPFs, 232, 233
  trends, 233, 234
  unprotected, 231
Short-term partnership, 257
Silent alarms, 241
Social disorganisation, 109
Social housing, 113, 135
Social rented accommodation, 133
Social rented housing, 133

Social renters, 111, 115, 121, 122, 125, 130,
        133–135, 143
Socio-economic and routine activity
        information, 119
Socio-economic characteristics, 114
Socio-economic factors, 109–111, 114
Socio-economic groups, 16
Solar lights, 177, 179
Street lighting, 35, 40
Surface area, 217
Surveillance, 54, 61–63, 267, 270
Surveillance cameras, 206

T
Target displacement, 169
Target hardening initiative
    crime prevention, 166, 185
    evaluation data, 183–185
    evaluation design, 185
    evaluation of burglary reduction
        methodological issues, 167–169
        theory of repeat and near repeat
            victimisation, 167
    house security upgrades, 166
    NCDP, 166
    Nottingham (see Nottingham)
    OPCC, 166
    pilot data
        activity log, 180–182
        CSEW, 182
        data capture forms, 180
        description, 182
        implementation, 181–182
        PCSOs, 180, 181
        properties, 182
        Security Surveys, 180, 181
        timers, 182
        WIDE-related target hardening
            upgrades, 180
    pilot target hardening initiative, 166
    population groups, 165
    selection, 166
    test areas, 185
    WIDE, 166, 185, 186
Territoriality, 53
Time-limited project, 246

U
Unequal burglary risks and security
        availability
    area types, 108
    crime areas, 107

crime drop (see Uneven crime drop)
crime prevention, 144, 145
criminological theories of lifestyle, 109
domestic burglary, 108
features, 109, 110
household types, 108
influences, 110, 111
population groups, 109
quantity and quality, 110
regions provide, 108
social disorganisation, 109
theory and policy, 108
victimisation, 110
WD, 108
Uneven crime drop
    and distributive justice
        distribution, 142
        equitable, 142
        horizontal equity, 142
        HRP, 144
        population groups, 141
        social renters, 143
        stagnant trends, 141
        vertical equity, 142
        victimisation divides, 142, 143
        WDE, 141
        WIDE, 141, 143, 144
    crime prevention, 144, 145
    data and methodology, 117–119
    income population groups, 116
    population groups and national average
        CSEW, 122
    security availability and burglary risks,
        119–121
        annual household income, 135–137
        by area of residence, 139–141
        ethnic groups, 127–129
        household car ownership, 138–139
        household composition, 129–133
        household tenure, 133–135
        national average, 125, 127
        population groups, 121, 122, 124, 125
United Nations Office on Drugs and Crime
        (UNODC), 5
Univariate response models, 157

V
Variable trajectories test, 226
Variables, 220–221
Vehicle thefts, 228, 229
Vertical equity, 142
Victimisation, 22, 28
Victimisation divides, 142, 143

Victimisation survey, 80–81, 195, 198, 214
Victims, 107
Video surveillance cameras, 219

W
'What works', 32, 37–40
W(I)DE pilot target hardening initiative, 183
Window and door locks (WD), 108

Window and door locks plus internal and
    external security lighting (WIDE),
    81, 108, 117–122, 125, 127–134,
    136–141, 143–145
Window and door locks, security chains and
    CCTV cameras (WDSC), 89, 108, 117
Window locks and security chains (WS), 102
Window locks, door locks and security chains
    (WDS), 102